Essentials of Geometry

.

Essentials of Geometry

Edited by
Josue Richards

Larsen & Keller
www.larsen-keller.com

Essentials of Geometry
Edited by Josue Richards
ISBN: 978-1-63549-135-7 (Hardback)

目 Larsen & Keller

Published by Larsen and Keller Education,
5 Penn Plaza,
19th Floor,
New York, NY 10001, USA

Cataloging-in-Publication Data

Essentials of geometry / edited by Josue Richards.
     p. cm.
Includes bibliographical references and index.
ISBN 978-1-63549-135-7
1. Geometry. 2. Mathematics. I. Richards, Josue.
QA445 .E87 2017
516--dc23

The publisher's policy is to use permanent paper from mills that operate a sustainable forestry policy. Furthermore, the publisher ensures that the text paper and cover boards used have met acceptable environmental accreditation standards.

Printed and bound in the United States of America.

For more information regarding Larsen and Keller Education and its products, please visit the publisher's website www.larsen-keller.com

Table of Contents

Permissions

Index

Preface

This textbook provides deep information about the basic concepts and fundamental theories of geometry. As a branch of mathematics, geometry studies abstract space and its properties. It deals with lengths, volumes and areas. This text elucidates new techniques and their applications in a multidisciplinary approach. It picks up individual branches and explains their need and contribution in the context of the growth of this field. Different approaches, evaluation and methodologies of geometric theory have been included in this book. Those in search of information to further their knowledge will be greatly assisted by this textbook.

Given below is the chapter wise description of the book:

Chapter 1- Geometry is a branch of mathematics. It concerns itself with shapes, sizes and figures. It has played an important role in a number of cultures and has progressed drastically throughout the years. This chapter helps the reader in understanding the basic concepts of geometry.

Chapter 2- Absolute geometry is a branch of geometry that is based on an axiom system. Absolute geometry is also referred to as neutral geometry. Some of the other branches of geometry that have been discussed in this section are algebraic geometry, analytic geometry, differential geometry, projective geometry and Euclidean geometry. This chapter is a compilation of the various branches of geometry that form an integral part of the broader subject matter.

Chapter 3- Geometry has a number of important concepts. Some of these concepts are lines, line segment, plane, angle, similarity and congruence. A line is a straight path that joins points and has no endpoints whereas a plane is a two-dimensional surface that lengthens infinitely. This text elucidates the key theories of geometry.

Chapter 4- Triangles are plain figures which have three angles and the three angles always add to 180°. Triangles on the bases of their sides can be categorized into three types, equilateral, isosceles and scalene. Equilateral triangles have all sides of the same length, an isosceles triangle has two sides of the same length and a scalene triangle has 3 different lengths.

Chapter 5- Circles have a center and are closed in shape. The distance from the center to any point of the circle is known as radius. Some of the basic aspects of a circle are circumference, diameter, arc and radius. The information provided in this section helps the reader to delve deep into the topics related to it.

Chapter 6- A quadrilateral is any figure that has four sides and four corners. They can be simple or complex and simple quadrilaterals can further be divided into convex or concave. This section will provide an integrated study of quadrilateral.

Chapter 7- Polyhedrons have three dimensions with straight edges and vertices. Some of the examples of polyhedrons are pyramids, prisms, cylinders, cones and spheres. The chapter strategically encompasses and incorporates the major components and key concepts of polyhedron, providing a complete understanding.

Chapter 8- The branch of mathematics that studies the relation between lengths and angles is termed as trigonometry. Some of the fields that use trigonometry are astronomy, electronics, chemistry, seismology and oceanography. Trigonometry in today's times has immense number of purposes.

Indeed, my job was extremely crucial and challenging as I had to ensure that every chapter is informative and structured in a student-friendly manner. I am thankful for the support provided by my family and colleagues during the completion of this book.

Editor

Introduction to Geometry

Geometry is a branch of mathematics. It concerns itself with shapes, sizes and figures. It has played an important role in a number of cultures and has progressed drastically throughout the years. This chapter helps the reader in understanding the basic concepts of geometry.

Geometry is a branch of mathematics concerned with questions of shape, size, relative position of figures, and the properties of space. A mathematician who works in the field of geometry is called a geometer.

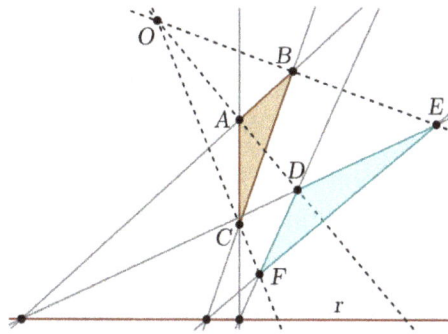

An illustration of Desargues' theorem, an important result in Euclidean and projective geometry

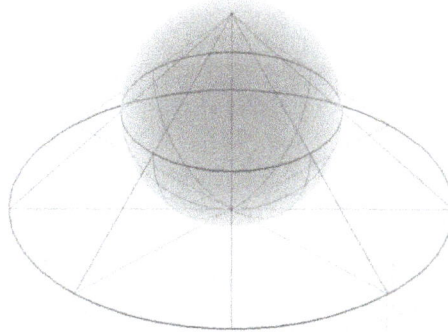

Projecting a sphere to a plane.

Geometry arose independently in a number of early cultures as a practical way for dealing with lengths, areas, and volumes. Geometry began to see elements of formal mathematical science emerging in the West as early as the 6th century BC. By the 3rd century BC, geometry was put into an axiomatic form by Euclid, whose treatment Euclid's Elements set a standard for many centuries to follow. Geometry arose independently in India, with texts providing rules for geometric constructions appearing as early as the 3rd century BC. Islamic scientists preserved Greek ideas and expanded on them during the Middle Ages. By the early 17th century, geometry had been put on a solid analytic footing by mathematicians such as René Descartes and Pierre de Fermat. Since then, and into modern times, geometry has expanded into non-Euclidean geometry and manifolds, describing spaces that lie beyond the normal range of human experience.

While geometry has evolved significantly throughout the years, there are some general concepts that are more or less fundamental to geometry. These include the concepts of points, lines, planes, surfaces, angles, and curves, as well as the more advanced notions of manifolds and topology or metric.

Contemporary geometry has many subfields:

- Euclidean geometry is geometry in its classical sense. The majority of nations includes the study of points, lines, planes, angles, triangles, congruence, similarity, solid figures, circles, and analytic geometry in their mandatory educational curriculum. Euclidean geometry also has applications in computer science, crystallography, and various branches of modern mathematics.

- Differential geometry uses techniques of calculus and linear algebra to study problems in geometry. It has applications in physics, including in general relativity.

- Topology is the field dealing with the properties of geometry that are unchanged by continuous functions. In practice, this means that it deals with large-scale properties of a spaces such as connectedness and compactness.

- Algebraic geometry studies geometry through the use of multivariate polynomials and other algebraic techniques. It has applications in many areas, including cryptography and string theory.

Geometry has applications to many fields, including art, architecture, physics, and other fields of mathematics.

History

A European and an Arab practicing geometry in the 15th century.

The earliest recorded beginnings of geometry can be traced to ancient Mesopotamia and Egypt in the 2nd millennium BC. Early geometry was a collection of empirically discovered principles concerning lengths, angles, areas, and volumes, which were developed to meet some practical need in surveying, construction, astronomy, and various crafts. The earliest known texts on geometry are the Egyptian *Rhind Papyrus* (2000–1800 BC) and *Moscow Papyrus* (c. 1890 BC), the Babylonian clay tablets such as Plimpton 322 (1900 BC). For example, the Moscow Papyrus gives a formula for calculating the volume of a truncated pyramid, or frustum. Later clay tablets (350–50 BC) demonstrate that Babylonian astronomers implemented trapezoid procedures for computing Jupiter's position and motion within time-velocity space. These geometric procedures anticipated the Oxford Calculators, including the mean speed theorem, by 14 centuries. South of Egypt the ancient Nubians established a system of geometry including early versions of sun clocks.

In the 7th century BC, the Greek mathematician Thales of Miletus used geometry to solve problems such as calculating the height of pyramids and the distance of ships from the shore. He is credited with the first use of deductive reasoning applied to geometry, by deriving four corollaries to Thales' Theorem. Pythagoras established the Pythagorean School, which is credited with the first proof of the Pythagorean theorem, though the statement of the theorem has a long history Eudoxus (408–c. 355 BC) developed the method of exhaustion, which allowed the calculation of areas and volumes of curvilinear figures, as well as a theory of ratios that avoided the problem of incommensurable magnitudes, which enabled subsequent geometers to make significant advances. Around 300 BC, geometry was revolutionized by Euclid, whose *Elements*, widely considered the most successful and influential textbook of all time, introduced mathematical rigor through the axiomatic method and is the earliest example of the format still used in mathematics today, that of definition, axiom, theorem, and proof. Although most of the contents of the *Elements* were already known, Euclid arranged them into a single, coherent logical framework. The *Elements* was known to all educated people in the West until the middle of the 20th century and its contents are still taught in geometry classes today. Archimedes (c. 287–212 BC) of Syracuse used the method of exhaustion to calculate the area under the arc of a parabola with the summation of an infinite series, and gave remarkably accurate approximations of Pi. He also studied the spiral bearing his name and obtained formulas for the volumes of surfaces of revolution.

Line art drawing of parallel lines and curves.

Indian mathematicians also made many important contributions in geometry. The *Satapatha Brahmana* (3rd century BC) contains rules for ritual geometric constructions that are similar to the *Sulba Sutras*. According to (Hayashi 2005, p. 363), the *Śulba Sūtras* contain "the earliest extant verbal expression of the Pythagorean Theorem in the world, although it had already been known to the Old Babylonians. They contain lists of Pythagorean triples, which are particular cases of Diophantine equations. In the Bakhshali manuscript, there is a handful of geometric problems (including problems about volumes of irregular solids). The Bakhshali manuscript also "employs a decimal place value system with a dot for zero." Aryabhata's *Aryabhatiya* (499) includes the computation of areas and volumes. Brahmagupta wrote his astronomical work *Brāhma Sphuṭa Siddhānta* in 628. Chapter 12, containing 66 Sanskrit verses, was divided into two sections: "basic operations" (including cube roots, fractions, ratio and proportion, and barter) and "practical mathematics" (including mixture, mathematical series, plane figures, stacking bricks, sawing of timber, and piling of grain). In the latter section, he stated his famous theorem on the diagonals of a cyclic quadrilateral. Chapter 12 also included a formula for the area of a cyclic quadrilateral (a generalization of Heron's formula), as well as a complete description of rational triangles (*i.e.* triangles with rational sides and rational areas).

In the Middle Ages, mathematics in medieval Islam contributed to the development of geometry, especially algebraic geometry. Al-Mahani (b. 853) conceived the idea of reducing geometrical problems such as duplicating the cube to problems in algebra. Thābit ibn Qurra (known as Thebit in Latin) (836–901) dealt with arithmetic operations applied to ratios of geometrical quantities, and contributed to the development of analytic geometry. Omar Khayyám (1048–1131) found geometric solutions to cubic equations. The theorems of Ibn al-Haytham (Alhazen), Omar Khayyam and Nasir al-Din al-Tusi on quadrilaterals, including the Lambert quadrilateral and Saccheri quadrilateral, were early results in hyperbolic geometry, and along with their alternative postulates, such as Playfair's axiom, these works had a considerable influence on the development of non-Euclidean geometry among later European geometers, including Witelo (c. 1230–c. 1314), Gersonides (1288–1344), Alfonso, John Wallis, and Giovanni Girolamo Saccheri.

In the early 17th century, there were two important developments in geometry. The first was the creation of analytic geometry, or geometry with coordinates and equations, by René Descartes (1596–1650) and Pierre de Fermat (1601–1665). This was a necessary precursor to the development of calculus and a precise quantitative science of physics. The second geometric development of this period was the systematic study of projective geometry by Girard Desargues (1591–1661). Projective geometry is a geometry without measurement or parallel lines, just the study of how points are related to each other.

Two developments in geometry in the 19th century changed the way it had been studied previously. These were the discovery of non-Euclidean geometries by Nikolai Ivanovich Lobachevsky, János Bolyai and Carl Friedrich Gauss and of the formulation of symmetry as the central consideration in the Erlangen Programme of Felix Klein (which generalized the Euclidean and non-Euclidean geometries). Two of the master geometers of the time were Bernhard Riemann (1826–1866), working primarily with tools from mathematical analysis, and introducing the Riemann surface, and Henri Poincaré, the founder of algebraic topology and the geometric theory of dynamical systems. As a consequence of these major changes in the conception of geometry, the concept of "space" became something rich and varied, and the natural background for theories as different as complex analysis and classical mechanics.

Important Concepts in Geometry

The following are some of the most important concepts in geometry.

Axioms

Euclid took an abstract approach to geometry in his Elements, one of the most influential books ever written. Euclid introduced certain axioms, or postulates, expressing primary or self-evident properties of points, lines, and planes. He proceeded to rigorously deduce other properties by mathematical reasoning. The characteristic feature of Euclid's approach to geometry was its rigor, and it has come to be known as *axiomatic* or *synthetic* geometry. At the start of the 19th century, the discovery of non-Euclidean geometries by Nikolai Ivanovich Lobachevsky (1792–1856), János Bolyai (1802–1860), Carl Friedrich Gauss (1777–1855) and others led to a revival of interest in this discipline, and in the 20th century, David Hilbert (1862–1943) employed axiomatic reasoning in an attempt to provide a modern foundation of geometry.

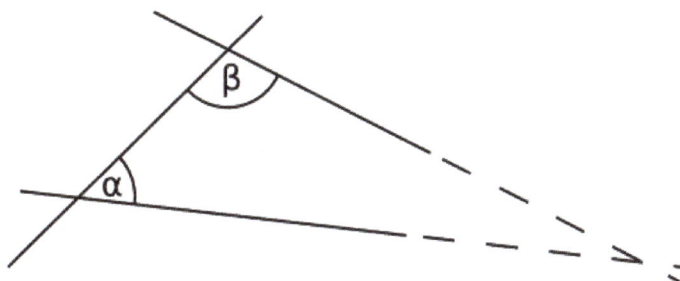

An illustration of Euclid's parallel postulate

Points

Points are considered fundamental objects in Euclidean geometry. They have been defined in a variety of ways, including Euclid's definition as 'that which has no part' and through the use of algebra or nested sets. In many areas of geometry, such as analytic geometry, differential geometry, and topology, all objects are considered to be built up from points. However, there has been some study of geometry without reference to points.

Lines

Euclid described a line as "breadthless length" which "lies equally with respect to the points on itself". In modern mathematics, given the multitude of geometries, the concept of a line is closely tied to the way the geometry is described. For instance, in analytic geometry, a line in the plane is often defined as the set of points whose coordinates satisfy a given linear equation, but in a more abstract setting, such as incidence geometry, a line may be an independent object, distinct from the set of points which lie on it. In differential geometry, a geodesic is a generalization of the notion of a line to curved spaces.

Planes

A plane is a flat, two-dimensional surface that extends infinitely far. Planes are used in every area of geometry. For instance, planes can be studied as a topological surface without reference to distances or angles; it can be studied as an affine space, where collinearity and ratios can be studied but not distances; it can be studied as the complex plane using techniques of complex analysis; and so on.

Angles

Euclid defines a plane angle as the inclination to each other, in a plane, of two lines which meet each other, and do not lie straight with respect to each other. In modern terms, an angle is the figure formed by two rays, called the *sides* of the angle, sharing a common endpoint, called the *vertex* of the angle.

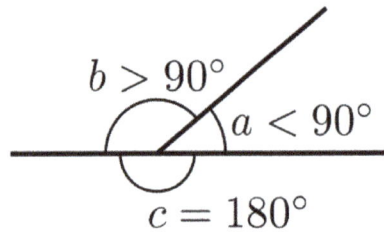

Acute (a), obtuse (b), and straight (c) angles. The acute and obtuse angles are also known as oblique angles.

In Euclidean geometry, angles are used to study polygons and triangles, as well as forming an object of study in their own right. The study of the angles of a triangle or of angles in a unit circle forms the basis of trigonometry.

In differential geometry and calculus, the angles between plane curves or space curves or surfaces can be calculated using the derivative.

Curves

A curve is a 1-dimensional object that may be straight (like a line) or not; curves in 2-dimensional space are called plane curves and those in 3-dimensional space are called space curves.

In topology, a curve is defined by a function from an interval of the real numbers to another space. In differential geometry, the same definition is used, but the defining function is required to be differentiable Algebraic geometry studies algebraic curves, which are defined as algebraic varieties of dimension one.

Surfaces

A surface is a two-dimensional object, such as a sphere or paraboloid. In differential geometry and topology, surfaces are described by two-dimensional 'patches' (or neighborhoods) that are assembled by diffeomorphisms or homeomorphisms, respectively. In algebraic geometry, surfaces are described by polynomial equations.

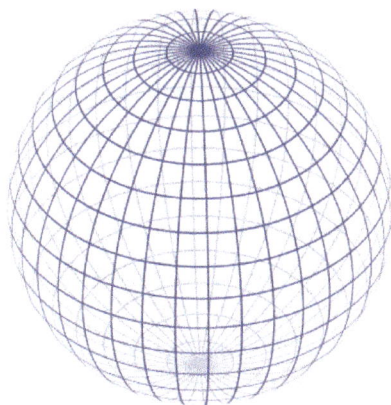

A sphere is a surface that can be defined parametrically (by $x = r \sin \theta \cos \varphi$, $y = r \sin \theta \sin \varphi$, $z = r \cos \theta$) or implicitly (by $x^2 + y^2 + z^2 - r^2 = 0$.)

Manifolds

A manifold is a generalization of the concepts of curve and surface. In topology, a manifold is a topological space where every point has a neighborhood that is homeomorphic to Euclidean space. In differential geometry, a differentiable manifold is a space where each neighborhood is diffeomorphic to Euclidean space.

Manifolds are used extensively in physics, including in general relativity and string theory

Topologies and Metrics

A topology is a mathematical structure on a set that tells how elements of the set relate spatially to each other. The best-known examples of topologies come from metrics, which are ways of measuring distances between points. For instance, the Euclidean metric measures the distance between points in the Euclidean plane, while the hyperbolic metric measures the distance in the hyperbolic plane. Other important examples of metrics include the Lorentz metric of special relativity and the semi-Riemannian metrics of general relativity.

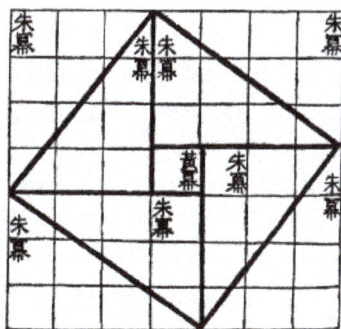

Visual checking of the Pythagorean theorem for the (3, 4, 5) triangle as in the Chou Pei Suan Ching 500–200 BC. The Pythagorean theorem is a consequence of the Euclidean metric.

Compass and Straightedge Constructions

Classical geometers paid special attention to constructing geometric objects that had been described in some other way. Classically, the only instruments allowed in geometric constructions are the compass and straightedge. Also, every construction had to be complete in a finite number of steps. However, some problems turned out to be difficult or impossible to solve by these means alone, and ingenious constructions using parabolas and other curves, as well as mechanical devices, were found.

Dimension

Where the traditional geometry allowed dimensions 1 (a line), 2 (a plane) and 3 (our ambient world conceived of as three-dimensional space), mathematicians have used higher dimensions for nearly two centuries. Dimension has gone through stages of being any natural number n, possibly infinite with the introduction of Hilbert space, and any positive real number in fractal geometry. Dimension theory is a technical area, initially within general topology, that discusses *definitions*; in common with most mathematical ideas, dimension is now defined rather than an intuition. Connected topological manifolds have a well-defined dimension; this is a theorem (invariance of domain) rather than anything *a priori*.

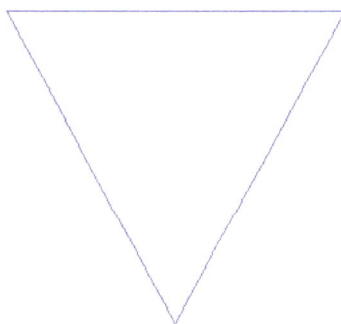

The Koch snowflake, with fractal dimension=log4/log3 and topological dimension=1

The issue of dimension still matters to geometry, in the absence of complete answers to classic questions. Dimensions 3 of space and 4 of space-time are special cases in geometric topology. Dimension 10 or 11 is a key number in string theory. Research may bring a satisfactory *geometric* reason for the significance of 10 and 11 dimensions.

Symmetry

The theme of symmetry in geometry is nearly as old as the science of geometry itself. Symmetric shapes such as the circle, regular polygons and platonic solids held deep significance for many ancient philosophers and were investigated in detail before the time of Euclid. Symmetric patterns occur in nature and were artistically rendered in a multitude of forms, including the graphics of M. C. Escher. Nonetheless, it was not until the second half of 19th century that the unifying role of symmetry in foundations of geometry was recognized. Felix Klein's Erlangen program proclaimed that, in a very precise sense, symmetry, expressed via the notion of a transformation group, determines what geometry *is*. Symmetry in classical Euclidean geometry is represented by congruences and rigid motions, whereas in projective geometry an analogous role is played by collineations, geometric transformations that take straight lines into straight lines. However it was in the new geometries of Bolyai and Lobachevsky, Riemann, Clifford and Klein, and Sophus Lie that Klein's idea to 'define a geometry via its symmetry group' proved most influential. Both discrete and continuous symmetries play prominent roles in geometry, the former in topology and geometric group theory, the latter in Lie theory and Riemannian geometry.

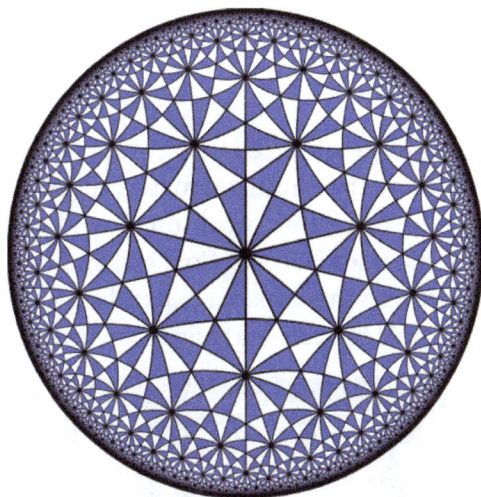

A tiling of the hyperbolic plane

A different type of symmetry is the principle of duality in projective geometry (Duality (projective geometry)) among other fields. This meta-phenomenon can roughly be described as follows: in any theorem, exchange *point* with *plane*, *join* with *meet*, *lies in* with *contains*, and you will get an equally true theorem. A similar and closely related form of duality exists between a vector space and its dual space.

Non-Euclidean Geometry

In the nearly two thousand years since Euclid, while the range of geometrical questions asked and answered inevitably expanded, the basic understanding of space remained essentially the same. Immanuel Kant argued that there is only one, *absolute*, geometry, which is known to be true *a priori* by an inner faculty of mind: Euclidean geometry was synthetic a priori. This dominant view was overturned by the revolutionary discovery of non-Euclidean geometry in the works of Bolyai, Lobachevsky, and Gauss (who never published his theory). They demonstrated that ordinary Euclidean space is only one possibility for development of geometry. A broad vision of the subject of geometry was then expressed by Riemann in his 1867 inauguration lecture *Über die Hypothesen, welche der Geometrie zu Grunde liegen* (*On the hypotheses on which geometry is based*), published only after his death. Riemann's new idea of space proved crucial in Einstein's general relativity theory, and Riemannian geometry, that considers very general spaces in which the notion of length is defined, is a mainstay of modern geometry.

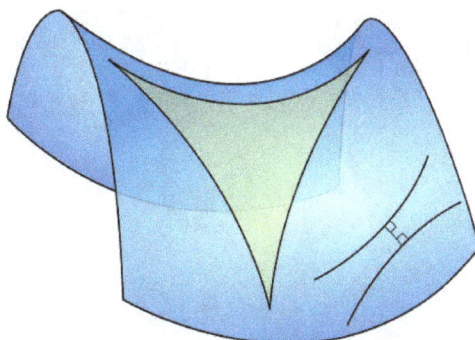

Differential geometry uses tools from calculus to study problems involving curvature.

Contemporary Geometry

Euclidean Geometry

Euclidean geometry has become closely connected with computational geometry, computer graphics, convex geometry, incidence geometry, finite geometry, discrete geometry, and some areas of combinatorics. Attention was given to further work on Euclidean geometry and the Euclidean groups by crystallography and the work of H. S. M. Coxeter, and can be seen in theories of Coxeter groups and polytopes. Geometric group theory is an expanding area of the theory of more general discrete groups, drawing on geometric models and algebraic techniques.

Geometry lessons in the 20th century

Differential Geometry

Differential geometry has been of increasing importance to mathematical physics due to Einstein's general relativity postulation that the universe is curved. Contemporary differential geometry is *intrinsic*, meaning that the spaces it considers are smooth manifolds whose geometric structure is governed by a Riemannian metric, which determines how distances are measured near each point, and not *a priori* parts of some ambient flat Euclidean space.

Topology and Geometry

A thickening of the trefoil knot

The field of topology, which saw massive development in the 20th century, is in a technical sense a type of transformation geometry, in which transformations are homeomorphisms. This has often been expressed in the form of the dictum 'topology is rubber-sheet geometry'. Contemporary geometric topology and differential topology, and particular subfields such as Morse theory, would be counted by most mathematicians as part of geometry. Algebraic topology and general topology have gone their own ways.

Algebraic Geometry

The field of algebraic geometry is the modern incarnation of the Cartesian geometry of co-ordinates. From late 1950s through mid-1970s it had undergone major foundational development, largely due to work of Jean-Pierre Serre and Alexander Grothendieck. This led to the introduction of schemes and greater emphasis on topological methods, including various cohomology theories. One of seven Millennium Prize problems, the Hodge conjecture, is a question in algebraic geometry.

Quintic Calabi–Yau threefold

The study of low-dimensional algebraic varieties, algebraic curves, algebraic surfaces and algebraic varieties of dimension 3 ("algebraic threefolds"), has been far advanced. Gröbner basis theory and real algebraic geometry are among more applied subfields of modern algebraic geometry. Arithmetic geometry is an active field combining algebraic geometry and number theory. Other directions of research involve moduli spaces and complex geometry. Algebro-geometric methods are commonly applied in string and brane theory.

Applications

Geometry has found applications in many fields, some of which are described below.

Art

Mathematics and art are related in a variety of ways. For instance, the theory of perspective showed that there is more to geometry than just the metric properties of figures: perspective is the origin of projective geometry.

Architecture

Mathematics and architecture are related, since, as with other arts, architects use mathematics for several reasons. Apart from the mathematics needed when engineering buildings, architects use geometry: to define the spatial form of a building; from the Pythagoreans of the sixth century BC onwards, to create forms considered harmonious, and thus to lay out buildings and their surroundings according to mathematical, aesthetic and sometimes religious principles; to decorate buildings with mathematical objects such as tessellations; and to meet environmental goals, such as to minimise wind speeds around the bases of tall buildings.

Physics

The field of astronomy, especially as it relates to mapping the positions of stars and planets on the celestial sphere and describing the relationship between movements of celestial bodies, have served as an important source of geometric problems throughout history.

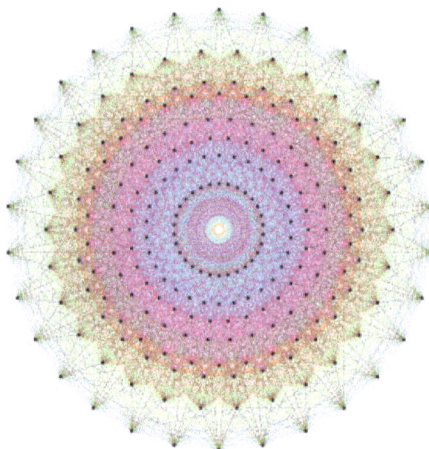

The 4_{21} polytope, orthogonally projected into the E_8 Lie group Coxeter plane. Lie groups have several applications in physics.

Modern geometry has many ties to physics as is exemplified by the links between pseudo-Riemannian geometry and general relativity. One of the youngest physical theories, string theory, is also very geometric in flavour.

Other Fields of Mathematics

Geometry has also had a large effect on other areas of mathematics. For instance, the introduction of coordinates by René Descartes and the concurrent developments of algebra marked a new stage for geometry, since geometric figures such as plane curves could now be represented analytically in the form of functions and equations. This played a key role in the emergence of infinitesimal calculus in the 17th century. The subject of geometry was further enriched by the study of the intrinsic structure of geometric objects that originated with Euler and Gauss and led to the creation of topology and differential geometry.

An important area of application is number theory. In ancient Greece the Pythagoreans considered the role of numbers in geometry. However, the discovery of incommensurable lengths, which contradicted their philosophical views, made them abandon abstract numbers in favor of concrete geometric quantities, such as length and area of figures. Since the 19th century, geometry has been used for solving problems in number theory, for example through the geometry of numbers or, more recently, scheme theory, which is used in Wiles's proof of Fermat's Last Theorem.

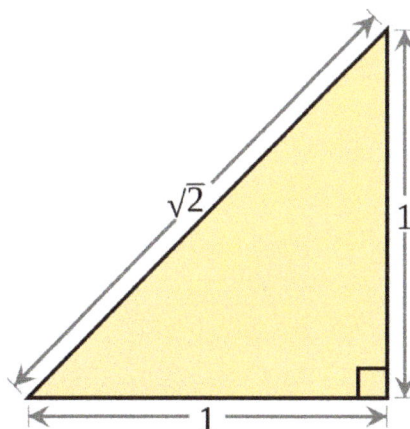

The Pythagoreans discovered that the sides of a triangle could have incommensurable lengths.

While the visual nature of geometry makes it initially more accessible than other mathematical areas such as algebra or number theory, geometric language is also used in contexts far removed from its traditional, Euclidean provenance (for example, in fractal geometry and algebraic geometry).

Analytic geometry applies methods of algebra to geometric questions, typically by relating geometric curves to algebraic equations. These ideas played a key role in the development of calculus in the 17th century and led to the discovery of many new properties of plane curves. Modern algebraic geometry considers similar questions on a vastly more abstract level.

Leonhard Euler, in studying problems like the Seven Bridges of Königsberg, considered the most fundamental properties of geometric figures based solely on shape, independent of their metric properties. Euler called this new branch of geometry *geometria situs* (geometry of place), but it is now known as topology. Topology grew out of geometry, but turned into a large independent discipline. It does not differentiate between objects that can be continuously deformed into each other. The objects may nevertheless retain some geometry, as in the case of hyperbolic knots.

References

- Martin J. Turner,Jonathan M. Blackledge,Patrick R. Andrews (1998). Fractal geometry in digital imaging. Academic Press. p. 1. ISBN 0-12-703970-8

- Euclid's Elements – All thirteen books in one volume, Based on Heath's translation, Green Lion Press ISBN 1-888009-18-7.

- Sidorov, L.A. (2001), "Angle", in Hazewinkel, Michiel, Encyclopedia of Mathematics, Springer, ISBN 978-1-55608-010-4

- Mumford, David (1999). The Red Book of Varieties and Schemes Includes the Michigan Lectures on Curves and Their Jacobians (2nd ed.). Springer-Verlag. ISBN 3-540-63293-X. Zbl 0945.14001.

- Dmitri Burago, Yu D Burago, Sergei Ivanov, A Course in Metric Geometry, American Mathematical Society, 2001, ISBN 0-8218-2129-6.

- Lamb, Evelyn (2015-11-08). "By Solving the Mysteries of Shape-Shifting Spaces, Mathematician Wins $3-Million Prize". Scientific American. Retrieved 2016-08-29.

- Ossendrijver, Mathieu (29 Jan 2016). "Ancient Babylonian astronomers calculated Jupiter's position from the area under a time-velocity graph". Science. 351 (6272): 482–484. doi:10.1126/science.aad8085. Retrieved 29 January 2016.

- Clark, Bowman L. (Jan 1985). "Individuals and Points". Notre Dame Journal of Formal Logic. 26 (1): 61–75. doi:10.1305/ndjfl/1093870761. Retrieved 29 August 2016.

- "geodesic – definition of geodesic in English from the Oxford dictionary". OxfordDictionaries.com. Retrieved 2016-01-20.

Branches of Geometry

Absolute geometry is a branch of geometry that is based on an axiom system. Absolute geometry is also referred to as neutral geometry. Some of the other branches of geometry that have been discussed in this section are algebraic geometry, analytic geometry, differential geometry, projective geometry and Euclidean geometry. This chapter is a compilation of the various branches of geometry that form an integral part of the broader subject matter.

Absolute Geometry

Absolute geometry is a geometry based on an axiom system for Euclidean geometry with the parallel postulate removed and none of its alternatives used in place of it. The term was introduced by János Bolyai in 1832. It is sometimes referred to as neutral geometry, as it is neutral with respect to the parallel postulate.

Properties

It might be imagined that absolute geometry is a rather weak system, but that is not the case. Indeed, in Euclid's *Elements*, the first 28 Propositions and Proposition 31 avoid using the parallel postulate, and therefore are valid in absolute geometry. One can also prove in absolute geometry the exterior angle theorem (an exterior angle of a triangle is larger than either of the remote angles), as well as the Saccheri–Legendre theorem, which states that the sum of the measures of the angles in a triangle has at most 180°.

Proposition 31 is the construction of a parallel line to a given line through a point not on the given line. As the proof only requires the use of Proposition 27 (the Alternate Interior Angle Theorem), it is a valid construction in absolute geometry. More precisely, given any line *l* and any point *P* not on *l*, there is *at least* one line through *P* which is parallel to *l*. This can be proved using a familiar construction: given a line *l* and a point *P* not on *l*, drop the perpendicular *m* from *P* to *l*, then erect a perpendicular *n* to *m* through *P*. By the alternate interior angle theorem, *l* is parallel to *n*. (The alternate interior angle theorem states that if lines a and b are cut by a transversal t such that there is a pair of congruent alternate interior angles, then a and b are parallel.) The foregoing construction, and the alternate interior angle theorem, do not depend on the parallel postulate and are therefore valid in absolute geometry.

In absolute geometry it is also provable that two lines perpendicular to the same line cannot intersect (which makes the two lines parallel by definition of parallel lines), proving that the summit angles of an Saccheri quadrilateral cannot be obtuse , and that spherical geometry is not an absolute geometry.

Relation to other Geometries

The theorems of absolute geometry hold in hyperbolic geometry, which is a non-Euclidean geometry, as well as in Euclidean geometry.

Absolute geometry is inconsistent with elliptic geometry: in that theory, there are no parallel lines at all, but it is a theorem of absolute geometry that parallel lines do exist.

Absolute geometry is an extension of ordered geometry, and thus, all theorems in ordered geometry hold in absolute geometry. The converse is not true. Absolute geometry assumes the first four of Euclid's Axioms (or their equivalents), to be contrasted with affine geometry, which does not assume Euclid's third and fourth axioms. (3: "To describe a circle with any centre and distance radius.", 4: "That all right angles are equal to one another.") Ordered geometry is a common foundation of both absolute and affine geometry.

The geometry of special relativity has been developed starting with nine axioms and eleven propositions of absolute geometry. The authors Edwin B. Wilson and Gilbert N. Lewis then proceed beyond absolute geometry when they introduce hyperbolic rotation as the transformation relating two frames of reference.

Incompleteness

Absolute geometry is an incomplete axiomatic system, in the sense that one can add extra independent axioms without making the axiom system inconsistent. One can extend absolute geometry by adding different axioms about parallel lines and get incompatible but consistent axiom systems, giving rise to Euclidean or hyperbolic geometry. Thus every theorem of absolute geometry is a theorem of hyperbolic geometry and Euclidean geometry. However the converse is not true.

Algebraic Geometry

Algebraic geometry is a branch of mathematics, classically studying zeros of multivariate polynomials. Modern algebraic geometry is based on the use of abstract algebraic techniques, mainly from commutative algebra, for solving geometrical problems about these sets of zeros.

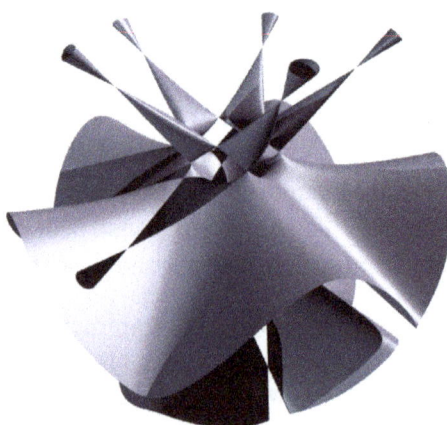

This Togliatti surface is an algebraic surface of degree five. The picture represents a portion of its real locus.

The fundamental objects of study in algebraic geometry are algebraic varieties, which are geometric manifestations of solutions of systems of polynomial equations. Examples of the most studied classes of algebraic varieties are: plane algebraic curves, which include lines, circles, parabolas, ellipses, hyperbolas, cubic curves like elliptic curves and quartic curves like lemniscates, and Cassini ovals. A point of the plane belongs to an algebraic curve if its coordinates satisfy a given polynomial equation. Basic questions involve the study of the points of special interest like the singular points, the inflection points and the points at infinity. More advanced questions involve the topology of the curve and relations between the curves given by different equations.

Algebraic geometry occupies a central place in modern mathematics and has multiple conceptual connections with such diverse fields as complex analysis, topology and number theory. Initially a study of systems of polynomial equations in several variables, the subject of algebraic geometry starts where equation solving leaves off, and it becomes even more important to understand the intrinsic properties of the totality of solutions of a system of equations, than to find a specific solution; this leads into some of the deepest areas in all of mathematics, both conceptually and in terms of technique.

In the 20th century, algebraic geometry split into several subareas.

- The mainstream of algebraic geometry is devoted to the study of the complex points of the algebraic varieties and more generally to the points with coordinates in an algebraically closed field.

- The study of the points of an algebraic variety with coordinates in the field of the rational numbers or in a number field became arithmetic geometry (or more classically Diophantine geometry), a subfield of algebraic number theory.

- The study of the real points of an algebraic variety is the subject of real algebraic geometry.

- A large part of singularity theory is devoted to the singularities of algebraic varieties.

- With the rise of the computers, a computational algebraic geometry area has emerged, which lies at the intersection of algebraic geometry and computer algebra. It consists essentially in developing algorithms and software for studying and finding the properties of explicitly given algebraic varieties.

Much of the development of the mainstream of algebraic geometry in the 20th century occurred within an abstract algebraic framework, with increasing emphasis being placed on "intrinsic" properties of algebraic varieties not dependent on any particular way of embedding the variety in an ambient coordinate space; this parallels developments in topology, differential and complex geometry. One key achievement of this abstract algebraic geometry is Grothendieck's scheme theory which allows one to use sheaf theory to study algebraic varieties in a way which is very similar to its use in the study of differential and analytic manifolds. This is obtained by extending the notion of point: In classical algebraic geometry, a point of an affine variety may be identified, through Hilbert's Nullstellensatz, with a maximal ideal of the coordinate ring, while the points of the corresponding affine scheme are all prime ideals of this ring. This means that a point of such a scheme may be either a usual point or a subvariety. This approach also enables a unification of the language and the tools of classical algebraic geometry, mainly concerned with complex points, and of algebraic number theory. Wiles's proof of the longstanding conjecture called Fermat's last theorem is an example of the power of this approach.

Basic Notions

Zeros of Simultaneous Polynomials

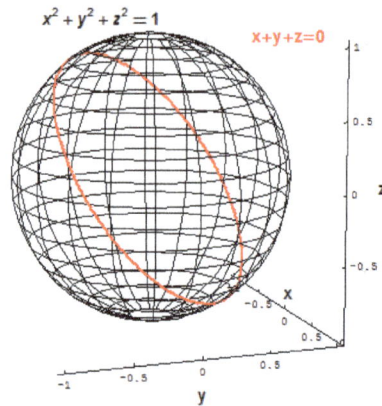

$$x^2 + y^2 + z^2 = 1$$
x+y+z=0

Sphere and slanted circle

In classical algebraic geometry, the main objects of interest are the vanishing sets of collections of polynomials, meaning the set of all points that simultaneously satisfy one or more polynomial equations. For instance, the two-dimensional sphere in three-dimensional Euclidean space R^3 could be defined as the set of all points (x,y,z) with

$$x^2 + y^2 + z^2 - 1 = 0.$$

A "slanted" circle in R^3 can be defined as the set of all points (x,y,z) which satisfy the two polynomial equations

$$x^2 + y^2 + z^2 - 1 = 0,$$

$$x + y + z = 0.$$

Affine Varieties

First we start with a field k. In classical algebraic geometry, this field was always the complex numbers C, but many of the same results are true if we assume only that k is algebraically closed. We consider the affine space of dimension n over k, denoted $A^n(k)$ (or more simply A^n, when k is clear from the context). When one fixes a coordinates system, one may identify $A^n(k)$ with k^n. The purpose of not working with k^n is to emphasize that one "forgets" the vector space structure that k^n carries.

A function $f : A^n \to A^1$ is said to be *polynomial* (or *regular*) if it can be written as a polynomial, that is, if there is a polynomial p in $k[x_1,...,x_n]$ such that $f(M) = p(t_1,...,t_n)$ for every point M with coordinates $(t_1,...,t_n)$ in A^n. The property of a function to be polynomial (or regular) does not depend on the choice of a coordinate system in A^n.

When a coordinate system is chosen, the regular functions on the affine n-space may be identified with the ring of polynomial functions in n variables over k. Therefore, the set of the regular functions on A^n is a ring, which is denoted $k[A^n]$.

We say that a polynomial *vanishes* at a point if evaluating it at that point gives zero. Let S be a set of polynomials in $k[A^n]$. The *vanishing set of S* (or *vanishing locus* or *zero set*) is the set $V(S)$ of all points in A^n where every polynomial in S vanishes. In other words,

$$V(S) = \{(t_1,\ldots,t_n) \mid \forall p \in S,\, p(t_1,\ldots,t_n) = 0\}.$$

A subset of A^n which is $V(S)$, for some S, is called an *algebraic set*. The V stands for *variety* (a specific type of algebraic set to be defined below).

Given a subset U of A^n, can one recover the set of polynomials which generate it? If U is *any* subset of A^n, define $I(U)$ to be the set of all polynomials whose vanishing set contains U. The I stands for ideal: if two polynomials f and g both vanish on U, then $f+g$ vanishes on U, and if h is any polynomial, then hf vanishes on U, so $I(U)$ is always an ideal of the polynomial ring $k[A^n]$.

Two natural questions to ask are:

- Given a subset U of A^n, when is $U = V(I(U))$?

- Given a set S of polynomials, when is $S = I(V(S))$?

The answer to the first question is provided by introducing the Zariski topology, a topology on A^n whose closed sets are the algebraic sets, and which directly reflects the algebraic structure of $k[A^n]$. Then $U = V(I(U))$ if and only if U is an algebraic set or equivalently a Zariski-closed set. The answer to the second question is given by Hilbert's Nullstellensatz. In one of its forms, it says that $I(V(S))$ is the radical of the ideal generated by S. In more abstract language, there is a Galois connection, giving rise to two closure operators; they can be identified, and naturally play a basic role in the theory; the example is elaborated at Galois connection.

For various reasons we may not always want to work with the entire ideal corresponding to an algebraic set U. Hilbert's basis theorem implies that ideals in $k[A^n]$ are always finitely generated.

An algebraic set is called *irreducible* if it cannot be written as the union of two smaller algebraic sets. Any algebraic set is a finite union of irreducible algebraic sets and this decomposition is unique. Thus its elements are called the *irreducible components* of the algebraic set. An irreducible algebraic set is also called a *variety*. It turns out that an algebraic set is a variety if and only if it may be defined as the vanishing set of a prime ideal of the polynomial ring.

Some authors do not make a clear distinction between algebraic sets and varieties and use *irreducible variety* to make the distinction when needed.

Regular Functions

Just as continuous functions are the natural maps on topological spaces and smooth functions are the natural maps on differentiable manifolds, there is a natural class of functions on an algebraic set, called *regular functions* or *polynomial functions*. A regular function on an algebraic set V contained in A^n is the restriction to V of a regular function on A^n. For an algebraic set defined on the field of the complex numbers, the regular functions are smooth and even analytic.

It may seem unnaturally restrictive to require that a regular function always extend to the ambient space, but it is very similar to the situation in a normal topological space, where the Tietze extension theorem guarantees that a continuous function on a closed subset always extends to the ambient topological space.

Just as with the regular functions on affine space, the regular functions on V form a ring, which we denote by $k[V]$. This ring is called the *coordinate ring of V*.

Since regular functions on V come from regular functions on A^n, there is a relationship between the coordinate rings. Specifically, if a regular function on V is the restriction of two functions f and g in $k[A^n]$, then $f - g$ is a polynomial function which is null on V and thus belongs to $I(V)$. Thus $k[V]$ may be identified with $k[A^n]/I(V)$.

Morphism of Affine Varieties

Using regular functions from an affine variety to A^1, we can define regular maps from one affine variety to another. First we will define a regular map from a variety into affine space: Let V be a variety contained in A^n. Choose m regular functions on V, and call them $f_1, ..., f_m$. We define a *regular map f* from V to A^m by letting $f = (f_1, ..., f_m)$. In other words, each f_i determines one coordinate of the range of f.

If V' is a variety contained in A^m, we say that f is a *regular map* from V to V' if the range of f is contained in V'.

The definition of the regular maps apply also to algebraic sets. The regular maps are also called *morphisms*, as they make the collection of all affine algebraic sets into a category, where the objects are the affine algebraic sets and the morphisms are the regular maps. The affine varieties is a subcategory of the category of the algebraic sets.

Given a regular map g from V to V' and a regular function f of $k[V']$, then $f \circ g \in k[V]$. The map $f \rightarrow f \circ g$ is a ring homomorphism from $k[V']$ to $k[V]$. Conversely, every ring homomorphism from $k[V']$ to $k[V]$ defines a regular map from V to V'. This defines an equivalence of categories between the category of algebraic sets and the opposite category of the finitely generated reduced k-algebras. This equivalence is one of the starting points of scheme theory.

Rational Function and Birational Equivalence

Contrarily to the preceding ones, this section concerns only varieties and not algebraic sets. On the other hand, the definitions extend naturally to projective varieties (next section), as an affine variety and its projective completion have the same field of functions.

If V is an affine variety, its coordinate ring is an integral domain and has thus a field of fractions which is denoted $k(V)$ and called the *field of the rational functions* on V or, shortly, the *function field* of V. Its elements are the restrictions to V of the rational functions over the affine space containing V. The domain of a rational function f is not V but the complement of the subvariety (a hypersurface) where the denominator of f vanishes.

Like for regular maps, one may define a *rational map* from a variety V to a variety V'. Like for the regular maps, the rational maps from V to V' may be identified to the field homomorphisms from $k(V')$ to $k(V)$.

Two affine varieties are *birationally equivalent* if there are two rational functions between them which are inverse one to the other in the regions where both are defined. Equivalently, they are birationally equivalent if their function fields are isomorphic.

An affine variety is a *rational variety* if it is birationally equivalent to an affine space. This means that the variety admits a rational parameterization. For example, the circle of equation $x^2 + y^2 - 1 = 0$ is a rational curve, as it has the parameterization

$$x = \frac{2t}{1+t^2}$$

$$y = \frac{1-t^2}{1+t^2},$$

which may also be viewed as a rational map from the line to the circle.

The problem of resolution of singularities is to know if every algebraic variety is birationally equivalent to a variety whose projective completion is nonsingular. It has been positively solved in characteristic 0 by Heisuke Hironaka in 1964 and is yet unsolved in finite characteristic.

Projective Variety

Just as the formulas for the roots of 2nd, 3rd and 4th degree polynomials suggest extending real numbers to the more algebraically complete setting of the complex numbers, many properties of algebraic varieties suggest extending affine space to a more geometrically complete projective space. Whereas the complex numbers are obtained by adding the number i, a root of the polynomial x^2 + 1, projective space is obtained by adding in appropriate points "at infinity", points where parallel lines may meet.

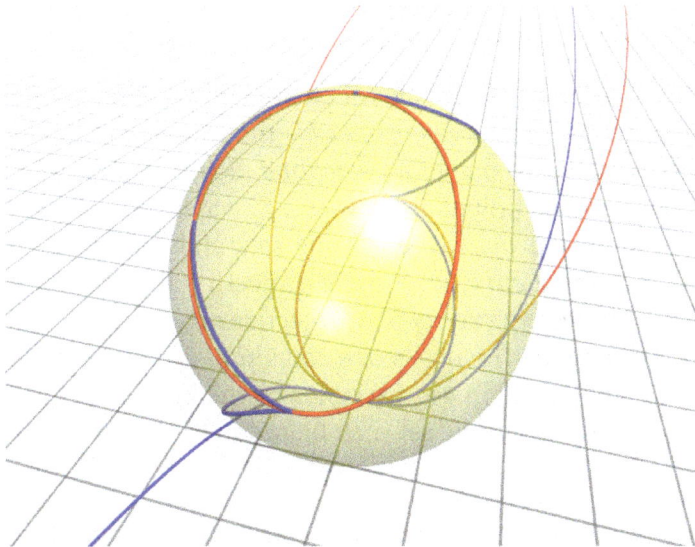

parabola ($y = x^2$, red) and cubic ($y = x^3$, blue) in projective space

To see how this might come about, consider the variety $V(y - x^2)$. If we draw it, we get a parabola. As x goes to positive infinity, the slope of the line from the origin to the point (x, x^2)

also goes to positive infinity. As x goes to negative infinity, the slope of the same line goes to negative infinity.

Compare this to the variety $V(y - x^3)$. This is a cubic curve. As x goes to positive infinity, the slope of the line from the origin to the point (x, x^3) goes to positive infinity just as before. But unlike before, as x goes to negative infinity, the slope of the same line goes to positive infinity as well; the exact opposite of the parabola. So the behavior "at infinity" of $V(y - x^3)$ is different from the behavior "at infinity" of $V(y - x^2)$.

The consideration of the *projective completion* of the two curves, which is their prolongation "at infinity" in the projective plane, allows to quantify this difference: the point at infinity of the parabola is a regular point, whose tangent is the line at infinity, while the point at infinity of the cubic curve is a cusp. Also, both curves are rational, as they are parameterized by x, and Riemann-Roch theorem implies that the cubic curve must have a singularity, which must be at infinity, as all its points in the affine space are regular.

Thus many of the properties of algebraic varieties, including birational equivalence and all the topological properties, depend on the behavior "at infinity" and so it is natural to study the varieties in projective space. Furthermore, the introduction of projective techniques made many theorems in algebraic geometry simpler and sharper: For example, Bézout's theorem on the number of intersection points between two varieties can be stated in its sharpest form only in projective space. For these reasons, projective space plays a fundamental role in algebraic geometry.

Nowadays, the *projective space* P^n of dimension n is usually defined as the set of the lines passing through a point, considered as the origin, in the affine space of dimension $n+1$, or equivalently to the set of the vector lines in a vector space of dimension $n+1$. When a coordinate system has been chosen in the space of dimension $n+1$, all the points of a line have the same set of coordinates, up to the multiplication by an element of k. This defines the homogeneous coordinates of a point of P^n as a sequence of $n+1$ elements of the base field k, defined up to the multiplication by a nonzero element of k (the same for the whole sequence).

Given a polynomial in $n+1$ variables, it vanishes at all the point of a line passing through the origin if and only if it is homogeneous. In this case, one says that the polynomial *vanishes* at the corresponding point of P^n. This allows to define a *projective algebraic set* in P^n as the set $V(f_1, ..., f_k)$ where a finite set of homogeneous polynomials $\{f_1, ..., f_k\}$ vanishes. Like for affine algebraic sets, there is a bijection between the projective algebraic sets and the reduced homogeneous ideals which define them. The *projective varieties* are the projective algebraic sets whose defining ideal is prime. In other words, a projective variety is a projective algebraic set, whose homogeneous coordinate ring is an integral domain, the *projective coordinates ring* being defined as the quotient of the graded ring or the polynomials in $n+1$ variables by the homogeneous (reduced) ideal defining the variety. Every projective algebraic set may be uniquely decomposed into a finite union of projective varieties.

The only regular functions which may be defined properly on a projective variety are the constant functions. Thus this notion is not used in projective situations. On the other hand, the *field of the rational functions* or *function field* is a useful notion, which, similarly as in the affine case, is defined as the set of the quotients of two homogeneous elements of the same degree in the homogeneous coordinate ring.

Real Algebraic Geometry

The real algebraic geometry is the study of the real points of the algebraic geometry.

The fact that the field of the reals number is an ordered field should not be ignored in such a study. For example, the curve of equation $x^2 + y^2 - a = 0$ is a circle if $a > 0$, but does not have any real point if $a < 0$. It follows that real algebraic geometry is not only the study of the real algebraic varieties, but has been generalized to the study of the *semi-algebraic sets*, which are the solutions of systems of polynomial equations and polynomial inequalities. For example, a branch of the hyperbola of equation $xy - 1 = 0$ is not an algebraic variety, but is a semi-algebraic set defined by $xy - 1 = 0$ and $x > 0$ or by $xy - 1 = 0$ and $x + y > 0$.

One of the challenging problems of real algebraic geometry is the unsolved Hilbert's sixteenth problem: Decide which respective positions are possible for the ovals of a nonsingular plane curve of degree 8.

Computational Algebraic Geometry

One may date the origin of computational algebraic geometry to meeting EUROSAM'79 (International Symposium on Symbolic and Algebraic Manipulation) held at Marseille, France in June 1979. At this meeting,

- Dennis S. Arnon showed that George E. Collins's Cylindrical algebraic decomposition (CAD) allows the computation of the topology of semi-algebraic sets,

- Bruno Buchberger presented the Gröbner bases and his algorithm to compute them,

- Daniel Lazard presented a new algorithm for solving systems of homogeneous polynomial equations with a computational complexity which is essentially polynomial in the expected number of solutions and thus simply exponential in the number of the unknowns. This algorithm is strongly related with Macaulay's multivariate resultant.

Since then, most results in this area are related to one or several of these items either by using or improving one of these algorithms, or by finding algorithms whose complexity is simply exponential in the number of the variables.

Gröbner Basis

A Gröbner basis is a system of generators of a polynomial ideal whose computation allows the deduction of many properties of the affine algebraic variety defined by the ideal.

Given an ideal I defining an algebraic set V:

- V is empty (over an algebraically closed extension of the basis field), if and only if the Gröbner basis for any monomial ordering is reduced to {1}.

- By means of the Hilbert series one may compute the dimension and the degree of V from any Gröbner basis of I for a monomial ordering refining the total degree.

- If the dimension of V is 0, one may compute the points (finite in number) of V from any Gröbner basis of I.

- A Gröbner basis computation allows to remove from V all irreducible components which are contained in a given hyper surface.

- A Gröbner basis computation allows to compute the Zariski closure of the image of V by the projection on the k first coordinates, and the subset of the image where the projection is not proper.

- More generally Gröbner basis computations allows to compute the Zariski closure of the image and the critical points of a rational function of V into another affine variety.

Gröbner basis computations do not allow to compute directly the primary decomposition of I nor the prime ideals defining the irreducible components of V, but most algorithms for this involve Gröbner basis computation. The algorithms which are not based on Gröbner bases use regular chains but may need Gröbner bases in some exceptional situations.

Gröbner base are deemed to be difficult to compute. In fact they may contain, in the worst case, polynomials whose degree is doubly exponential in the number of variables and a number of polynomials which is also doubly exponential. However, this is only a worst case complexity, and the complexity bound of Lazard's algorithm of 1979 may frequently apply. Faugère's F4 and F5 algorithms realize this complexity, as F5 algorithm may be viewed as an improvement of Lazard's 1979 algorithm. It follows that the best implementations allow to compute almost routinely with algebraic sets of degree more than 100. This means that, presently, the difficulty of computing a Gröbner basis is strongly related to the intrinsic difficulty of the problem.

Cylindrical Algebraic Decomposition (CAD)

CAD is an algorithm which had been introduced in 1973 by G. Collins to implement with an acceptable complexity the Tarski–Seidenberg theorem on quantifier elimination over the real numbers.

This theorem concerns the formulas of the first-order logic whose atomic formulas are polynomial equalities or inequalities between polynomials with real coefficients. These formulas are thus the formulas which may be constructed from the atomic formulas by the logical operators *and* ($\wedge$), *or* ($\vee$), *not* ($\neg$), *for all* ($\forall$) and *exists* ($\exists$). Tarski's theorem asserts that, from such a formula, one may compute an equivalent formula without quantifier ($\forall, \exists$).

The complexity of CAD is doubly exponential in the number of variables. This means that CAD allow, in theory, to solve every problem of real algebraic geometry which may be expressed by such a formula, that is almost every problem concerning explicitly given varieties and semi-algebraic sets.

While Gröbner basis computation has doubly exponential complexity only in rare cases, CAD has almost always this high complexity. This implies that, unless if most polynomials appearing in the input are linear, it may not solve problems with more than four variables.

Since 1973, most of the research on this subject is devoted either to improve CAD or to find alternate algorithms in special cases of general interest.

As an example of the state of art, there are efficient algorithms to find at least a point in every connected component of a semi-algebraic set, and thus to test if a semi-algebraic set is empty. On the other hand, CAD is yet, in practice, the best algorithm to count the number of connected components.

Asymptotic Complexity vs. Practical Efficiency

The basic general algorithms of computational geometry have a double exponential worst case complexity. More precisely, if d is the maximal degree of the input polynomials and n the number of variables, their complexity is at most $d^{2^{cn}}$ for some constant c, and, for some inputs, the complexity is at least $d^{2^{c'n}}$ for another constant c'.

During the last 20 years of 20th century, various algorithms have been introduced to solve specific subproblems with a better complexity. Most of these algorithms have a complexity $d^{O(n^2)}$.

Among these algorithms which solve a sub problem of the problems solved by Gröbner bases, one may cite *testing if an affine variety is empty* and *solving nonhomogeneous polynomial systems which have a finite number of solutions*. Such algorithms are rarely implemented because, on most entries Faugère's F4 and F5 algorithms have a better practical efficiency and probably a similar or better complexity (*probably* because the evaluation of the complexity of Gröbner basis algorithms on a particular class of entries is a difficult task which has been done only in a few special cases).

The main algorithms of real algebraic geometry which solve a problem solved by CAD are related to the topology of semi-algebraic sets. One may cite *counting the number of connected components, testing if two points are in the same components* or *computing a Whitney stratification of a real algebraic set*. They have a complexity of $d^{O(n^2)}$, but the constant involved by O notation is so high that using them to solve any nontrivial problem effectively solved by CAD, is impossible even if one could use all the existing computing power in the world. Therefore, these algorithms have never been implemented and this is an active research area to search for algorithms with have together a good asymptotic complexity and a good practical efficiency.

Abstract Modern Viewpoint

The modern approaches to algebraic geometry redefine and effectively extend the range of basic objects in various levels of generality to schemes, formal schemes, ind-schemes, algebraic spaces, algebraic stacks and so on. The need for this arises already from the useful ideas within theory of varieties, e.g. the formal functions of Zariski can be accommodated by introducing nilpotent elements in structure rings; considering spaces of loops and arcs, constructing quotients by group actions and developing formal grounds for natural intersection theory and deformation theory lead to some of the further extensions.

Most remarkably, in late 1950s, algebraic varieties were subsumed into Alexander Grothendieck's concept of a scheme. Their local objects are affine schemes or prime spectra which are locally ringed spaces which form a category which is antiequivalent to the category of commutative unital rings, extending the duality between the category of affine algebraic varieties over a field k, and the category of finitely generated reduced k-algebras. The gluing is along Zariski topology; one can glue within the category of locally ringed spaces, but also, using the Yoneda embedding, within the more abstract category of presheaves of sets over the category of affine schemes. The Zariski topology in the set theoretic sense is then replaced by a Grothendieck topology. Grothendieck introduced Grothendieck topologies having in mind more exotic but geometrically finer and more sensitive examples than the crude Zariski topology, namely the étale topology, and the two flat

Grothendieck topologies: fppf and fpqc; nowadays some other examples became prominent including Nisnevich topology. Sheaves can be furthermore generalized to stacks in the sense of Grothendieck, usually with some additional representability conditions leading to Artin stacks and, even finer, Deligne-Mumford stacks, both often called algebraic stacks.

Sometimes other algebraic sites replace the category of affine schemes. For example, Nikolai Durov has introduced commutative algebraic monads as a generalization of local objects in a generalized algebraic geometry. Versions of a tropical geometry, of an absolute geometry over a field of one element and an algebraic analogue of Arakelov's geometry were realized in this setup.

Another formal generalization is possible to Universal algebraic geometry in which every variety of algebras has its own algebraic geometry. The term *variety of algebras* should not be confused with *algebraic variety*.

The language of schemes, stacks and generalizations has proved to be a valuable way of dealing with geometric concepts and became cornerstones of modern algebraic geometry.

Algebraic stacks can be further generalized and for many practical questions like deformation theory and intersection theory, this is often the most natural approach. One can extend the Grothendieck site of affine schemes to a higher categorical site of derived affine schemes, by replacing the commutative rings with an infinity category of differential graded commutative algebras, or of simplicial commutative rings or a similar category with an appropriate variant of a Grothendieck topology. One can also replace presheaves of sets by presheaves of simplicial sets (or of infinity groupoids). Then, in presence of an appropriate homotopic machinery one can develop a notion of derived stack as such a presheaf on the infinity category of derived affine schemes, which is satisfying certain infinite categorical version of a sheaf axiom (and to be algebraic, inductively a sequence of representability conditions). Quillen model categories, Segal categories and quasicategories are some of the most often used tools to formalize this yielding the *derived algebraic geometry*, introduced by the school of Carlos Simpson, including Andre Hirschowitz, Bertrand Toën, Gabrielle Vezzosi, Michel Vaquié and others; and developed further by Jacob Lurie, Bertrand Toën, and Gabrielle Vezzosi. Another (noncommutative) version of derived algebraic geometry, using A-infinity categories has been developed from early 1990s by Maxim Kontsevich and followers.

History

Prehistory: Before the 16th Century

Some of the roots of algebraic geometry date back to the work of the Hellenistic Greeks from the 5th century BC. The Delian problem, for instance, was to construct a length x so that the cube of side x contained the same volume as the rectangular box a^2b for given sides a and b. Menaechmus (circa 350 BC) considered the problem geometrically by intersecting the pair of plane conics $ay = x^2$ and $xy = ab$. The later work, in the 3rd century BC, of Archimedes and Apollonius studied more systematically problems on conic sections, and also involved the use of coordinates. The Arab mathematicians were able to solve by purely algebraic means certain cubic equations, and then to interpret the results geometrically. This was done, for instance, by Ibn al-Haytham in the 10th century AD. Subsequently, Persian mathematician Omar Khayyám (born 1048 A.D.) discovered the general method of solving cubic equations by intersecting a parabola with a circle. Each of these early developments in algebraic geometry dealt with questions of finding and describing the intersections of algebraic curves.

Renaissance

Such techniques of applying geometrical constructions to algebraic problems were also adopted by a number of Renaissance mathematicians such as Gerolamo Cardano and Niccolò Fontana "Tartaglia" on their studies of the cubic equation. The geometrical approach to construction problems, rather than the algebraic one, was favored by most 16th and 17th century mathematicians, notably Blaise Pascal who argued against the use of algebraic and analytical methods in geometry. The French mathematicians Franciscus Vieta and later René Descartes and Pierre de Fermat revolutionized the conventional way of thinking about construction problems through the introduction of coordinate geometry. They were interested primarily in the properties of *algebraic curves*, such as those defined by Diophantine equations (in the case of Fermat), and the algebraic reformulation of the classical Greek works on conics and cubics (in the case of Descartes).

During the same period, Blaise Pascal and Gérard Desargues approached geometry from a different perspective, developing the synthetic notions of projective geometry. Pascal and Desargues also studied curves, but from the purely geometrical point of view: the analog of the Greek *ruler and compass construction*. Ultimately, the analytic geometry of Descartes and Fermat won out, for it supplied the 18th century mathematicians with concrete quantitative tools needed to study physical problems using the new calculus of Newton and Leibniz. However, by the end of the 18th century, most of the algebraic character of coordinate geometry was subsumed by the *calculus of infinitesimals* of Lagrange and Euler.

19th and Early 20th Century

It took the simultaneous 19th century developments of non-Euclidean geometry and Abelian integrals in order to bring the old algebraic ideas back into the geometrical fold. The first of these new developments was seized up by Edmond Laguerre and Arthur Cayley, who attempted to ascertain the generalized metric properties of projective space. Cayley introduced the idea of *homogeneous polynomial forms*, and more specifically quadratic forms, on projective space. Subsequently, Felix Klein studied projective geometry (along with other types of geometry) from the viewpoint that the geometry on a space is encoded in a certain class of transformations on the space. By the end of the 19th century, projective geometers were studying more general kinds of transformations on figures in projective space. Rather than the projective linear transformations which were normally regarded as giving the fundamental Kleinian geometry on projective space, they concerned themselves also with the higher degree birational transformations. This weaker notion of congruence would later lead members of the 20th century Italian school of algebraic geometry to classify algebraic surfaces up to birational isomorphism.

The second early 19th century development, that of Abelian integrals, would lead Bernhard Riemann to the development of Riemann surfaces.

In the same period began the algebraization of the algebraic geometry through commutative algebra. The prominent results in this direction are Hilbert's basis theorem and Hilbert's Nullstellensatz, which are the basis of the connexion between algebraic geometry and commutative algebra, and Macaulay's multivariate resultant, which is the basis of elimination theory. Probably because of the size of the computation which is implied by multivariate resultants, elimination theory was forgotten during the middle of the 20th century until it was renewed by singularity theory and computational algebraic geometry.

20th Century

B. L. van der Waerden, Oscar Zariski and André Weil developed a foundation for algebraic geometry based on contemporary commutative algebra, including valuation theory and the theory of ideals. One of the goals was to give a rigorous framework for proving the results of Italian school of algebraic geometry. In particular, this school used systematically the notion of generic point without any precise definition, which was first given by these authors during the 1930s.

In the 1950s and 1960s Jean-Pierre Serre and Alexander Grothendieck recast the foundations making use of sheaf theory. Later, from about 1960, and largely led by Grothendieck, the idea of schemes was worked out, in conjunction with a very refined apparatus of homological techniques. After a decade of rapid development the field stabilized in the 1970s, and new applications were made, both to number theory and to more classical geometric questions on algebraic varieties, singularities and moduli.

An important class of varieties, not easily understood directly from their defining equations, are the abelian varieties, which are the projective varieties whose points form an abelian group. The prototypical examples are the elliptic curves, which have a rich theory. They were instrumental in the proof of Fermat's last theorem and are also used in elliptic curve cryptography.

In parallel with the abstract trend of the algebraic geometry, which is concerned with general statements about varieties, methods for effective computation with concretely-given varieties have also been developed, which lead to the new area of computational algebraic geometry. One of the founding methods of this area is the theory of Gröbner bases, introduced by Bruno Buchberger in 1965. Another founding method, more specially devoted to real algebraic geometry, is the cylindrical algebraic decomposition, introduced by George E. Collins in 1973.

Analytic Geometry

An analytic variety is defined locally as the set of common solutions of several equations involving analytic functions. It is analogous to the included concept of real or complex algebraic variety. Any complex manifold is an analytic variety. Since analytic varieties may have singular points, not all analytic varieties are manifolds.

Modern analytic geometry is essentially equivalent to real and complex algebraic geometry, as has been shown by Jean-Pierre Serre in his paper *GAGA*, the name of which is French for *Algebraic geometry and analytic geometry*. Nevertheless, the two fields remain distinct, as the methods of proof are quite different and algebraic geometry includes also geometry in finite characteristic.

Applications

Algebraic geometry now finds applications in statistics, control theory, robotics, error-correcting codes, phylogenetics and geometric modelling. There are also connections to string theory, game theory, graph matchings, solitons and integer programming.

Birational Geometry

In mathematics, birational geometry is a field of algebraic geometry the goal of which is to determine when two algebraic varieties are isomorphic outside lower-dimensional subsets. This amounts to studying mappings that are given by rational functions rather than polynomials; the map may fail to be defined where the rational functions have poles.

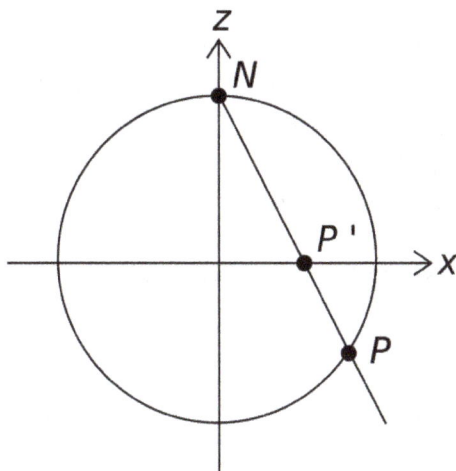

The circle is birationally equivalent to the line. One birational map between them is stereographic projection, pictured here.

Birational Maps

A rational map from one variety (understood to be irreducible) X to another variety Y, written as a dashed arrow $X \dashrightarrow Y$, is defined as a morphism from a nonempty open subset U of X to Y. By definition of the Zariski topology used in algebraic geometry, a nonempty open subset U is always the complement of a lower-dimensional subset of X. Concretely, a rational map can be written in coordinates using rational functions.

A birational map from X to Y is a rational map $f: X \dashrightarrow Y$ such that there is a rational map $Y \dashrightarrow X$ inverse to f. A birational map induces an isomorphism from a nonempty open subset of X to a nonempty open subset of Y. In this case, X and Y are said to be birational, or birationally equivalent. In algebraic terms, two varieties over a field k are birational if and only if their function fields are isomorphic as extension fields of k.

A special case is a birational morphism $f: X \rightarrow Y$, meaning a morphism which is birational. That is, f is defined everywhere, but its inverse may not be. Typically, this happens because a birational morphism contracts some subvarieties of X to points in Y.

A variety X is said to be rational if it is birational to affine space (or equivalently, to projective space) of some dimension. Rationality is a very natural property: it means that X minus some lower-dimensional subset can be identified with affine space minus some lower-dimensional subset. For example, the circle with equation $x^2 + y^2 - 1 = 0$ is a rational curve, because the formulas

$$x = \frac{2t}{1+t^2}$$

$$y = \frac{1-t^2}{1+t^2},$$

define a birational map from the affine line to the circle. (Applying this map with t a rational number gives a systematic construction of Pythagorean triples.) The inverse map sends (x,y) to $(1-y)/x$.

More generally, a smooth quadric (degree 2) hypersurface X of any dimension n is rational, by stereographic projection. (For X a quadric over a field k, X must be assumed to have a k-rational point; this is automatic if k is algebraically closed.) To define stereographic projection, let p be a point in X. Then a birational map from X to the projective space P^n of lines through p is given by sending a point q in X to the line through p and q. This is a birational equivalence but not an isomorphism of varieties, because it fails to be defined where $q = p$ (and the inverse map fails to be defined at those lines through p which are contained in X).

Minimal Models and Resolution of Singularities

Every algebraic variety is birational to a projective variety (Chow's lemma). So, for the purposes of birational classification, it is enough to work only with projective varieties, and this is usually the most convenient setting.

Much deeper is Hironaka's 1964 theorem on resolution of singularities: over a field of characteristic 0 (such as the complex numbers), every variety is birational to a smooth projective variety. Given that, it is enough to classify smooth projective varieties up to birational equivalence.

In dimension 1, if two smooth projective curves are birational, then they are isomorphic. But that fails in dimension at least 2, by the blowing up construction. By blowing up, every smooth projective variety of dimension at least 2 is birational to infinitely many "bigger" varieties, for example with bigger Betti numbers.

This leads to the idea of minimal models: is there a unique simplest variety in each birational equivalence class? The modern definition is that a projective variety X is minimal if the canonical line bundle K_X has nonnegative degree on every curve in X; in other words, K_X is nef. It is easy to check that blown-up varieties are never minimal.

This notion works perfectly for algebraic surfaces (varieties of dimension 2). In modern terms, one central result of the Italian school of algebraic geometry from 1890–1910, part of the classification of surfaces, is that every surface X is birational either to a product $\mathrm{P}^1 \times C$ for some curve C or to a minimal surface Y. The two cases are mutually exclusive, and Y is unique if it exists. When Y exists, it is called the minimal model of X.

Birational Invariants

At first, it is not clear how to show that there are any algebraic varieties which are not rational. In order to prove this, some birational invariants of algebraic varieties are needed.

One useful set of birational invariants are the plurigenera. The canonical bundle of a smooth variety X of dimension n means the line bundle of n-forms $K_X = \Omega^n$, which is the nth exterior power of the cotangent bundle of X. For an integer d, the dth tensor power of K_X is again a line bundle. For $d \geq 0$, the vector space of global sections $H^0(X, K_X^d)$ has the remarkable property that a birational map $f: X \dashrightarrow Y$ between smooth projective varieties induces an isomorphism $H^0(X, K_X^d) \cong H^0(Y, K_Y^d)$.

For $d \geq 0$, define the dth plurigenus P_d as the dimension of the vector space $H^0(X, K_X^d)$; then the plurigenera are birational invariants for smooth projective varieties. In particular, if any plurigenus P_d with $d > 0$ is not zero, then X is not rational.

A fundamental birational invariant is the Kodaira dimension, which measures the growth of the plurigenera P_d as d goes to infinity. The Kodaira dimension divides all varieties of dimension n into $n + 2$ types, with Kodaira dimension $-\infty$, 0, 1, ..., or n. This is a measure of the complexity of a variety, with projective space having Kodaira dimension $-\infty$. The most complicated varieties are those with Kodaira dimension equal to their dimension n, called varieties of general type.

More generally, for any natural summand $E(\Omega^1)$ of the rth tensor power of the cotangent bundle Ω^1 with $r \geq 0$, the vector space of global sections $H^0(X, E(\Omega^1))$ is a birational invariant for smooth projective varieties. In particular, the Hodge numbers $h^{r,0} = \dim H^0(X, \Omega^r)$ are birational invariants of X. (Most other Hodge numbers $h^{p,q}$ are not birational invariants, as shown by blowing up.)

The fundamental group $\pi_1(X)$ is a birational invariant for smooth complex projective varieties.

The "Weak factorization theorem", proved by Abramovich, Karu, Matsuki, and Włodarczyk (2002), says that any birational map between two smooth complex projective varieties can be decomposed into finitely many blow-ups or blow-downs of smooth subvarieties. This is important to know, but it can still be very hard to determine whether two smooth projective varieties are birational.

Minimal Models in Higher Dimensions

A projective variety X is called minimal if the canonical bundle K_X is nef. For X of dimension 2, it is enough to consider smooth varieties in this definition. In dimensions at least 3, minimal varieties must be allowed to have certain mild singularities, for which K_X is still well-behaved; these are called terminal singularities.

That being said, the minimal model conjecture would imply that every variety X is either covered by rational curves or birational to a minimal variety Y. When it exists, Y is called a minimal model of X.

Minimal models are not unique in dimensions at least 3, but any two minimal varieties which are birational are very close. For example, they are isomorphic outside subsets of codimension at least 2, and more precisely they are related by a sequence of flops. So the minimal model conjecture would give strong information about the birational classification of algebraic varieties.

The conjecture was proved in dimension 3 by Mori (1988). There has been great progress in higher dimensions, although the general problem remains open. In particular, Birkar, Cascini, Hacon, and McKernan (2010) proved that every variety of general type over a field of characteristic zero has a minimal model.

Uniruled Varieties

A variety is called uniruled if it is covered by rational curves. A uniruled variety does not have a minimal model, but there is a good substitute: Birkar, Cascini, Hacon, and McKernan showed that every uniruled variety over a field of characteristic zero is birational to a Fano fiber space. This leads to the problem of the birational classification of Fano fiber spaces and (as the most interesting special case) Fano varieties. By definition, a projective variety X is Fano if the anticanonical bundle K_X^* is ample. Fano varieties can be considered the algebraic varieties which are most similar to projective space.

In dimension 2, every Fano variety (known as a Del Pezzo surface) over an algebraically closed field is rational. A major discovery in the 1970s was that starting in dimension 3, there are many Fano varieties which are not rational. In particular, smooth cubic 3-folds are not rational by Clemens–Griffiths (1972), and smooth quartic 3-folds are not rational by Iskovskikh–Manin (1971). Nonetheless, the problem of determining exactly which Fano varieties are rational is far from solved. For example, it is not known whether there is any smooth cubic hypersurface in P^{n+1} with $n \geq 4$ which is not rational.

Birational Automorphism Groups

Algebraic varieties differ widely in how many birational automorphisms they have. Every variety of general type is extremely rigid, in the sense that its birational automorphism group is finite. At the other extreme, the birational automorphism group of projective space P^n over a field k, known as the Cremona group $Cr_n(k)$, is large (in a sense, infinite-dimensional) for $n \geq 2$. For $n = 2$, the complex Cremona group $Cr_2(C)$ is generated by the "quadratic transformation"

$$[x,y,z] \rightarrow [1/x, 1/y, 1/z]$$

together with the group $PGL(3,C)$ of automorphisms of P^2, by Max Noether and Castelnuovo. By contrast, the Cremona group in dimensions $n \geq 3$ is very much a mystery: no explicit set of generators is known.

Iskovskikh–Manin (1971) showed that the birational automorphism group of a smooth quartic 3-fold is equal to its automorphism group, which is finite. In this sense, quartic 3-folds are far from being rational, since the birational automorphism group of a rational variety is enormous. This phenomenon of "birational rigidity" has since been discovered in many other Fano fiber spaces.

Analytic Geometry

In classical mathematics, analytic geometry, also known as coordinate geometry, or Cartesian geometry, is the study of geometry using a coordinate system. This contrasts with synthetic geometry.

Analytic geometry is widely used in physics and engineering, and is the foundation of most modern fields of geometry, including algebraic, differential, discrete and computational geometry.

Usually the Cartesian coordinate system is applied to manipulate equations for planes, straight lines, and squares, often in two and sometimes in three dimensions. Geometrically, one studies the Euclidean plane (two dimensions) and Euclidean space (three dimensions). As taught in school books, analytic geometry can be explained more simply: it is concerned with defining and representing geometrical shapes in a numerical way and extracting numerical information from shapes' numerical definitions and representations. The numerical output, however, might also be a vector or a shape. That the algebra of the real numbers can be employed to yield results about the linear continuum of geometry relies on the Cantor–Dedekind axiom.

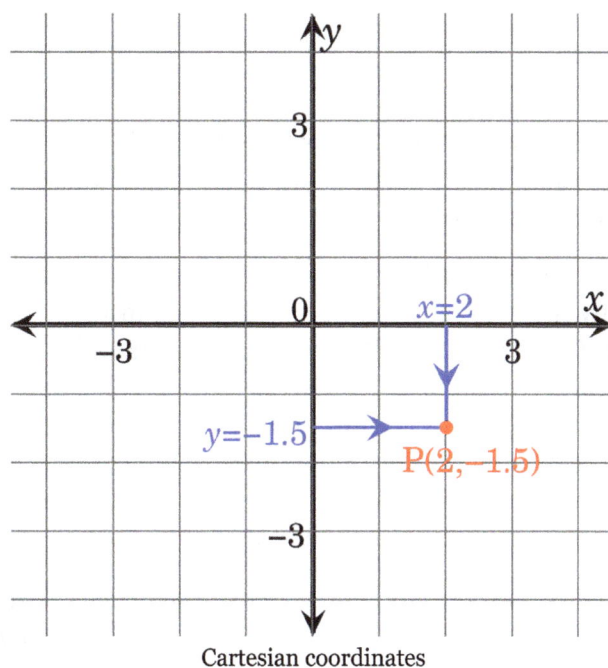

Cartesian coordinates

History

Ancient Greece

The Greek mathematician Menaechmus solved problems and proved theorems by using a method that had a strong resemblance to the use of coordinates and it has sometimes been maintained that he had introduced analytic geometry.

Apollonius of Perga, in *On Determinate Section*, dealt with problems in a manner that may be called an analytic geometry of one dimension; with the question of finding points on a line that were in a ratio to the others. Apollonius in the *Conics* further developed a method that is so similar to analytic geometry that his work is sometimes thought to have anticipated the work of Descartes by some 1800 years. His application of reference lines, a diameter and a tangent is essentially no different from our modern use of a coordinate frame, where the distances measured along the diameter from the point of tangency are the abscissas, and the segments parallel to the tangent and intercepted between the axis and the curve are the ordinates. He further developed relations between the abscissas and the corresponding ordinates that are equivalent to rhetorical equations of curves. However, although Apollonius came close to developing analytic geometry, he did not manage to do so since he did not take into account negative magnitudes and in every case the coordinate system was superimposed upon a given curve *a posteriori* instead of *a priori*. That is, equations were determined by curves, but curves were not determined by equations. Coordinates, variables, and equations were subsidiary notions applied to a specific geometric situation.

Persia

The eleventh century Persian mathematician Omar Khayyám saw a strong relationship between geometry and algebra, and was moving in the right direction when he helped to close the gap between numerical and geometric algebra with his geometric solution of the general cubic equations, but the decisive step came later with Descartes.

Western Europe

Analytic geometry was independently invented by René Descartes and Pierre de Fermat, although Descartes is sometimes given sole credit. *Cartesian geometry*, the alternative term used for analytic geometry, is named after Descartes.

Descartes made significant progress with the methods in an essay titled *La Geometrie (Geometry)*, one of the three accompanying essays (appendices) published in 1637 together with his *Discourse on the Method for Rightly Directing One's Reason and Searching for Truth in the Sciences*, commonly referred to as *Discourse on Method*. This work, written in his native French tongue, and its philosophical principles, provided a foundation for calculus in Europe. Initially the work was not well received, due, in part, to the many gaps in arguments and complicated equations. Only after the translation into Latin and the addition of commentary by van Schooten in 1649 (and further work thereafter) did Descartes's masterpiece receive due recognition.

Pierre de Fermat also pioneered the development of analytic geometry. Although not published in his lifetime, a manuscript form of *Ad locos planos et solidos isagoge* (Introduction to Plane and Solid Loci) was circulating in Paris in 1637, just prior to the publication of Descartes' *Discourse*. Clearly written and well received, the *Introduction* also laid the groundwork for analytical geometry. The key difference between Fermat's and Descartes' treatments is a matter of viewpoint: Fermat always started with an algebraic equation and then described the geometric curve which satisfied it, whereas Descartes started with geometric curves and produced their equations as one of several properties of the curves. As a consequence of this approach, Descartes had to deal with more complicated equations and he had to develop the methods to work with polynomial equations of higher degree. It was Leonard Euler who first applied the coordinate method in a systematic study of space curves and surfaces.

Coordinates

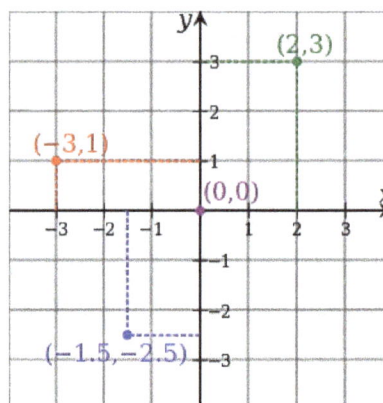

Illustration of a Cartesian coordinate plane. Four points are marked and labeled with their coordinates: (2,3) in green, (−3,1) in red, (−1.5,−2.5) in blue, and the origin (0,0) in purple.

In analytic geometry, the plane is given a coordinate system, by which every point has a pair of real number coordinates. Similarly, Euclidean space is given coordinates where every point has three coordinates. The value of the coordinates depends on the choice of the initial point of origin. There are a variety of coordinate systems used, but the most common are the following:

Cartesian Coordinates (in a Plane or Space)

The most common coordinate system to use is the Cartesian coordinate system, where each point has an x-coordinate representing its horizontal position, and a y-coordinate representing its vertical position. These are typically written as an ordered pair (x, y). This system can also be used for three-dimensional geometry, where every point in Euclidean space is represented by an ordered triple of coordinates (x, y, z).

Polar Coordinates (in a Plane)

In polar coordinates, every point of the plane is represented by its distance r from the origin and its angle θ from the polar axis.

Cylindrical Coordinates (in a Space)

In cylindrical coordinates, every point of space is represented by its height z, its radius r from the z-axis and the angle θ its projection on the xy-plane makes with respect to the horizontal axis.

Spherical Coordinates (in a Space)

In spherical coordinates, every point in space is represented by its distance ρ from the origin, the angle θ its projection on the xy-plane makes with respect to the horizontal axis, and the angle φ that it makes with respect to the z-axis. The names of the angles are often reversed in physics.

Equations and Curves

In analytic geometry, any equation involving the coordinates specifies a subset of the plane, namely the solution set for the equation, or locus. For example, the equation $y = x$ corresponds to the set of all the points on the plane whose x-coordinate and y-coordinate are equal. These points form a line, and $y = x$ is said to be the equation for this line. In general, linear equations involving x and y specify lines, quadratic equations specify conic sections, and more complicated equations describe more complicated figures.

Usually, a single equation corresponds to a curve on the plane. This is not always the case: the trivial equation $x = x$ specifies the entire plane, and the equation $x^2 + y^2 = 0$ specifies only the single point $(0, 0)$. In three dimensions, a single equation usually gives a surface, and a curve must be specified as the intersection of two surfaces, or as a system of parametric equations. The equation $x^2 + y^2 = r^2$ is the equation for any circle centered at the origin $(0, 0)$ with a radius of r.

Lines and Planes

Lines in a Cartesian plane or, more generally, in affine coordinates, can be described algebraically by *linear* equations. In two dimensions, the equation for non-vertical lines is often given in the *slope-intercept form*:

$$y = mx + b$$

where:

m is the slope or gradient of the line.

b is the y-intercept of the line.

x is the independent variable of the function $y = f(x)$.

In a manner analogous to the way lines in a two-dimensional space are described using a point-slope form for their equations, planes in a three dimensional space have a natural description using a point in the plane and a vector orthogonal to it (the normal vector) to indicate its "inclination".

Specifically, let $\mathbf{r}_0$ be the position vector of some point P_0 (x_0, y_0, z_0), and let $\mathbf{n} = (a, b, c)$ be a nonzero vector. The plane determined by this point and vector consists of those points P, with position vector $\mathbf{r}$, such that the vector drawn from P_0 to P is perpendicular to $\mathbf{n}$. Recalling that two vectors are perpendicular if and only if their dot product is zero, it follows that the desired plane can be described as the set of all points $\mathbf{r}$ such that

$$\mathbf{n} \cdot (\mathbf{r} - \mathbf{r}_0) = 0.$$

(The dot here means a dot product, not scalar multiplication.) Expanded this becomes

$$a(x - x_0) + b(y - y_0) + c(z - z_0) = 0,$$

which is the *point-normal* form of the equation of a plane. This is just a linear equation:

$$ax + by + cz + d = 0, \text{ where } d = -(ax_0 + by_0 + cz_0).$$

Conversely, it is easily shown that if a, b, c and d are constants and a, b, and c are not all zero, then the graph of the equation

$$ax + by + cz + d = 0,$$

is a plane having the vector $\mathbf{n} = (a, b, c)$ as a normal. This familiar equation for a plane is called the *general form* of the equation of the plane.

In three dimensions, lines can *not* be described by a single linear equation, so they are frequently described by parametric equations:

$$x = x_0 + at$$

$$y = y_0 + bt$$

$$z = z_0 + ct$$

where:

> x, y, and z are all functions of the independent variable t which ranges over the real numbers.

> (x_o, y_o, z_o) is any point on the line.

> a, b, and c are related to the slope of the line, such that the vector (a, b, c) is parallel to the line.

Conic Sections

In the Cartesian coordinate system, the graph of a quadratic equation in two variables is always a conic section – though it may be degenerate, and all conic sections arise in this way. The equation will be of the form

$$Ax^2 + Bxy + Cy^2 + Dx + Ey + F = 0 \text{ with } A, B, C \text{ not all zero.}$$

As scaling all six constants yields the same locus of zeros, one can consider conics as points in the five-dimensional projective space $\mathbf{P}^5$.

The conic sections described by this equation can be classified using the discriminant

$$B^2 - 4AC.$$

If the conic is non-degenerate, then:

- if $B^2 - 4AC < 0$, the equation represents an ellipse;
  - if $A = C$ and $B = 0$, the equation represents a circle, which is a special case of an ellipse;
- if $B^2 - 4AC = 0$, the equation represents a parabola;
- if $B^2 - 4AC > 0$, the equation represents a hyperbola;
  - if we also have $A + C = 0$, the equation represents a rectangular hyperbola.

Quadric Surfaces

A quadric, or quadric surface, is a 2-dimensional surface in 3-dimensional space defined as the locus of zeros of a quadratic polynomial. In coordinates x_1, x_2, x_3, the general quadric is defined by the algebraic equation

$$\sum_{i,j=1}^{3} x_i Q_{ij} x_j + \sum_{i=1}^{3} P_i x_i + R = 0.$$

Quadric surfaces include ellipsoids (including the sphere), paraboloids, hyperboloids, cylinders, cones, and planes.

Distance and Angle

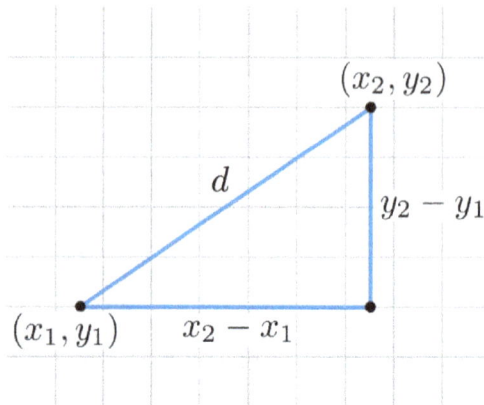

The distance formula on the plane follows from the Pythagorean theorem.

In analytic geometry, geometric notions such as distance and angle measure are defined using formulas. These definitions are designed to be consistent with the underlying Euclidean geometry. For example, using Cartesian coordinates on the plane, the distance between two points (x_1, y_1) and (x_2, y_2) is defined by the formula

$$d = \sqrt{(x_2 - x_1)^2 + (y_2 - y_1)^2},$$

which can be viewed as a version of the Pythagorean theorem. Similarly, the angle that a line makes with the horizontal can be defined by the formula

$$\theta = \arctan(m),$$

where m is the slope of the line.

In three dimensions, distance is given by the generalization of the Pythagorean theorem:

$$d = \sqrt{(x_2 - x_1)^2 + (y_2 - y_1)^2 + (z_2 - z_1)^2},$$

while the angle between two vectors is given by the dot product. The dot product of two Euclidean vectors A and B is defined by

$$\mathbf{A} \cdot \mathbf{B} = \|\mathbf{A}\| \|\mathbf{B}\| \cos\theta,$$

where θ is the angle between A and B.

Transformations

Transformations are applied to a parent function to turn it into a new function with similar characteristics.

The graph of $R(x, y)$ is changed by standard transformations as follows:

- Changing x to $x - h$ moves the graph to the right h units.

- Changing y to $y-k$ moves the graph up k units.

- Changing x to $x >$ stretches the graph horizontally by a factor of b. (think of the x as being dilated)

- Changing y to y/a stretches the graph vertically.

- Changing x to $x\cos A + y\sin A$ and changing y to $-x\sin A + y\cos A$ rotates the graph by an angle A.

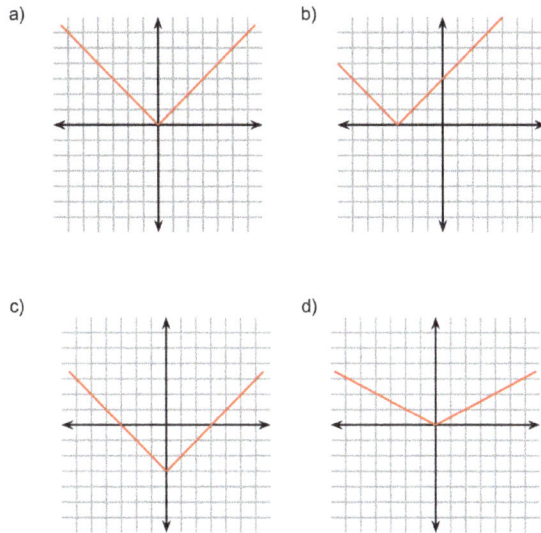

a) y = f(x) = |x| b) y = f(x+3)   c) y = f(x)-3   d) y = 1/2 f(x)

There are other standard transformation not typically studied in elementary analytic geometry because the transformations change the shape of objects in ways not usually considered. Skewing is an example of a transformation not usually considered.

For example, the parent function $y = 1/x$ has a horizontal and a vertical asymptote, and occupies the first and third quadrant, and all of its transformed forms have one horizontal and vertical asymptote, and occupies either the 1st and 3rd or 2nd and 4th quadrant. In general, if $y = f(x)$, then it can be transformed into $y = af(b(x-k))+h$. In the new transformed function, a is the factor that vertically stretches the function if it is greater than 1 or vertically compresses the function if it is less than 1, and for negative a values, the function is reflected in the x-axis. The b value compresses the graph of the function horizontally if greater than 1 and stretches the function horizontally if less than 1, and like a, reflects the function in the y-axis when it is negative. The k and h values introduce translations, h, vertical, and k horizontal. Positive h and k values mean the function is translated to the positive end of its axis and negative meaning translation towards the negative end.

Transformations can be applied to any geometric equation whether or not the equation represents a function. Transformations can be considered as individual transactions or in combinations.

Suppose that $R(x, y)$ is a relation in the xy plane. For example,

$$x^2 + y^2 - 1 = 0$$

is the relation that describes the unit circle.

Finding Intersections of Geometric Objects

For two geometric objects P and Q represented by the relations $P(x, y)$ and $Q(x, y)$ the intersection is the collection of all points (x, y) which are in both relations.

For example, P might be the circle with radius 1 and center P: $P = \{(x, y) \mid x^2 + y^2 = 1\}$ and Q might be the circle with radius 1 and center $(1, 0)$: $Q = \{(x, y) \mid (x - 1)^2 + y^2 = 1\}$. The intersection of these two circles is the collection of points which make both equations true. Does the point $(0, 0)$ make both equations true? Using $(0, 0)$ for (x, y), the equation for Q becomes $(0 - 1)^2 + 0^2 = 1$ or $(-1)^2 = 1$ which is true, so $(0, 0)$ is in the relation Q. On the other hand, still using $(0, 0)$ for (x, y) the equation for P becomes $0^2 + 0^2 = 1$ or $0 = 1$ which is false. $(0, 0)$ is not in P so it is not in the intersection.

The intersection of P and Q can be found by solving the simultaneous equations:

$$x^2 + y^2 = 1$$

$$(x - 1)^2 + y^2 = 1.$$

Traditional methods for finding intersections include substitution and elimination.

Substitution: Solve the first equation for y in terms of x and then substitute the expression for y into the second equation:

$$x^2 + y^2 = 1$$

$$y^2 = 1 - x^2.$$

We then substitute this value for y^2 into the other equation and proceed to solve for x:

$$(x - 1)^2 + (1 - x^2) = 1$$

$$x^2 - 2x + 1 + 1 - x^2 = 1$$

$$-2x = -1$$

$$x = 1/2.$$

Next, we place this value of x in either of the original equations and solve for y:

$$(1/2)^2 + y^2 = 1$$

$$y^2 = 3/4$$

$$y = \frac{\pm\sqrt{3}}{2}.$$

So our intersection has two points:

$$\left(1/2, \frac{+\sqrt{3}}{2}\right) \text{ and } \left(1/2, \frac{-\sqrt{3}}{2}\right).$$

Elimination: Add (or subtract) a multiple of one equation to the other equation so that one of the variables is eliminated. For our current example, if we subtract the first equation from the second we get $(x-1)^2 - x^2 = 0$. The y^2 in the first equation is subtracted from the y^2 in the second equation leaving no y term. The variable y has been eliminated. We then solve the remaining equation for x, in the same way as in the substitution method:

$$x^2 - 2x + 1 + 1 - x^2 = 1$$

$$-2x = -1$$

$$x = 1/2.$$

We then place this value of x in either of the original equations and solve for y:

$$(1/2)^2 + y^2 = 1$$

$$y^2 = 3/4$$

$$y = \frac{\pm\sqrt{3}}{2}.$$

So our intersection has two points:

$$\left(1/2, \frac{+\sqrt{3}}{2}\right) \text{ and } \left(1/2, \frac{-\sqrt{3}}{2}\right).$$

For conic sections, as many as 4 points might be in the intersection.

Finding Intercepts

One type of intersection which is widely studied is the intersection of a geometric object with the x and y coordinate axes.

The intersection of a geometric object and the y-axis is called the y-intercept of the object. The intersection of a geometric object and the x-axis is called the x-intercept of the object.

For the line $y = mx + b$, the parameter b specifies the point where the line crosses the y axis. Depending on the context, either b or the point $(0, b)$ is called the y-intercept.

Tangents and Normals

Tangent Lines and Planes

In geometry, the tangent line (or simply tangent) to a plane curve at a given point is the straight line that "just touches" the curve at that point. Informally, it is a line through a pair of infinitely close points on the curve. More precisely, a straight line is said to be a tangent of a curve $y = f(x)$ at a point $x = c$ on the curve if the line passes through the point $(c, f(c))$ on the curve and has slope $f'(c)$ where f is the derivative of f. A similar definition applies to space curves and curves in n-dimensional Euclidean space.

As it passes through the point where the tangent line and the curve meet, called the point of tangency, the tangent line is "going in the same direction" as the curve, and is thus the best straight-line approximation to the curve at that point.

Similarly, the tangent plane to a surface at a given point is the plane that "just touches" the surface at that point. The concept of a tangent is one of the most fundamental notions in differential geometry and has been extensively generalized.

Normal Line and Vector

In geometry, a normal is an object such as a line or vector that is perpendicular to a given object. For example, in the two-dimensional case, the normal line to a curve at a given point is the line perpendicular to the tangent line to the curve at the point.

In the three-dimensional case a surface normal, or simply normal, to a surface at a point P is a vector that is perpendicular to the tangent plane to that surface at P. The word "normal" is also used as an adjective: a line normal to a plane, the normal component of a force, the normal vector, etc. The concept of normality generalizes to orthogonality.

Differential Geometry

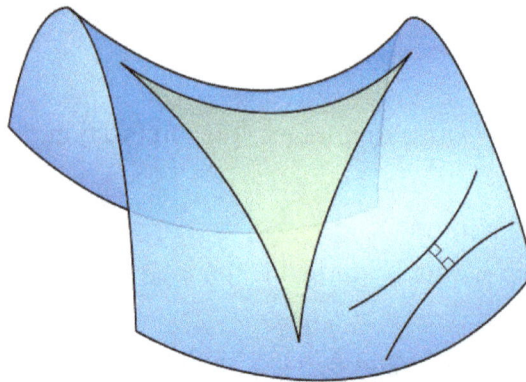

A triangle immersed in a saddle-shape plane (a hyperbolic paraboloid), as well as two diverging ultraparallel lines.

Differential geometry is a mathematical discipline that uses the techniques of differential calculus, integral calculus, linear algebra and multilinear algebra to study problems in geometry. The theory of plane and space curves and surfaces in the three-dimensional Euclidean space formed the basis for development of differential geometry during the 18th century and the 19th century.

Since the late 19th century, differential geometry has grown into a field concerned more generally with the geometric structures on differentiable manifolds. Differential geometry is closely related to differential topology and the geometric aspects of the theory of differential equations. The differential geometry of surfaces captures many of the key ideas and techniques characteristic of this field.

History of Development

Differential geometry arose and developed as a result of and in connection to the mathematical analysis of curves and surfaces. Mathematical analysis of curves and surfaces had been developed to answer some of the nagging and unanswered questions that appeared in Calculus, like the reasons for relationships between complex shapes and curves, series and analytic functions. These unanswered questions indicated greater, hidden relationships and symmetries in nature, which the standard methods of analysis could not address.

When curves, surfaces enclosed by curves, and points on curves were found to be quantitatively, and generally, related by mathematical forms the formal study of the nature of curves and surfaces became a field of study in its own right, with Monge's paper in 1795, and especially, with Gauss's publication of his article, titled 'Disquisitiones Generales Circa Superficies Curvas', in *Commentationes Societatis Regiae Scientiarum Gottingesis Recentiores* in 1827.

Initially applied to the Euclidean space, further explorations led to non-Euclidean space, and metric and topological spaces.

Branches of Differential Geometry

Riemannian Geometry

Riemannian geometry studies Riemannian manifolds, smooth manifolds with a *Riemannian metric*. This is a concept of distance expressed by means of a smooth positive definite symmetric bilinear form defined on the tangent space at each point. Riemannian geometry generalizes Euclidean geometry to spaces that are not necessarily flat, although they still resemble the Euclidean space at each point infinitesimally, i.e. in the first order of approximation. Various concepts based on length, such as the arc length of curves, area of plane regions, and volume of solids all possess natural analogues in Riemannian geometry. The notion of a directional derivative of a function from multivariable calculus is extended in Riemannian geometry to the notion of a covariant derivative of a tensor. Many concepts and techniques of analysis and differential equations have been generalized to the setting of Riemannian manifolds.

A distance-preserving diffeomorphism between Riemannian manifolds is called an isometry. This notion can also be defined *locally*, i.e. for small neighborhoods of points. Any two regular curves are locally isometric. However, the Theorema Egregium of Carl Friedrich Gauss showed that for surfaces, the existence of a local isometry imposes strong compatibility conditions on their metrics: the Gaussian curvatures at the corresponding points must be the same. In higher dimensions, the Riemann curvature tensor is an important pointwise invariant associated with a Riemannian manifold that measures how close it is to being flat. An important class of Riemannian manifolds is the Riemannian symmetric spaces, whose curvature is not necessarily constant. These are the closest analogues to the "ordinary" plane and space considered in Euclidean and non-Euclidean geometry.

Pseudo-Riemannian Geometry

Pseudo-Riemannian geometry generalizes Riemannian geometry to the case in which the metric tensor need not be positive-definite. A special case of this is a Lorentzian manifold, which is the mathematical basis of Einstein's general relativity theory of gravity.

Finsler Geometry

Finsler geometry has the *Finsler manifold* as the main object of study. This is a differential manifold with a Finsler metric, i.e. a Banach norm defined on each tangent space. Riemannian manifolds are special cases of the more general Finsler manifolds. A Finsler structure on a manifold M is a function $F : TM \to [0,\infty)$ such that:

1. $F(x, my) = |m|F(x,y)$ for all x, y in TM,

2. F is infinitely differentiable in $TM - \{0\}$,

3. The vertical Hessian of F^2 is positive definite.

Symplectic Geometry

Symplectic geometry is the study of symplectic manifolds. An almost symplectic manifold is a differentiable manifold equipped with a smoothly varying non-degenerate skew-symmetric bilinear form on each tangent space, i.e., a nondegenerate 2-form ω, called the *symplectic form*. A symplectic manifold is an almost symplectic manifold for which the symplectic form ω is closed: $d\omega = 0$.

A diffeomorphism between two symplectic manifolds which preserves the symplectic form is called a symplectomorphism. Non-degenerate skew-symmetric bilinear forms can only exist on even-dimensional vector spaces, so symplectic manifolds necessarily have even dimension. In dimension 2, a symplectic manifold is just a surface endowed with an area form and a symplectomorphism is an area-preserving diffeomorphism. The phase space of a mechanical system is a symplectic manifold and they made an implicit appearance already in the work of Joseph Louis Lagrange on analytical mechanics and later in Carl Gustav Jacobi's and William Rowan Hamilton's formulations of classical mechanics.

By contrast with Riemannian geometry, where the curvature provides a local invariant of Riemannian manifolds, Darboux's theorem states that all symplectic manifolds are locally isomorphic. The only invariants of a symplectic manifold are global in nature and topological aspects play a prominent role in symplectic geometry. The first result in symplectic topology is probably the Poincaré-Birkhoff theorem, conjectured by Henri Poincaré and then proved by G.D. Birkhoff in 1912. It claims that if an area preserving map of an annulus twists each boundary component in opposite directions, then the map has at least two fixed points.

Contact Geometry

Contact geometry deals with certain manifolds of odd dimension. It is close to symplectic geometry and like the latter, it originated in questions of classical mechanics. A *contact structure* on a (2n + 1) - dimensional manifold M is given by a smooth hyperplane field H in the tangent bundle that is as far as possible from being associated with the level sets of a differentiable function on M (the technical term is "completely nonintegrable tangent hyperplane distribution"). Near each point p, a hyperplane distribution is determined by a nowhere vanishing 1-form α, which is unique up to multiplication by a nowhere vanishing function:

$$H_p = \ker \alpha_p \subset T_p M.$$

A local 1-form on M is a *contact form* if the restriction of its exterior derivative to H is a non-degenerate two-form and thus induces a symplectic structure on H_p at each point. If the distribution H can be defined by a global one-form α then this form is contact if and only if the top-dimensional form

$$\alpha \wedge (d\alpha)^n$$

is a volume form on M, i.e. does not vanish anywhere. A contact analogue of the Darboux theorem holds: all contact structures on an odd-dimensional manifold are locally isomorphic and can be brought to a certain local normal form by a suitable choice of the coordinate system.

Complex and Kähler Geometry

Complex differential geometry is the study of complex manifolds. An almost complex manifold is a *real* manifold M, endowed with a tensor of type (1, 1), i.e. a vector bundle endomorphism (called an *almost complex structure*)

$$J : TM \to TM \text{ , such that } J^2 = -1.$$

It follows from this definition that an almost complex manifold is even-dimensional.

An almost complex manifold is called *complex* if $N_J = 0$, where N_J is a tensor of type (2, 1) related to J, called the Nijenhuis tensor (or sometimes the *torsion*). An almost complex manifold is complex if and only if it admits a holomorphic coordinate atlas. An *almost Hermitian structure* is given by an almost complex structure J, along with a Riemannian metric g, satisfying the compatibility condition

$$g(JX, JY) = g(X, Y).$$

An almost Hermitian structure defines naturally a differential two-form

$$\omega_{J,g}(X, Y) := g(JX, Y).$$

The following two conditions are equivalent:

$$N_J = 0 \text{ and } d\omega = 0$$

$$\nabla J = 0$$

where ∇ is the Levi-Civita connection of g. In this case, (J, g) is called a *Kähler structure*, and a *Kähler manifold* is a manifold endowed with a Kähler structure. In particular, a Kähler manifold is both a complex and a symplectic manifold. A large class of Kähler manifolds (the class of Hodge manifolds) is given by all the smooth complex projective varieties.

CR geometry

CR geometry is the study of the intrinsic geometry of boundaries of domains in complex manifolds.

Differential Topology

Differential topology is the study of (global) geometric invariants without a metric or symplectic form. It starts from the natural operations such as Lie derivative of natural vector bundles and de Rham differential of forms. Beside Lie algebroids, also Courant algebroids start playing a more important role.

Lie Groups

A Lie group is a group in the category of smooth manifolds. Beside the algebraic properties this enjoys also differential geometric properties. The most obvious construction is that of a Lie algebra which is the tangent space at the unit endowed with the Lie bracket between left-invariant vector fields. Beside the structure theory there is also the wide field of representation theory.

Bundles and Connections

The apparatus of vector bundles, principal bundles, and connections on bundles plays an extraordinarily important role in modern differential geometry. A smooth manifold always carries a natural vector bundle, the tangent bundle. Loosely speaking, this structure by itself is sufficient only for developing analysis on the manifold, while doing geometry requires, in addition, some way to relate the tangent spaces at different points, i.e. a notion of parallel transport. An important example is provided by affine connections. For a surface in R³, tangent planes at different points can be identified using a natural path-wise parallelism induced by the ambient Euclidean space, which has a well-known standard definition of metric and parallelism. In Riemannian geometry, the Levi-Civita connection serves a similar purpose. (The Levi-Civita connection defines path-wise parallelism in terms of a given arbitrary Riemannian metric on a manifold.) More generally, differential geometers consider spaces with a vector bundle and an arbitrary affine connection which is not defined in terms of a metric. In physics, the manifold may be the space-time continuum and the bundles and connections are related to various physical fields.

Intrinsic Versus Extrinsic

From the beginning and through the middle of the 18th century, differential geometry was studied from the *extrinsic* point of view: curves and surfaces were considered as lying in a Euclidean space of higher dimension (for example a surface in an ambient space of three dimensions). The simplest results are those in the differential geometry of curves and differential geometry of surfaces.

Starting with the work of Riemann, the *intrinsic* point of view was developed, in which one cannot speak of moving "outside" the geometric object because it is considered to be given in a free-standing way. The fundamental result here is Gauss's theorema egregium, to the effect that Gaussian curvature is an intrinsic invariant.

The intrinsic point of view is more flexible. For example, it is useful in relativity where space-time cannot naturally be taken as extrinsic (what would be "outside" of it?). However, there is a price to pay in technical complexity: the intrinsic definitions of curvature and connections become much less visually intuitive.

These two points of view can be reconciled, i.e. the extrinsic geometry can be considered as a structure additional to the intrinsic one. In the formalism of geometric calculus both extrinsic and intrinsic geometry of a manifold can be characterized by a single bivector-valued one-form called the shape operator.

Applications

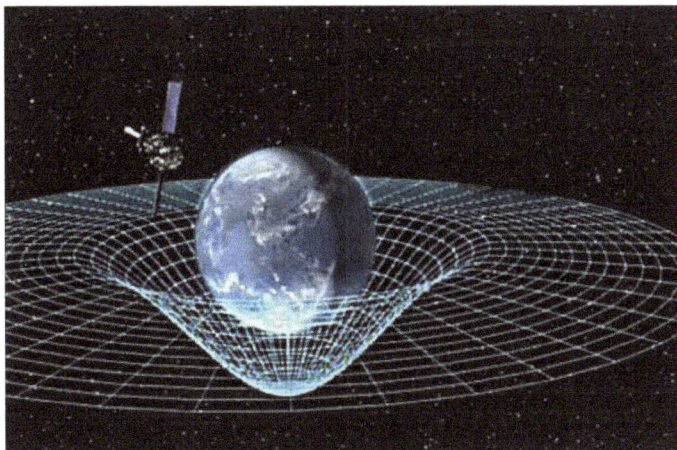

Part of a series on Spacetime

Below are some examples of how differential geometry is applied to other fields of science and mathematics.

- In physics, four uses will be mentioned:

    o Differential geometry is the language in which Einstein's general theory of relativity is expressed. According to the theory, the universe is a smooth manifold equipped with a pseudo-Riemannian metric, which describes the curvature of space-time. Understanding this curvature is essential for the positioning of satellites into orbit around the earth. Differential geometry is also indispensable in the study of gravitational lensing and black holes.

    o Differential forms are used in the study of electromagnetism.

    o Differential geometry has applications to both Lagrangian mechanics and Hamiltonian mechanics. Symplectic manifolds in particular can be used to study Hamiltonian systems.

- o Riemannian geometry and contact geometry have been used to construct the formalism of geometrothermodynamics which has found applications in classical equilibrium thermodynamics.

- In chemistry and biophysics when modelling cell membrane structure under varying pressure.

- In economics, differential geometry has applications to the field of econometrics.

- Geometric modeling (including computer graphics) and computer-aided geometric design draw on ideas from differential geometry.

- In engineering, differential geometry can be applied to solve problems in digital signal processing.

- In control theory, differential geometry can be used to analyze nonlinear controllers, particularly geometric control

- In probability, statistics, and information theory, one can interpret various structures as Riemannian manifolds, which yields the field of information geometry, particularly via the Fisher information metric.

- In structural geology, differential geometry is used to analyze and describe geologic structures.

- In computer vision, differential geometry is used to analyze shapes.

- In image processing, differential geometry is used to process and analyse data on non-flat surfaces.

- Grigori Perelman's proof of the Poincaré conjecture using the techniques of Ricci flows demonstrated the power of the differential-geometric approach to questions in topology and it highlighted the important role played by its analytic methods.

- In wireless communications, Grassmannian manifolds are used for beamforming techniques in multiple antenna systems.

Riemannian Geometry

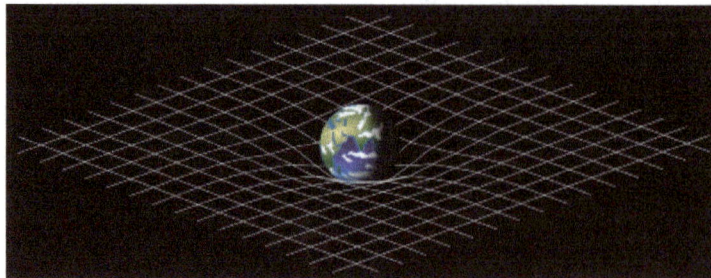

General relativity

$$G_{\mu\nu} + \Lambda g_{\mu\nu} = \frac{8\pi G}{c^4} T_{\mu\nu}$$

Riemannian geometry is the branch of differential geometry that studies Riemannian manifolds, smooth manifolds with a *Riemannian metric*, i.e. with an inner product on the tangent space at each point that varies smoothly from point to point. This gives, in particular, local notions of angle, length of curves, surface area, and volume. From those some other global quantities can be derived by integrating local contributions.

Riemannian geometry originated with the vision of Bernhard Riemann expressed in his inaugural lecture *Ueber die Hypothesen, welche der Geometrie zu Grunde liegen* (*On the Hypotheses on which Geometry is based*). It is a very broad and abstract generalization of the differential geometry of surfaces in R³. Development of Riemannian geometry resulted in synthesis of diverse results concerning the geometry of surfaces and the behavior of geodesics on them, with techniques that can be applied to the study of differentiable manifolds of higher dimensions. It enabled Einstein's general relativity theory, made profound impact on group theory and representation theory, as well as analysis, and spurred the development of algebraic and differential topology.

Introduction

Riemannian geometry was first put forward in generality by Bernhard Riemann in the 19th century. It deals with a broad range of geometries whose metric properties vary from point to point, including the standard types of Non-Euclidean geometry.

Bernhard Riemann

Any smooth manifold admits a Riemannian metric, which often helps to solve problems of differential topology. It also serves as an entry level for the more complicated structure of pseudo-Riemannian manifolds, which (in four dimensions) are the main objects of the theory of general relativity. Other generalizations of Riemannian geometry include Finsler geometry.

There exists a close analogy of differential geometry with the mathematical structure of defects in regular crystals. Dislocations and Disclinations produce torsions and curvature.

The following articles provide some useful introductory material:

- Metric tensor

- Riemannian manifold

- Levi-Civita connection

- Curvature

- Curvature tensor

- List of differential geometry topics

- Glossary of Riemannian and metric geometry

Classical Theorems in Riemannian Geometry

What follows is an incomplete list of the most classical theorems in Riemannian geometry. The choice is made depending on its importance, beauty, and simplicity of formulation. Most of the results can be found in the classic monograph by Jeff Cheeger and D. Ebin.

The formulations given are far from being very exact or the most general. This list is oriented to those who already know the basic definitions and want to know what these definitions are about.

General Theorems

1. Gauss–Bonnet theorem The integral of the Gauss curvature on a compact 2-dimensional Riemannian manifold is equal to $2\pi\chi(M)$ where $\chi(M)$ denotes the Euler characteristic of M. This theorem has a generalization to any compact even-dimensional Riemannian manifold.

2. Nash embedding theorems also called fundamental theorems of Riemannian geometry. They state that every Riemannian manifold can be isometrically embedded in a Euclidean space R^n.

Geometry in Large

In all of the following theorems we assume some local behavior of the space (usually formulated using curvature assumption) to derive some information about the global structure of the space, including either some information on the topological type of the manifold or on the behavior of points at "sufficiently large" distances.

Pinched Sectional Curvature

1. Sphere theorem. If M is a simply connected compact n-dimensional Riemannian manifold with sectional curvature strictly pinched between 1/4 and 1 then M is diffeomorphic to a sphere.

2. Cheeger's finiteness theorem. Given constants C, D and V, there are only finitely many (up to diffeomorphism) compact n-dimensional Riemannian manifolds with sectional curvature $|K| \leq C$, diameter $\leq D$ and volume $\geq V$.

3. Gromov's almost flat manifolds. There is an $\varepsilon_n > 0$ such that if an n-dimensional Riemannian manifold has a metric with sectional curvature $|K| \leq \varepsilon_n$ and diameter ≤ 1 then its finite cover is diffeomorphic to a nil manifold.

Sectional Curvature Bounded Below

1. Cheeger-Gromoll's Soul theorem. If M is a non-compact complete non-negatively curved n-dimensional Riemannian manifold, then M contains a compact, totally geodesic submanifold S such that M is diffeomorphic to the normal bundle of S (S is called the soul of M.) In particular, if M has strictly positive curvature everywhere, then it is diffeomorphic to R^n. G. Perelman in 1994 gave an astonishingly elegant/short proof of the Soul Conjecture: M is diffeomorphic to R^n if it has positive curvature at only one point.

2. Gromov's Betti number theorem. There is a constant $C = C(n)$ such that if M is a compact connected n-dimensional Riemannian manifold with positive sectional curvature then the sum of its Betti numbers is at most C.

3. Grove–Petersen's finiteness theorem. Given constants C, D and V, there are only finitely many homotopy types of compact n-dimensional Riemannian manifolds with sectional curvature $K \geq C$, diameter $\leq D$ and volume $\geq V$.

Sectional Curvature Bounded Above

1. The Cartan–Hadamard theorem states that a complete simply connected Riemannian manifold M with nonpositive sectional curvature is diffeomorphic to the Euclidean space R^n with $n = \dim M$ via the exponential map at any point. It implies that any two points of a simply connected complete Riemannian manifold with nonpositive sectional curvature are joined by a unique geodesic.

2. The geodesic flow of any compact Riemannian manifold with negative sectional curvature is ergodic.

3. If M is a complete Riemannian manifold with sectional curvature bounded above by a strictly negative constant k then it is a CAT(k) space. Consequently, its fundamental group $\Gamma = \pi_1(M)$ is Gromov hyperbolic. This has many implications for the structure of the fundamental group:

   - it is finitely presented;

   - the word problem for Γ has a positive solution;

   - the group Γ has finite virtual cohomological dimension;

   - it contains only finitely many conjugacy classes of elements of finite order;

   - the abelian subgroups of Γ are virtually cyclic, so that it does not contain a subgroup isomorphic to $Z \times Z$.

Ricci Curvature Bounded Below

1. Myers theorem. If a compact Riemannian manifold has positive Ricci curvature then its fundamental group is finite.

2. Splitting theorem. If a complete n-dimensional Riemannian manifold has nonnegative Ricci curvature and a straight line (i.e. a geodesic that minimizes distance on each interval)

then it is isometric to a direct product of the real line and a complete (n-1)-dimensional Riemannian manifold that has nonnegative Ricci curvature.

3. Bishop–Gromov inequality. The volume of a metric ball of radius r in a complete n-dimensional Riemannian manifold with positive Ricci curvature has volume at most that of the volume of a ball of the same radius r in Euclidean space.

4. Gromov's compactness theorem. The set of all Riemannian manifolds with positive Ricci curvature and diameter at most D is pre-compact in the Gromov-Hausdorff metric.

Negative Ricci curvature

1. The isometry group of a compact Riemannian manifold with negative Ricci curvature is discrete.

2. Any smooth manifold of dimension $n \geq 3$ admits a Riemannian metric with negative Ricci curvature. (*This is not true for surfaces.*)

Positive scalar curvature

1. The n-dimensional torus does not admit a metric with positive scalar curvature.

2. If the injectivity radius of a compact n-dimensional Riemannian manifold is $\geq \pi$ then the average scalar curvature is at most $n(n$-1$)$.

Projective Geometry

Projective geometry is a topic of mathematics. It is the study of geometric properties that are invariant with respect to projective transformations. This means that, compared to elementary geometry, projective geometry has a different setting, projective space, and a selective set of basic geometric concepts. The basic intuitions are that projective space has *more* points than Euclidean space, for a given dimension, and that geometric transformations are permitted that transform the extra points (called "points at infinity") to Euclidean points, and vice versa.

Properties meaningful for projective geometry are respected by this new idea of transformation, which is more radical in its effects than expressible by a transformation matrix and translations (the affine transformations). The first issue for geometers is what kind of geometry is adequate for a novel situation. It is not possible to refer to angles in projective geometry as it is in Euclidean geometry, because angle is an example of a concept not invariant with respect to projective transformations, as is seen in perspective drawing. One source for projective geometry was indeed the theory of perspective. Another difference from elementary geometry is the way in which parallel lines can be said to meet in a point at infinity, once the concept is translated into projective geometry's terms. Again this notion has an intuitive basis, such as railway tracks meeting at the horizon in a perspective drawing.

While the ideas were available earlier, projective geometry was mainly a development of the 19th century. This included the theory of complex projective space, the coordinates used (homogeneous coordinates) being complex numbers. Several major types of more abstract mathematics (including invariant theory, the Italian school of algebraic geometry, and Felix Klein's Erlangen programme resulting in the study of the classical groups) were based on projective geometry. It was also a subject with a large number of practitioners for its own sake, as synthetic geometry. Another topic that developed from axiomatic studies of projective geometry is finite geometry.

The topic of projective geometry is itself now divided into many research subtopics, two examples of which are projective algebraic geometry (the study of projective varieties) and projective differential geometry (the study of differential invariants of the projective transformations).

Overview

Projective geometry is an elementary non-metrical form of geometry, meaning that it is not based on a concept of distance. In two dimensions it begins with the study of configurations of points and lines. That there is indeed some geometric interest in this sparse setting was seen as projective geometry was developed by Desargues and others in their exploration of the principles of perspective art. In higher dimensional spaces there are considered hyperplanes (that always meet), and other linear subspaces, which exhibit the principle of duality. The simplest illustration of duality is in the projective plane, where the statements "two distinct points determine a unique line" (i.e. the line through them) and "two distinct lines determine a unique point" (i.e. their point of intersection) show the same structure as propositions. Projective geometry can also be seen as a geometry of constructions with a straight-edge alone. Since projective geometry excludes compass constructions, there are no circles, no angles, no measurements, no parallels, and no concept of intermediacy. It was realised that the theorems that do apply to projective geometry are simpler statements. For example, the different conic sections are all equivalent in (complex) projective geometry, and some theorems about circles can be considered as special cases of these general theorems.

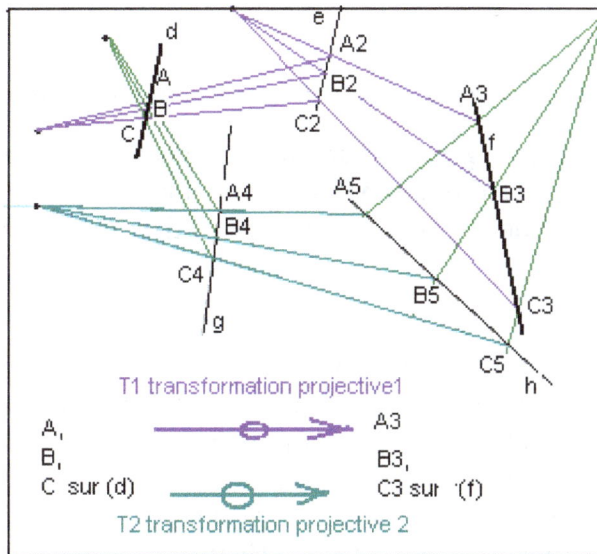

The Fundamental Theory of Projective Geometry

During the early 19th century the work of Jean-Victor Poncelet, Lazare Carnot and others established projective geometry as an independent field of mathematics . Its rigorous foundations were addressed by Karl von Staudt and perfected by Italians Giuseppe Peano, Mario Pieri, Alessandro Padoa and Gino Fano during the late 19th century. Projective geometry, like affine and Euclidean geometry, can also be developed from the Erlangen program of Felix Klein; projective geometry is characterized by invariants under transformations of the projective group.

After much work on the very large number of theorems in the subject, therefore, the basics of projective geometry became understood. The incidence structure and the cross-ratio are fundamental invariants under projective transformations. Projective geometry can be modeled by the affine plane (or affine space) plus a line (hyperplane) "at infinity" and then treating that line (or hyperplane) as "ordinary". An algebraic model for doing projective geometry in the style of analytic geometry is given by homogeneous coordinates. On the other hand, axiomatic studies revealed the existence of non-Desarguesian planes, examples to show that the axioms of incidence can be modelled (in two dimensions only) by structures not accessible to reasoning through homogeneous coordinate systems.

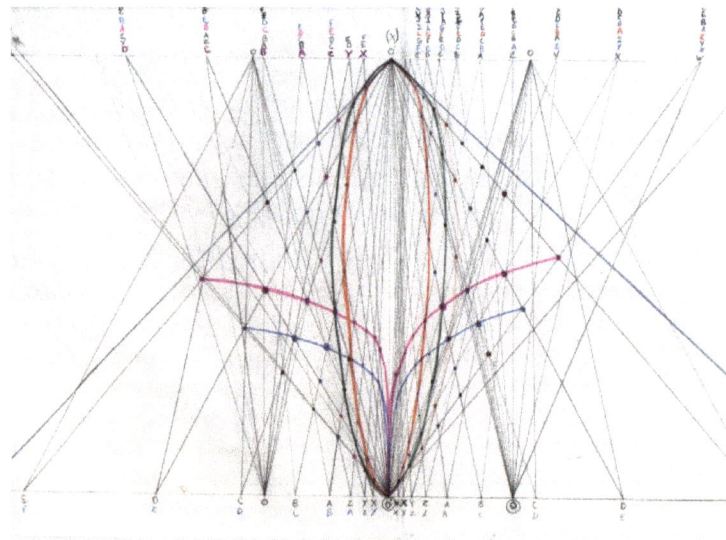

Growth measure and the polar vortices. Based on the work of Lawrence Edwards

In a foundational sense, projective geometry and ordered geometry are elementary since they involve a minimum of axioms and either can be used as the foundation for affine and Euclidean geometry. Projective geometry is not "ordered" and so it is a distinct foundation for geometry.

History

The first geometrical properties of a projective nature were discovered during the 3rd century by Pappus of Alexandria. Filippo Brunelleschi (1404–1472) started investigating the geometry of perspective during 1425. Johannes Kepler (1571–1630) and Gérard Desargues (1591–1661) independently developed the concept of the "point at infinity". Desargues developed an alternative way of constructing perspective drawings by generalizing the use of vanishing points to include the case when these are infinitely far away. He made Euclidean geometry, where parallel lines are truly parallel, into a special case of an all-encompassing geometric system. Desargues's study on conic sections drew the attention of 16-year-old Blaise Pascal and helped him formulate Pascal's theorem. The works of Gaspard Monge at the end of 18th and beginning of 19th century were important for the subsequent development of projective geometry. The work of Desargues was ignored until Michel Chasles chanced upon a handwritten copy during 1845. Meanwhile, Jean-Victor Poncelet had published the foundational treatise on projective geometry during 1822. Poncelet separated the projective properties of objects in individual class and establishing a relationship between metric and projective properties. The non-Euclidean geometries discovered soon thereafter were eventually demonstrated to have models, such as the Klein model of hyperbolic space, relating to projective geometry.

This early 19th century projective geometry was intermediate from analytic geometry to algebraic geometry. When treated in terms of homogeneous coordinates, projective geometry seems like an extension or technical improvement of the use of coordinates to reduce geometric problems to algebra, an extension reducing the number of special cases. The detailed study of quadrics and the "line geometry" of Julius Plücker still form a rich set of examples for geometers working with more general concepts.

The work of Poncelet, Jakob Steiner and others was not intended to extend analytic geometry. Techniques were supposed to be *synthetic*: in effect projective space as now understood was to be introduced axiomatically. As a result, reformulating early work in projective geometry so that it satisfies current standards of rigor can be somewhat difficult. Even in the case of the projective plane alone, the axiomatic approach can result in models not describable via linear algebra.

This period in geometry was overtaken by research on the general algebraic curve by Clebsch, Riemann, Max Noether and others, which stretched existing techniques, and then by invariant theory. Towards the end of the century, the Italian school of algebraic geometry (Enriques, Segre, Severi) broke out of the traditional subject matter into an area demanding deeper techniques.

During the later part of the 19th century, the detailed study of projective geometry became less fashionable, although the literature is voluminous. Some important work was done in enumerative geometry in particular, by Schubert, that is now considered as anticipating the theory of Chern classes, taken as representing the algebraic topology of Grassmannians.

Paul Dirac studied projective geometry and used it as a basis for developing his concepts of Quantum Mechanics, although his published results were always in algebraic form. See a blog article referring to an article and a book on this subject, also to a talk Dirac gave to a general audience during 1972 in Boston about projective geometry, without specifics as to its application in his physics.

Description

Projective geometry is less restrictive than either Euclidean geometry or affine geometry. It is an intrinsically non-metrical geometry, whose facts are independent of any metric structure. Under the projective transformations, the incidence structure and the relation of projective harmonic conjugates are preserved. A projective range is the one-dimensional foundation. Projective geometry formalizes one of the central principles of perspective art: that parallel lines meet at infinity, and therefore are drawn that way. In essence, a projective geometry may be thought of as an extension of Euclidean geometry in which the "direction" of each line is subsumed within the line as an extra "point", and in which a "horizon" of directions corresponding to coplanar lines is regarded as a "line". Thus, two parallel lines meet on a horizon line in virtue of their possessing the same direction.

Idealized directions are referred to as points at infinity, while idealized horizons are referred to as lines at infinity. In turn, all these lines lie in the plane at infinity. However, infinity is a metric concept, so a purely projective geometry does not single out any points, lines or plane in this regard—those at infinity are treated just like any others.

Because a Euclidean geometry is contained within a projective geometry, with projective geometry having a simpler foundation, general results in Euclidean geometry may be derived in a more transparent manner, where separate but similar theorems of Euclidean geometry may be handled collectively within the framework of projective geometry. For example, parallel and nonparallel lines need not be treated as separate cases – we single out some arbitrary projective plane as the ideal plane and locate it "at infinity" using homogeneous coordinates.

Additional properties of fundamental importance include Desargues' Theorem and the Theorem of Pappus. In projective spaces of dimension 3 or greater there is a construction that allows one to prove Desargues' Theorem. But for dimension 2, it must be separately postulated.

Using Desargues' Theorem, combined with the other axioms, it is possible to define the basic operations of arithmetic, geometrically. The resulting operations satisfy the axioms of a field — except that the commutativity of multiplication requires Pappus's hexagon theorem. As a result, the points of each line are in one-to-one correspondence with a given field, F, supplemented by an additional element, ∞, such that $r \cdot \infty = \infty$, $-\infty = \infty$, $r + \infty = \infty$, $r / 0 = \infty$, $r / \infty = 0$, $\infty - r = r - \infty = \infty$. However, $0 / 0$, ∞ / ∞, $\infty + \infty$, $\infty - \infty$, $0 \cdot \infty$ and $\infty \cdot 0$ remain undefined.

Projective geometry also includes a full theory of conic sections, a subject already very well developed in Euclidean geometry. There are advantages in being able to think of a hyperbola and an ellipse as distinguished only by the way the hyperbola *lies across the line at infinity*; and that a parabola is distinguished only by being tangent to the same line. The whole family of circles can be considered as *conics passing through two given points on the line at infinity* — at the cost of requiring complex coordinates. Since coordinates are not "synthetic", one replaces them by fixing a line and two points on it, and considering the *linear system* of all conics passing through those points as the basic object of study. This method proved very attractive to talented geometers, and the topic was studied thoroughly. An example of this method is the multi-volume treatise by H. F. Baker.

There are many projective geometries, which may be divided into discrete and continuous: a *discrete* geometry comprises a set of points, which may or may not be *finite* in number, while a *continuous* geometry has infinitely many points with no gaps in between.

The only projective geometry of dimension 0 is a single point. A projective geometry of dimension 1 consists of a single line containing at least 3 points. The geometric construction of arithmetic operations cannot be performed in either of these cases. For dimension 2, there is a rich structure in virtue of the absence of Desargues' Theorem.

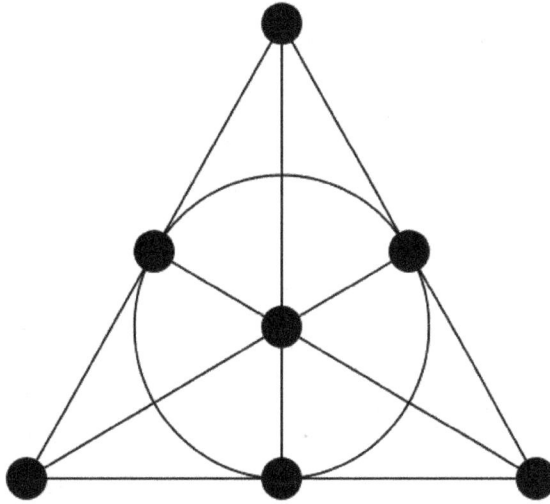

The Fano plane is the projective plane with the fewest points and lines.

According to Greenberg (1999) and others, the simplest 2-dimensional projective geometry is the Fano plane, which has 3 points on every line, with 7 points and 7 lines in all, having the following collinearities:

- [ABC]

- [ADE]

- [AFG]

- [BDG]

- [BEF]

- [CDF]

- [CEG]

with homogeneous coordinates A = (0,0,1), B = (0,1,1), C = (0,1,0), D = (1,0,1), E = (1,0,0), F = (1,1,1), G = (1,1,0), or, in affine coordinates, A = (0,0), B = (0,1), C = (∞), D = (1,0), E = (0), F = (1,1)and G = (1). The affine coordinates in a Desarguesian plane for the points designated to be the points at infinity (in this example: C, E and G) can be defined in several other ways.

In standard notation, a finite projective geometry is written PG(a, b) where:

   a is the projective (or geometric) dimension, and

   b is one less than the number of points on a line (called the *order* of the geometry).

Thus, the example having only 7 points is written PG(2, 2).

The term "projective geometry" is used sometimes to indicate the generalised underlying abstract geometry, and sometimes to indicate a particular geometry of wide interest, such as the metric geometry of flat space which we analyse through the use of homogeneous coordinates, and in which Euclidean geometry may be embedded (hence its name, Extended Euclidean plane).

The fundamental property that singles out all projective geometries is the *elliptic* incidence property that any two distinct lines *L* and *M* in the projective plane intersect at exactly one point *P*. The special case in analytic geometry of *parallel* lines is subsumed in the smoother form of a line *at infinity* on which *P* lies. The *line at infinity* is thus a line like any other in the theory: it is in no way special or distinguished. (In the later spirit of the Erlangen programme one could point to the way the group of transformations can move any line to the *line at infinity*).

The parallel properties of elliptic, Euclidean and hyperbolic geometries contrast as follows:

Given a line *l* and a point *P* not on the line,		
Elliptic	:	there exists no line through *P* that does not meet *l*
Euclidean	:	there exists exactly one line through *P* that does not meet *l*
Hyperbolic	:	there exists more than one line through *P* that does not meet *l*

The parallel property of elliptic geometry is the key idea that leads to the principle of projective duality, possibly the most important property that all projective geometries have in common.

Duality

In 1825, Joseph Gergonne noted the principle of duality characterizing projective plane geometry: given any theorem or definition of that geometry, substituting *point* for *line*, *lie on* for *pass through*, *collinear* for *concurrent*, *intersection* for *join*, or vice versa, results in another theorem or valid definition, the "dual" of the first. Similarly in 3 dimensions, the duality relation holds between points and planes, allowing any theorem to be transformed by swapping *point* and *plane, is contained by* and *contains*. More generally, for projective spaces of dimension N, there is a duality between the subspaces of dimension R and dimension N–R–1. For N = 2, this specializes to the most commonly known form of duality—that between points and lines. The duality principle was also discovered independently by Jean-Victor Poncelet.

To establish duality only requires establishing theorems which are the dual versions of the axioms for the dimension in question. Thus, for 3-dimensional spaces, one needs to show that (1*) every point lies in 3 distinct planes, (2*) every two planes intersect in a unique line and a dual version of (3*) to the effect: if the intersection of plane P and Q is coplanar with the intersection of plane R and S, then so are the respective intersections of planes P and R, Q and S (assuming planes P and S are distinct from Q and R).

In practice, the principle of duality allows us to set up a *dual correspondence* between two geometric constructions. The most famous of these is the polarity or reciprocity of two figures in a conic curve (in 2 dimensions) or a quadric surface (in 3 dimensions). A commonplace example is found in the reciprocation of a symmetrical polyhedron in a concentric sphere to obtain the dual polyhedron.

Another example is Brianchon's theorem, the dual of the already mentioned Pascal's theorem, and one of whose proofs simply consists of applying the principle of duality to Pascal's. Here are comparative statements of these two theorems (in both cases within the framework of the projective plane):

- Pascal: If all six vertices of a hexagon lie on a conic, then the intersections of its opposite sides *(regarded as full lines, since in the projective plane there is no such thing as a "line segment")* are three collinear points. The line joining them is then called the Pascal line of the hexagon.

- Brianchon: If all six sides of a hexagon are tangent to a conic, then its diagonals (i.e. the lines joining opposite vertices) are three concurrent lines. Their point of intersection is then called the Brianchon point of the hexagon.

  (If the conic degenerates into two straight lines, Pascal's becomes Pappus's theorem, which has no interesting dual, since the Brianchon point trivially becomes the two lines' intersection point.)

Axioms of Projective Geometry

Any given geometry may be deduced from an appropriate set of axioms. Projective geometries are characterised by the "elliptic parallel" axiom, that *any two planes always meet in just one line*, or in the plane, *any two lines always meet in just one point*. In other words, there are no such things as parallel lines or planes in projective geometry. Many alternative sets of axioms for projective geometry have been proposed (Hilbert & Cohn-Vossen 1999, Greenberg 1980).

Whitehead's Axioms

These axioms are based on Whitehead, "The Axioms of Projective Geometry". There are two types, points and lines, and one "incidence" relation between points and lines. The three axioms are:

- G1: Every line contains at least 3 points

- G2: Every two points, A and B, lie on a unique line, AB.

- G3: If lines AB and CD intersect, then so do lines AC and BD (where it is assumed that A and D are distinct from B and C).

The reason each line is assumed to contain at least 3 points is to eliminate some degenerate cases. The spaces satisfying these three axioms either have at most one line, or are projective spaces of some dimension over a division ring, or are non-Desarguesian planes.

One can add further axioms restricting the dimension or the coordinate ring. For example, Coxeter's *Projective Geometry*, references Veblen in the three axioms above, together with a further 5 axioms that make the dimension 3 and the coordinate ring a commutative field of characteristic not 2.

Axioms Using a Ternary Relation

One can pursue axiomatization by postulating a ternary relation, [ABC] to denote when three points (not all necessarily distinct) are collinear. An axiomatization may be written down in terms of this relation as well:

- C0: [ABA]

- C1: If A and B are two points such that [ABC] and [ABD] then [BDC]

- C2: If A and B are two points then there is a third point C such that [ABC]

- C3: If A and C are two points, B and D also, with [BCE], [ADE] but not [ABE] then there is a point F such that [ACF] and [BDF].

For two different points, A and B, the line AB is defined as consisting of all points C for which [ABC]. The axioms C0 and C1 then provide a formalization of G2; C2 for G1 and C3 for G3.

The concept of line generalizes to planes and higher-dimensional subspaces. A subspace, AB...XY may thus be recursively defined in terms of the subspace AB...X as that containing all the points of all lines YZ, as Z ranges over AB...X. Collinearity then generalizes to the relation of "independence". A set {A, B, ..., Z} of points is independent, [AB...Z] if {A, B, ..., Z} is a minimal generating subset for the subspace AB...Z.

The projective axioms may be supplemented by further axioms postulating limits on the dimension of the space. The minimum dimension is determined by the existence of an independent set of the required size. For the lowest dimensions, the relevant conditions may be stated in equivalent form as follows. A projective space is of:

- (L1) at least dimension 0 if it has at least 1 point,

- (L2) at least dimension 1 if it has at least 2 distinct points (and therefore a line),

- (L3) at least dimension 2 if it has at least 3 non-collinear points (or two lines, or a line and a point not on the line),

- (L4) at least dimension 3 if it has at least 4 non-coplanar points.

The maximum dimension may also be determined in a similar fashion. For the lowest dimensions, they take on the following forms. A projective space is of:

- (M1) at most dimension 0 if it has no more than 1 point,

- (M2) at most dimension 1 if it has no more than 1 line,

- (M3) at most dimension 2 if it has no more than 1 plane,

and so on. It is a general theorem (a consequence of axiom (3)) that all coplanar lines intersect—the very principle Projective Geometry was originally intended to embody. Therefore, property (M3) may be equivalently stated that all lines intersect one another.

It is generally assumed that projective spaces are of at least dimension 2. In some cases, if the focus is on projective planes, a variant of M3 may be postulated. The axioms of (Eves 1997: 111), for instance, include (1), (2), (L3) and (M3). Axiom (3) becomes vacuously true under (M3) and is therefore not needed in this context.

Axioms for Projective Planes

In incidence geometry, most authors give a treatment that embraces the Fano plane PG(2, 2) as the minimal finite projective plane. An axiom system that achieves this is as follows:

- (P1) Any two distinct points lie on a unique line.

- (P2) Any two distinct lines meet in a unique point.

- (P3) There exist at least four points of which no three are collinear.

Coxeter's *Introduction to Geometry* gives a list of five axioms for a more restrictive concept of a projective plane attributed to Bachmann, adding Pappus's theorem to the list of axioms above (which eliminates non-Desarguesian planes) and excluding projective planes over fields of characteristic 2 (those that don't satisfy Fano's axiom). The restricted planes given in this manner more closely resemble the real projective plane.

Discrete Geometry

Discrete geometry and combinatorial geometry are branches of geometry that study combinatorial properties and constructive methods of discrete geometric objects. Most questions in discrete geometry involve finite or discrete sets of basic geometric objects, such as points, lines, planes, circles, spheres, polygons, and so forth. The subject focuses on the combinatorial properties of these objects, such as how they intersect one another, or how they may be arranged to cover a larger object.

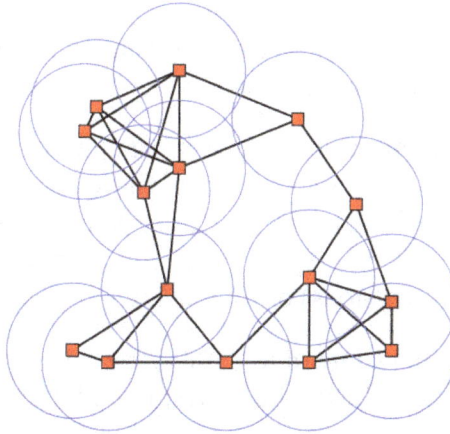

A collection of circles and the corresponding unit disk graph

Discrete geometry has large overlap with convex geometry and computational geometry, and is closely related to subjects such as finite geometry, combinatorial optimization, digital geometry, discrete differential geometry, geometric graph theory, toric geometry, and combinatorial topology.

History

Although polyhedra and tessellations had been studied for many years by people such as Kepler and Cauchy, modern discrete geometry has its origins in the late 19th century. Early topics studied were: the density of circle packings by Thue, projective configurations by Reye and Steinitz, the geometry of numbers by Minkowski, and map colourings by Tait, Heawood, and Hadwiger.

László Fejes Tóth, H.S.M. Coxeter and Paul Erdős, laid the foundations of *discrete geometry*.

Topics in Discrete Geometry

Polyhedra and Polytopes

A polytope is a geometric object with flat sides, which exists in any general number of dimensions. A polygon is a polytope in two dimensions, a polyhedron in three dimensions, and so on in higher dimensions (such as a 4-polytope in four dimensions). Some theories further generalize the idea to include such objects as unbounded polytopes (apeirotopes and tessellations), and abstract polytopes.

The following are some of the aspects of polytopes studied in discrete geometry:

- Polyhedral combinatorics
- Lattice polytopes
- Ehrhart polynomials
- Pick's theorem
- Hirsch conjecture

Packings, Coverings and Tilings

Packings, coverings, and tilings are all ways of arranging uniform objects (typically circles, spheres, or tiles) in a regular way on a surface or manifold.

A sphere packing is an arrangement of non-overlapping spheres within a containing space. The spheres considered are usually all of identical size, and the space is usually three-dimensional Euclidean space. However, sphere packing problems can be generalised to consider unequal spheres, n-dimensional Euclidean space (where the problem becomes circle packing in two dimensions, or hypersphere packing in higher dimensions) or to non-Euclidean spaces such as hyperbolic space.

A tessellation of a flat surface is the tiling of a plane using one or more geometric shapes, called tiles, with no overlaps and no gaps. In mathematics, tessellations can be generalized to higher dimensions.

Specific topics in this area include:

- Circle packings
- Sphere packings
- Kepler conjecture
- Quasicrystals
- Aperiodic tilings
- Periodic Graphs (Geometry)
- Finite subdivision rules

Structural Rigidity and Flexibility

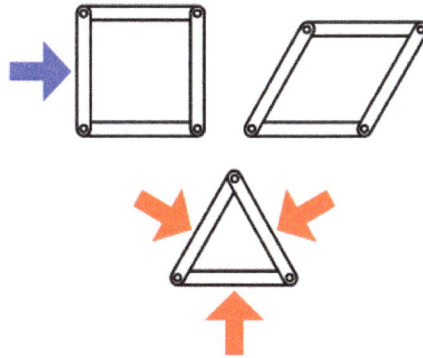

Graphs are drawn as rods connected by rotating hinges. The cycle graph C_4 drawn as a square can be tilted over by the blue force into a parallelogram, so it is a flexible graph. K_3, drawn as a triangle, cannot be altered by any force that is applied to it, so it is a rigid graph.

Structural rigidity is a combinatorial theory for predicting the flexibility of ensembles formed by rigid bodies connected by flexible linkages or hinges.

Topics in this area include:

- Cauchy's theorem

- Flexible polyhedra

Incidence Structures

Incidence structures generalize planes (such as affine, projective, and Möbius planes) as can be seen from their axiomatic definitions. Incidence structures also generalize the higher-dimensional analogs and the finite structures are sometimes called finite geometries.

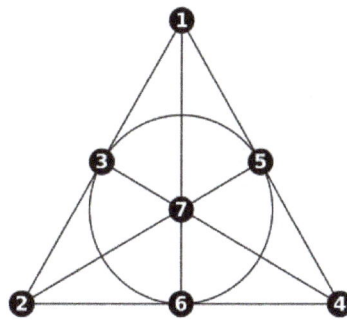

Seven points are elements of seven lines in the Fano plane, an example of an incidence structure.

Formally, an incidence structure is a triple

$$C = (P, L, I).$$

where P is a set of "points", L is a set of "lines" and $I \subseteq P \times L$ is the incidence relation. The elements of I are called flags. If

$$(p, l) \in I,$$

we say that point p "lies on" line l.

Topics in this area include:

- Configurations

- Line arrangements

- Hyperplane arrangements

- Buildings

Oriented Matroids

An oriented matroid is a mathematical structure that abstracts the properties of directed graphs and of arrangements of vectors in a vector space over an ordered field (particularly for partially ordered vector spaces). In comparison, an ordinary (i.e., non-oriented) matroid abstracts the dependence properties that are common both to graphs, which are not necessarily *directed*, and to arrangements of vectors over fields, which are not necessarily *ordered*.

Geometric Graph Theory

A geometric graph is a graph in which the vertices or edges are associated with geometric objects. Examples include Euclidean graphs, the 1-skeleton of a polyhedron or polytope, intersection graphs, and visibility graphs.

Topics in this area include:

- Graph drawing

- Polyhedral graphs

- Voronoi diagrams and Delaunay triangulations

Simplicial Complexes

A simplicial complex is a topological space of a certain kind, constructed by "gluing together" points, line segments, triangles, and their n-dimensional counterparts. Simplicial complexes should not be confused with the more abstract notion of a simplicial set appearing in modern simplicial homotopy theory. The purely combinatorial counterpart to a simplicial complex is an abstract simplicial complex.

Topological Combinatorics

The discipline of combinatorial topology used combinatorial concepts in topology and in the early 20th century this turned into the field of algebraic topology.

In 1978 the situation was reversed – methods from algebraic topology were used to solve a problem in combinatorics – when László Lovász proved the Kneser conjecture, thus beginning the new study of topological combinatorics. Lovász's proof used the Borsuk-Ulam theorem and this theorem retains a prominent role in this new field. This theorem has many equivalent versions and analogs and has been used in the study of fair division problems.

Topics in this are include:

- Sperner's lemma

- Regular maps

Lattices and Discrete Groups

A discrete group is a group G equipped with the discrete topology. With this topology, G becomes a topological group. A discrete subgroup of a topological group G is a subgroup H whose relative topology is the discrete one. For example, the integers, Z, form a discrete subgroup of the reals, R (with the standard metric topology), but the rational numbers, Q, do not.

A lattice in a locally compact topological group is a discrete subgroup with the property that the quotient space has finite invariant measure. In the special case of subgroups of R^n, this amounts to the usual geometric notion of a lattice, and both the algebraic structure of lattices and the geometry of the totality of all lattices are relatively well understood. Deep results of Borel, Harish-Chandra, Mostow, Tamagawa, M. S. Raghunathan, Margulis, Zimmer obtained from the 1950s through the 1970s provided examples and generalized much of the theory to the setting of nilpotent Lie groups and semisimple algebraic groups over a local field. In the 1990s, Bass and Lubotzky initiated the study of *tree lattices*, which remains an active research area.

Topics in this area include:

- Reflection groups

- Triangle groups

Digital Geometry

Digital geometry deals with discrete sets (usually discrete point sets) considered to be digitized models or images of objects of the 2D or 3D Euclidean space.

Simply put, digitizing is replacing an object by a discrete set of its points. The images we see on the TV screen, the raster display of a computer, or in newspapers are in fact digital images.

Its main application areas are computer graphics and image analysis. See Li Chen, Digital and discrete geometry: Theory and Algorithms, Springer, 2014. (http://www.springer.com/us/book/9783319120980)

Discrete Differential Geometry

Discrete differential geometry is the study of discrete counterparts of notions in differential geometry. Instead of smooth curves and surfaces, there are polygons, meshes, and simplicial complexes. It is used in the study of computer graphics and topological combinatorics.

Topics in this area include:

- Discrete Laplace operator

- Discrete exterior calculus

- Discrete Morse theory

- Topological combinatorics

- Spectral shape analysis

- Abstract differential geometry

- Analysis on fractals

Euclidean Geometry

Euclidean geometry is a mathematical system attributed to the Alexandrian Greek mathematician Euclid, which he described in his textbook on geometry: the Elements. Euclid's method consists in assuming a small set of intuitively appealing axioms, and deducing many other propositions (theorems) from these. Although many of Euclid's results had been stated by earlier mathematicians,[1] Euclid was the first to show how these propositions could fit into a comprehensive deductive and logical system.[2] The Elements begins with plane geometry, still taught in secondary school as the first axiomatic system and the first examples of formal proof. It goes on to the solid geometry of three dimensions. Much of the Elements states results of what are now called algebra and number theory, explained in geometrical language.

Detail from Raphael's *The School of Athens* featuring a Greek mathematician – perhaps representing Euclid or Archimedes – using a compass to draw a geometric construction.

For more than two thousand years, the adjective "Euclidean" was unnecessary because no other sort of geometry had been conceived. Euclid's axioms seemed so intuitively obvious (with the possible exception of the parallel postulate) that any theorem proved from them was deemed true in an absolute, often metaphysical, sense. Today, however, many other self-consistent non-Euclidean geometries are known, the first ones having been discovered in the early 19th century. An implication of Albert Einstein's theory of general relativity is that physical space itself is not Euclidean, and Euclidean space is a good approximation for it only where the gravitational field is weak.

Euclidean geometry is an example of synthetic geometry, in that it proceeds logically from axioms to propositions without the use of coordinates. This is in contrast to analytic geometry, which uses coordinates.

The Elements

The *Elements* is mainly a systematization of earlier knowledge of geometry. Its improvement over earlier treatments was rapidly recognized, with the result that there was little interest in preserving the earlier ones, and they are now nearly all lost.

There are 13 total books in the *Elements*:

Books I–IV and VI discuss plane geometry. Many results about plane figures are proved, for example *"In any triangle two angles taken together in any manner are less than two right angles."* (Book 1 proposition 17) and the Pythagorean theorem *"In right angled triangles the square on the side subtending the right angle is equal to the squares on the sides containing the right angle."* (Book I, proposition 47)

Books V and VII–X deal with number theory, with numbers treated geometrically via their representation as line segments with various lengths. Notions such as prime numbers and rational and irrational numbers are introduced. The infinitude of prime numbers is proved.

Books XI–XIII concern solid geometry. A typical result is the 1:3 ratio between the volume of a cone and a cylinder with the same height and base.

Axioms

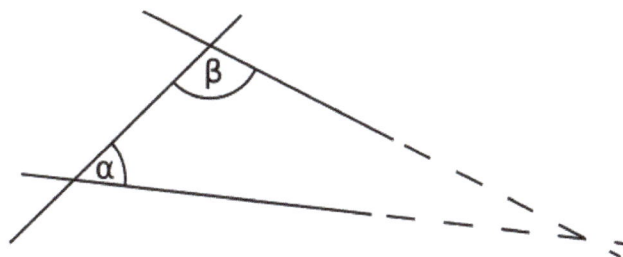

The parallel postulate (Postulate 5): If two lines intersect a third in such a way that the sum of the inner angles on one side is less than two right angles, then the two lines inevitably must intersect each other on that side if extended far enough.

Euclidean geometry is an axiomatic system, in which all theorems ("true statements") are derived from a small number of axioms. Near the beginning of the first book of the *Elements*, Euclid gives five postulates (axioms) for plane geometry, stated in terms of constructions (as translated by Thomas Heath):

"Let the following be postulated":

1. "To draw a straight line from any point to any point."

2. "To produce [extend] a finite straight line continuously in a straight line."

3. "To describe a circle with any centre and distance [radius]."

4. "That all right angles are equal to one another."

5. *The parallel postulate*: "That, if a straight line falling on two straight lines make the interior angles on the same side less than two right angles, the two straight lines, if produced indefinitely, meet on that side on which are the angles less than the two right angles."

Although Euclid's statement of the postulates only explicitly asserts the existence of the constructions, they are also taken to be unique.

The *Elements* also include the following five "common notions":

1. Things that are equal to the same thing are also equal to one another (formally the Euclidean property of equality, but may be considered a consequence of the transitivity property of equality).

2. If equals are added to equals, then the wholes are equal (Addition property of equality).

3. If equals are subtracted from equals, then the remainders are equal (Subtraction property of equality).

4. Things that coincide with one another are equal to one another (Reflexive Property).

5. The whole is greater than the part.

Parallel Postulate

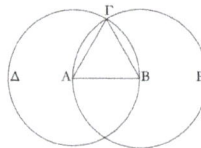

A proof from Euclid's *Elements* that, given a line segment, an equilateral triangle exists that includes the segment as one of its sides. The proof is by construction: an equilateral triangle ΑΒΓ is made by drawing circles Δ and Ε centered on the points A and B, and taking one intersection of the circles as the third vertex of the triangle.

To the ancients, the parallel postulate seemed less obvious than the others. They were concerned with creating a system which was absolutely rigorous and to them it seemed as if the parallel line postulate should have been able to be proven rather than simply accepted as a fact. It is now known that such a proof is impossible. Euclid himself seems to have considered it as being qualitatively different from the others, as evidenced by the organization of the *Elements*: the first 28 propositions he presents are those that can be proved without it.

Many alternative axioms can be formulated that have the same logical consequences as the parallel postulate. For example, Playfair's axiom states:

> In a plane, through a point not on a given straight line, at most one line can be drawn that never meets the given line.

Methods of Proof

Euclidean Geometry is *constructive*. Postulates 1, 2, 3, and 5 assert the existence and uniqueness of certain geometric figures, and these assertions are of a constructive nature: that is, we are not only told that certain things exist, but are also given methods for creating them with no more than a compass and an unmarked straightedge. In this sense, Euclidean geometry is more concrete than many modern axiomatic systems such as set theory, which often assert the existence of objects without saying how to construct them, or even assert the existence of objects that cannot be constructed within the theory. Strictly speaking, the lines on paper are *models* of the objects defined within the formal system, rather than instances of those objects. For example, a Euclidean straight line has no width, but any real drawn line will. Though nearly all modern mathematicians consider nonconstructive methods just as sound as constructive ones, Euclid's constructive proofs often supplanted fallacious nonconstructive ones—e.g., some of the Pythagoreans' proofs that involved irrational numbers, which usually required a statement such as "Find the greatest common measure of ..."

Euclid often used proof by contradiction. Euclidean geometry also allows the method of superposition, in which a figure is transferred to another point in space. For example, proposition I.4, side-angle-side congruence of triangles, is proved by moving one of the two triangles so that one of its sides coincides with the other triangle's equal side, and then proving that the other sides coincide as well. Some modern treatments add a sixth postulate, the rigidity of the triangle, which can be used as an alternative to superposition.

System of Measurement and Arithmetic

Euclidean geometry has two fundamental types of measurements: angle and distance. The angle scale is absolute, and Euclid uses the right angle as his basic unit, so that, e.g., a 45-degree angle would be referred to as half of a right angle. The distance scale is relative; one arbitrarily picks a line segment with a certain nonzero length as the unit, and other distances are expressed in relation to it. Addition of distances is represented by a construction in which one line segment is copied onto the end of another line segment to extend its length, and similarly for subtraction.

Measurements of area and volume are derived from distances. For example, a rectangle with a width of 3 and a length of 4 has an area that represents the product, 12. Because this geometrical interpretation of multiplication was limited to three dimensions, there was no direct way of interpreting the product of four or more numbers, and Euclid avoided such products, although they are implied, e.g., in the proof of book IX, proposition 20.

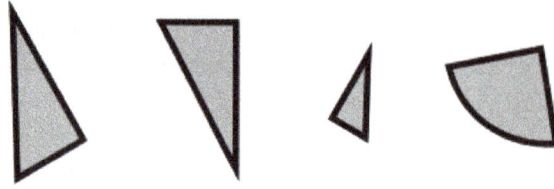

An example of congruence. The two figures on the left are congruent, while the third is similar to them. The last figure is neither. Note that congruences alter some properties, such as location and orientation, but leave others unchanged, like distance and angles. The latter sort of properties are called invariants and studying them is the essence of geometry.

Euclid refers to a pair of lines, or a pair of planar or solid figures, as "equal" their lengths, areas, or volumes are equal, and similarly for angles. The stronger term "congruent" refers to the idea that an entire figure is the same size and shape as another figure. Alternatively, two figures are congruent if one can be moved on top of the other so that it matches up with it exactly. (Flipping it over is allowed.) Thus, for example, a 2x6 rectangle and a 3x4 rectangle are equal but not congruent, and the letter R is congruent to its mirror image. Figures that would be congruent except for their differing sizes are referred to as similar. Corresponding angles in a pair of similar shapes are congruent and corresponding sides are in proportion to each other.

Notation and Terminology

Naming of Points and Figures

Points are customarily named using capital letters of the alphabet. Other figures, such as lines, triangles, or circles, are named by listing a sufficient number of points to pick them out unambiguously from the relevant figure, e.g., triangle ABC would typically be a triangle with vertices at points A, B, and C.

Complementary and Supplementary Angles

Angles whose sum is a right angle are called complementary. Complementary angles are formed when a ray shares the same vertex and is pointed in a direction that is in between the two original rays that form the right angle. The number of rays in between the two original rays is infinite.

Angles whose sum is a straight angle are supplementary. Supplementary angles are formed when a ray shares the same vertex and is pointed in a direction that is in between the two original rays that form the straight angle (180 degree angle). The number of rays in between the two original rays is infinite.

Modern Versions of Euclid's Notation

In modern terminology, angles would normally be measured in degrees or radians.

Modern school textbooks often define separate figures called lines (infinite), rays (semi-infinite), and line segments (of finite length). Euclid, rather than discussing a ray as an object that extends to infinity in one direction, would normally use locutions such as "if the line is extended to a sufficient length," although he occasionally referred to "infinite lines." A "line" in Euclid could be either straight or curved, and he used the more specific term "straight line" when necessary.

Some Important or Well Known Results

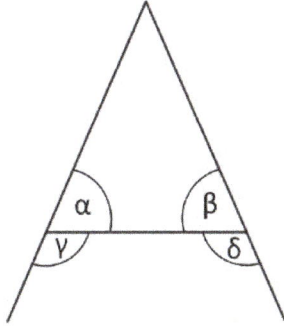

The Pons Asinorum or Bridge of Asses theorem states that in an isosceles triangle, $\alpha = \beta$ and $\gamma = \delta$.

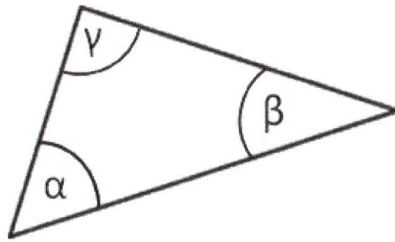

The Triangle Angle Sum theorem states that the sum of the three angles of any triangle, in this case angles α, β, and γ, will always equal 180 degrees.

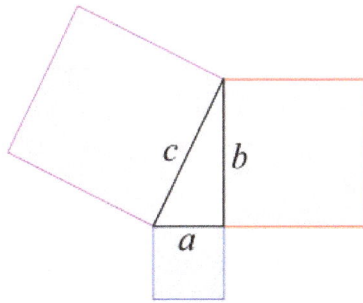

The Pythagorean theorem states that the sum of the areas of the two squares on the legs (a and b) of a right triangle equals the area of the square on the hypotenuse (c).

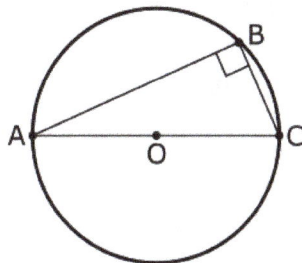

Thales' theorem states that if AC is a diameter, then the angle at B is a right angle.

Pons Asinorum

The Bridge of Asses (*Pons Asinorum*) states that *in isosceles triangles the angles at the base equal one another, and, if the equal straight lines are produced further, then the angles under the base equal one another*. Its name may be attributed to its frequent role as the first real test in the *Elements* of the intelligence of the reader and as a bridge to the harder propositions that followed. It might also be so named because of the geometrical figure's resemblance to a steep bridge that only a sure-footed donkey could cross.

Congruence of Triangles

Triangles are congruent if they have all three sides equal (SSS), two sides and the angle between them equal (SAS), or two angles and a side equal (ASA) (Book I, propositions 4, 8, and 26). Triangles with three equal angles (AAA) are similar, but not necessarily congruent. Also, triangles with two equal sides and an adjacent angle are not necessarily equal or congruent.

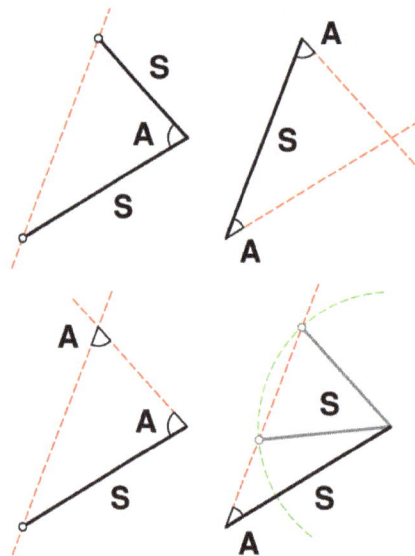

Congruence of triangles is determined by specifying two sides and the angle between them (SAS), two angles and the side between them (ASA) or two angles and a corresponding adjacent side (AAS). Specifying two sides and an adjacent angle (SSA), however, can yield two distinct possible triangles unless the angle specified is a right angle.

Triangle Angle Sum

The sum of the angles of a triangle is equal to a straight angle (180 degrees). This causes an equilateral triangle to have 3 interior angles of 60 degrees. Also, it causes every triangle to have at least 2 acute angles and up to 1 obtuse or right angle.

Pythagorean Theorem

The celebrated Pythagorean theorem (book I, proposition 47) states that in any right triangle, the area of the square whose side is the hypotenuse (the side opposite the right angle) is equal to the sum of the areas of the squares whose sides are the two legs (the two sides that meet at a right angle).

Thales' Theorem

Thales' theorem, named after Thales of Miletus states that if A, B, and C are points on a circle where the line AC is a diameter of the circle, then the angle ABC is a right angle. Cantor supposed that Thales proved his theorem by means of Euclid Book I, Prop. 32 after the manner of Euclid Book III, Prop. 31. Tradition has it that Thales sacrificed an ox to celebrate this theorem.

Scaling of Area and Volume

In modern terminology, the area of a plane figure is proportional to the square of any of its linear dimensions, $A \propto L^2$, and the volume of a solid to the cube, $V \propto L^3$. Euclid proved these results in various special cases such as the area of a circle and the volume of a parallelepipedal solid. Euclid determined some, but not all, of the relevant constants of proportionality. E.g., it was his successor Archimedes who proved that a sphere has 2/3 the volume of the circumscribing cylinder.

Applications

Because of Euclidean geometry's fundamental status in mathematics, it would be impossible to give more than a representative sampling of applications here.

A surveyor uses a level

Sphere packing applies to a stack of oranges.

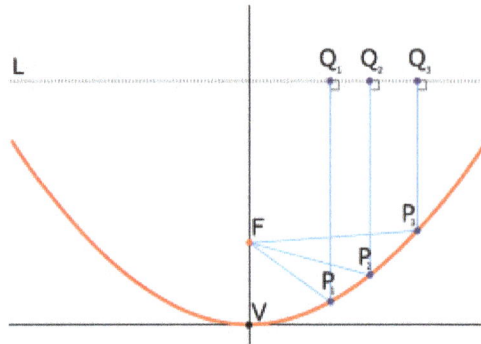

A parabolic mirror brings parallel rays of light to a focus.

As suggested by the etymology of the word, one of the earliest reasons for interest in geometry was surveying, and certain practical results from Euclidean geometry, such as the right-angle property of the 3-4-5 triangle, were used long before they were proved formally. The fundamental types of measurements in Euclidean geometry are distances and angles, and both of these quantities can be measured directly by a surveyor. Historically, distances were often measured by chains such as Gunter's chain, and angles using graduated circles and, later, the theodolite.

An application of Euclidean solid geometry is the determination of packing arrangements, such as the problem of finding the most efficient packing of spheres in n dimensions. This problem has applications in error detection and correction.

Geometric optics uses Euclidean geometry to analyze the focusing of light by lenses and mirrors.

Geometry is used in art and architecture.

The water tower consists of a cone, a cylinder, and a hemisphere. Its volume can be calculated using solid geometry.

Geometry can be used to design origami.

Geometry is used extensively in architecture.

Geometry can be used to design origami. Some classical construction problems of geometry are impossible using compass and straightedge, but can be solved using origami.

As a Description of the Structure of Space

Euclid believed that his axioms were self-evident statements about physical reality. Euclid's proofs depend upon assumptions perhaps not obvious in Euclid's fundamental axioms, in particular that certain movements of figures do not change their geometrical properties such as the lengths of sides and interior angles, the so-called *Euclidean motions*, which include translations, reflections and rotations of figures. Taken as a physical description of space, postulate 2 (extending a line) asserts that space does not have holes or boundaries (in other words, space is homogeneous and unbounded); postulate 4 (equality of right angles) says that space is isotropic and figures may be moved to any location while maintaining congruence; and postulate 5 (the parallel postulate) that space is flat (has no intrinsic curvature).

As discussed in more detail below, Einstein's theory of relativity significantly modifies this view.

The ambiguous character of the axioms as originally formulated by Euclid makes it possible for different commentators to disagree about some of their other implications for the structure of space, such as whether or not it is infinite and what its topology is. Modern, more rigorous reformulations of the system typically aim for a cleaner separation of these issues. Interpreting Euclid's axioms in the spirit of this more modern approach, axioms 1-4 are consistent with either infinite or finite space (as in elliptic geometry), and all five axioms are consistent with a variety of topologies (e.g., a plane, a cylinder, or a torus for two-dimensional Euclidean geometry).

Later Work

Archimedes and Apollonius

Archimedes (ca. 287 BCE – ca. 212 BCE), a colorful figure about whom many historical anecdotes are recorded, is remembered along with Euclid as one of the greatest of ancient mathematicians. Although the foundations of his work were put in place by Euclid, his work, unlike Euclid's, is believed to have been entirely original. He proved equations for the volumes and areas of various figures in two and three dimensions, and enunciated the Archimedean property of finite numbers.

A sphere has 2/3 the volume and surface area of its circumscribing cylinder. A sphere and cylinder were placed on the tomb of Archimedes at his request.

Apollonius of Perga (ca. 262 BCE–ca. 190 BCE) is mainly known for his investigation of conic sections.

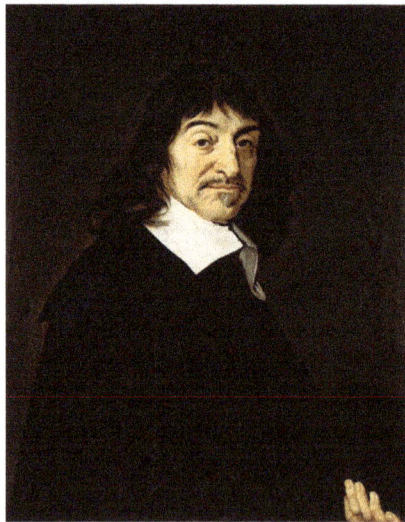

René Descartes. Portrait after Frans Hals, 1648.

17th Century: Descartes

René Descartes (1596–1650) developed analytic geometry, an alternative method for formalizing geometry which focused on turning geometry into algebra.

In this approach, a point on a plane is represented by its Cartesian (x, y) coordinates, a line is represented by its equation, and so on.

In Euclid's original approach, the Pythagorean theorem follows from Euclid's axioms. In the Cartesian approach, the axioms are the axioms of algebra, and the equation expressing the Pythagorean theorem is then a definition of one of the terms in Euclid's axioms, which are now considered theorems.

The equation

$$|PQ| = \sqrt{(p_x - q_x)^2 + (p_y - q_y)^2}$$

defining the distance between two points $P = (p_x, p_y)$ and $Q = (q_x, q_y)$ is then known as the *Euclidean metric*, and other metrics define non-Euclidean geometries.

In terms of analytic geometry, the restriction of classical geometry to compass and straightedge constructions means a restriction to first- and second-order equations, e.g., $y = 2x + 1$ (a line), or $x^2 + y^2 = 7$ (a circle).

Also in the 17th century, Girard Desargues, motivated by the theory of perspective, introduced the concept of idealized points, lines, and planes at infinity. The result can be considered as a type of generalized geometry, projective geometry, but it can also be used to produce proofs in ordinary Euclidean geometry in which the number of special cases is reduced.

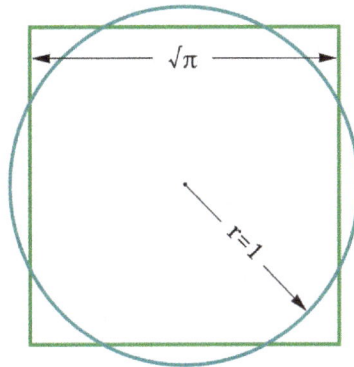

Squaring the circle: the areas of this square and this circle are equal. In 1882, it was proven that this figure cannot be constructed in a finite number of steps with an idealized compass and straightedge.

18th Century

Geometers of the 18th century struggled to define the boundaries of the Euclidean system. Many tried in vain to prove the fifth postulate from the first four. By 1763 at least 28 different proofs had been published, but all were found incorrect.

Leading up to this period, geometers also tried to determine what constructions could be accomplished in Euclidean geometry. For example, the problem of trisecting an angle with a compass and straightedge is one that naturally occurs within the theory, since the axioms refer to constructive operations that can be carried out with those tools. However, centuries of efforts failed to find a solution to this problem, until Pierre Wantzel published a proof in 1837 that such a construction was impossible. Other constructions that were proved impossible include doubling the cube and squaring the circle. In the case of doubling the cube, the impossibility of the construction originates from the fact that the compass and straightedge method involve equations whose order is an integral power of two, while doubling a cube requires the solution of a third-order equation.

Euler discussed a generalization of Euclidean geometry called affine geometry, which retains the fifth postulate unmodified while weakening postulates three and four in a way that eliminates the notions of angle (whence right triangles become meaningless) and of equality of length of line segments in general (whence circles become meaningless) while retaining the notions of parallelism as an equivalence relation between lines, and equality of length of parallel line segments (so line segments continue to have a midpoint).

19th Century and Non-Euclidean Geometry

In the early 19th century, Carnot and Möbius systematically developed the use of signed angles and line segments as a way of simplifying and unifying results.

The century's most significant development in geometry occurred when, around 1830, János Bolyai and Nikolai Ivanovich Lobachevsky separately published work on non-Euclidean geometry, in which the parallel postulate is not valid. Since non-Euclidean geometry is provably relatively consistent with Euclidean geometry, the parallel postulate cannot be proved from the other postulates.

In the 19th century, it was also realized that Euclid's ten axioms and common notions do not suffice to prove all of the theorems stated in the *Elements*. For example, Euclid assumed implicitly that any line contains at least two points, but this assumption cannot be proved from the other axioms, and therefore must be an axiom itself. The very first geometric proof in the *Elements,* shown in the figure above, is that any line segment is part of a triangle; Euclid constructs this in the usual way, by drawing circles around both endpoints and taking their intersection as the third vertex. His axioms, however, do not guarantee that the circles actually intersect, because they do not assert the geometrical property of continuity, which in Cartesian terms is equivalent to the completeness property of the real numbers. Starting with Moritz Pasch in 1882, many improved axiomatic systems for geometry have been proposed, the best known being those of Hilbert, George Birkhoff, and Tarski.

20th Century and General Relativity

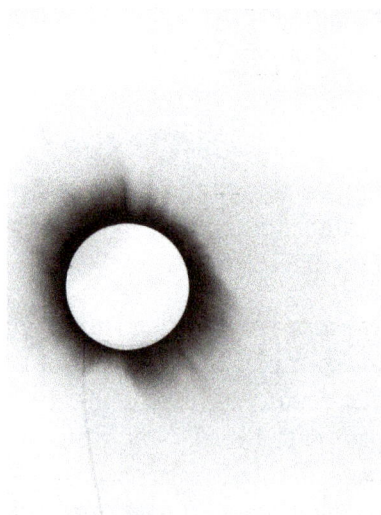

A disproof of Euclidean geometry as a description of physical space. In a 1919 test of the general theory of relativity, stars (marked with short horizontal lines) were photographed during a solar eclipse. The rays of starlight were bent by the Sun's gravity on their way to the earth. This is interpreted as evidence in favor of Einstein's prediction that gravity would cause deviations from Euclidean geometry.

Einstein's theory of general relativity shows that the true geometry of spacetime is not Euclidean geometry. For example, if a triangle is constructed out of three rays of light, then in general the interior angles do not add up to 180 degrees due to gravity. A relatively weak gravitational field, such as the Earth's or the sun's, is represented by a metric that is approximately, but not exactly, Euclidean. Until the 20th century, there was no technology capable of detecting the deviations from Euclidean geometry, but Einstein predicted that such deviations would exist. They were later verified by observations such as the slight bending of starlight by the Sun during a solar eclipse in 1919, and such considerations are now an integral part of the software that runs the GPS system. It is possible to object to this interpretation of general relativity on the grounds that light rays might be improper physical models of Euclid's lines, or that relativity could be rephrased so as to avoid the geometrical interpretations. However, one of the consequences of Einstein's theory is that there is no possible physical test that can distinguish between a beam of light as a model of a geometrical line and any other physical model. Thus, the only logical possibilities are to accept non-Euclidean geometry as physically real, or to reject the entire notion of physical tests of the axioms of geometry, which can then be imagined as a formal system without any intrinsic real-world meaning.

Treatment of Infinity

Infinite Objects

Euclid sometimes distinguished explicitly between "finite lines" (e.g., Postulate 2) and "infinite lines" (book I, proposition 12). However, he typically did not make such distinctions unless they were necessary. The postulates do not explicitly refer to infinite lines, although for example some commentators interpret postulate 3, existence of a circle with any radius, as implying that space is infinite.

The notion of infinitesimal quantities had previously been discussed extensively by the Eleatic School, but nobody had been able to put them on a firm logical basis, with paradoxes such as Zeno's paradox occurring that had not been resolved to universal satisfaction. Euclid used the method of exhaustion rather than infinitesimals.

Later ancient commentators such as Proclus (410–485 CE) treated many questions about infinity as issues demanding proof and, e.g., Proclus claimed to prove the infinite divisibility of a line, based on a proof by contradiction in which he considered the cases of even and odd numbers of points constituting it.

At the turn of the 20th century, Otto Stolz, Paul du Bois-Reymond, Giuseppe Veronese, and others produced controversial work on non-Archimedean models of Euclidean geometry, in which the distance between two points may be infinite or infinitesimal, in the Newton–Leibniz sense. Fifty years later, Abraham Robinson provided a rigorous logical foundation for Veronese's work.

Infinite Processes

One reason that the ancients treated the parallel postulate as less certain than the others is that verifying it physically would require us to inspect two lines to check that they never intersected, even at some very distant point, and this inspection could potentially take an infinite amount of time.

The modern formulation of proof by induction was not developed until the 17th century, but some later commentators consider it implicit in some of Euclid's proofs, e.g., the proof of the infinitude of primes.

Supposed paradoxes involving infinite series, such as Zeno's paradox, predated Euclid. Euclid avoided such discussions, giving, for example, the expression for the partial sums of the geometric series in IX.35 without commenting on the possibility of letting the number of terms become infinite.

Logical Basis

Classical Logic

Euclid frequently used the method of proof by contradiction, and therefore the traditional presentation of Euclidean geometry assumes classical logic, in which every proposition is either true or false, i.e., for any proposition P, the proposition "P or not P" is automatically true.

Modern Standards of Rigor

Placing Euclidean geometry on a solid axiomatic basis was a preoccupation of mathematicians for centuries. The role of primitive notions, or undefined concepts, was clearly put forward by Alessandro Padoa of the Peano delegation at the 1900 Paris conference:

...when we begin to formulate the theory, we can imagine that the undefined symbols are *completely devoid of meaning* and that the unproved propositions are simply *conditions* imposed upon the undefined symbols.

Then, the *system of ideas* that we have initially chosen is simply *one interpretation* of the undefined symbols; but..this interpretation can be ignored by the reader, who is free to replace it in his mind by *another interpretation*.. that satisfies the conditions...

Logical questions thus become completely independent of *empirical* or *psychological* questions...

The system of undefined symbols can then be regarded as the *abstraction* obtained from the *specialized theories* that result when...the system of undefined symbols is successively replaced by each of the interpretations...

— *Padoa, Essai d'une théorie algébrique des nombre entiers, avec une Introduction logique à une théorie déductive qulelconque*

That is, mathematics is context-independent knowledge within a hierarchical framework. As said by Bertrand Russell:

If our hypothesis is about *anything*, and not about some one or more particular things, then our deductions constitute mathematics. Thus, mathematics may be defined as the subject in which we never know what we are talking about, nor whether what we are saying is true.

— *Bertrand Russell, Mathematics and the metaphysicians*

Such foundational approaches range between foundationalism and formalism.

Axiomatic Formulations

Geometry is the science of correct reasoning on incorrect figures.

— George Polyá, How to Solve It, p. 208

- Euclid's axioms: In his dissertation to Trinity College, Cambridge, Bertrand Russell summarized the changing role of Euclid's geometry in the minds of philosophers up to that time. It was a conflict between certain knowledge, independent of experiment, and empiricism, requiring experimental input. This issue became clear as it was discovered that the parallel postulate was not necessarily valid and its applicability was an empirical matter, deciding whether the applicable geometry was Euclidean or non-Euclidean.

- Hilbert's axioms: Hilbert's axioms had the goal of identifying a *simple* and *complete* set of *independent* axioms from which the most important geometric theorems could be deduced. The outstanding objectives were to make Euclidean geometry rigorous (avoiding hidden assumptions) and to make clear the ramifications of the parallel postulate.

- Birkhoff's axioms: Birkhoff proposed four postulates for Euclidean geometry that can be confirmed experimentally with scale and protractor. This system relies heavily on the properties of the real numbers. The notions of *angle* and *distance* become primitive concepts.

- Tarski's axioms: Alfred Tarski (1902–1983) and his students defined *elementary* Euclidean geometry as the geometry that can be expressed in first-order logic and does not depend on set theory for its logical basis, in contrast to Hilbert's axioms, which involve point sets. Tarski proved that his axiomatic formulation of elementary Euclidean geometry is consistent and complete in a certain sense: there is an algorithm that, for every proposition, can be shown either true or false. (This doesn't violate Gödel's theorem, because Euclidean geometry cannot describe a sufficient amount of arithmetic for the theorem to apply.) This is equivalent to the decidability of real closed fields, of which elementary Euclidean geometry is a model.

Constructive Approaches and Pedagogy

The process of abstract axiomatization as exemplified by Hilbert's axioms reduces geometry to theorem proving or predicate logic. In contrast, the Greeks used construction postulates, and emphasized problem solving. For the Greeks, constructions are more primitive than existence propositions, and can be used to prove existence propositions, but not *vice versa*. To describe problem solving adequately requires a richer system of logical concepts. The contrast in approach may be summarized:

- Axiomatic proof: Proofs are deductive derivations of propositions from primitive premises that are 'true' in some sense. The aim is to justify the proposition.

- Analytic proof: Proofs are non-deductive derivations of hypotheses from problems. The aim is to find hypotheses capable of giving a solution to the problem. One can argue that Euclid's axioms were arrived upon in this manner. In particular, it is thought that Euclid felt the parallel postulate was forced upon him, as indicated by his reluctance to make use of it, and his arrival upon it by the method of contradiction.

Andrei Nicholaevich Kolmogorov proposed a problem solving basis for geometry. This work was a precursor of a modern formulation in terms of constructive type theory. This development has implications for pedagogy as well.

If proof simply follows conviction of truth rather than contributing to its construction and is only experienced as a demonstration of something already known to be true, it is likely to remain meaningless and purposeless in the eyes of students.

— Celia Hoyles, The curricular shaping of students' approach to proof

Non-Euclidean Geometry

In mathematics, non-Euclidean geometry consists of two geometries based on axioms closely related to those specifying Euclidean geometry. As Euclidean geometry lies at the intersection of metric geometry and affine geometry, non-Euclidean geometry arises when either the metric requirement is relaxed, or the parallel postulate is replaced with an alternative one. In the latter case one obtains hyperbolic geometry and elliptic geometry, the traditional non-Euclidean geometries. When the metric requirement is relaxed, then there are affine planes associated with the planar algebras which give rise to kinematic geometries that have also been called non-Euclidean geometry.

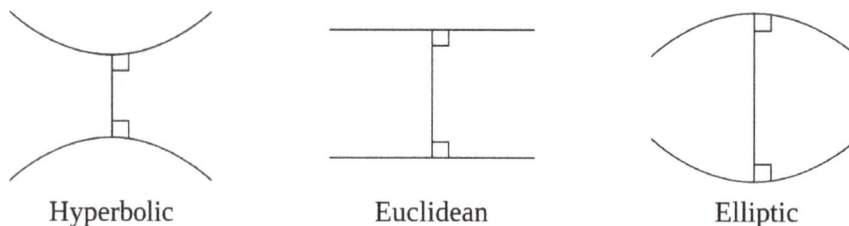

| Hyperbolic | Euclidean | Elliptic |

Behavior of lines with a common perpendicular in each of the three types of geometry

The essential difference between the metric geometries is the nature of parallel lines. Euclid's fifth postulate, the parallel postulate, is equivalent to Playfair's postulate, which states that, within a two-dimensional plane, for any given line ℓ and a point A, which is not on ℓ, there is exactly one line through A that does not intersect ℓ. In hyperbolic geometry, by contrast, there are infinitely many lines through A not intersecting ℓ, while in elliptic geometry, any line through A intersects ℓ.

Another way to describe the differences between these geometries is to consider two straight lines indefinitely extended in a two-dimensional plane that are both perpendicular to a third line:

- In Euclidean geometry the lines remain at a constant distance from each other (meaning that a line drawn perpendicular to one line at any point will intersect the other line and the length of the line segment joining the points of intersection remains constant) and are known as parallels.

- In hyperbolic geometry they "curve away" from each other, increasing in distance as one moves further from the points of intersection with the common perpendicular; these lines are often called ultraparallels.

- In elliptic geometry the lines "curve toward" each other and intersect.

History

Early History

While Euclidean geometry, named after the Greek mathematician Euclid, includes some of the oldest known mathematics, non-Euclidean geometries were not widely accepted as legitimate until the 19th century.

The debate that eventually led to the discovery of the non-Euclidean geometries began almost as soon as Euclid's work *Elements* was written. In the *Elements*, Euclid began with a limited number of assumptions (23 definitions, five common notions, and five postulates) and sought to prove all the other results (propositions) in the work. The most notorious of the postulates is often referred to as "Euclid's Fifth Postulate," or simply the "parallel postulate", which in Euclid's original formulation is:

If a straight line falls on two straight lines in such a manner that the interior angles on the same side are together less than two right angles, then the straight lines, if produced indefinitely, meet on that side on which are the angles less than the two right angles.

Other mathematicians have devised simpler forms of this property. Regardless of the form of the postulate, however, it consistently appears to be more complicated than Euclid's other postulates:

1. To draw a straight line from any point to any point.

2. To produce [extend] a finite straight line continuously in a straight line.

3. To describe a circle with any centre and distance [radius].

4. That all right angles are equal to one another.

For at least a thousand years, geometers were troubled by the disparate complexity of the fifth postulate, and believed it could be proved as a theorem from the other four. Many attempted to find a proof by contradiction, including Ibn al-Haytham (Alhazen, 11th century), Omar Khayyám (12th century), Nasīr al-Dīn al-Tūsī (13th century), and Giovanni Girolamo Saccheri (18th century).

The theorems of Ibn al-Haytham, Khayyam and al-Tusi on quadrilaterals, including the Lambert quadrilateral and Saccheri quadrilateral, were "the first few theorems of the hyperbolic and the elliptic geometries." These theorems along with their alternative postulates, such as Playfair's axiom, played an important role in the later development of non-Euclidean geometry. These early attempts at challenging the fifth postulate had a considerable influence on its development among later European geometers, including Witelo, Levi ben Gerson, Alfonso, John Wallis and Saccheri. All of these early attempts made at trying to formulate non-Euclidean geometry however provided flawed proofs of the parallel postulate, containing assumptions that were essentially equivalent to the parallel postulate. These early attempts did, however, provide some early properties of the hyperbolic and elliptic geometries.

Khayyam, for example, tried to derive it from an equivalent postulate he formulated from "the principles of the Philosopher" (Aristotle): *"Two convergent straight lines intersect and it is impossible for two convergent straight lines to diverge in the direction in which they converge."*

Khayyam then considered the three cases right, obtuse, and acute that the summit angles of a Saccheri quadrilateral can take and after proving a number of theorems about them, he correctly refuted the obtuse and acute cases based on his postulate and hence derived the classic postulate of Euclid which he didn't realize was equivalent to his own postulate. Another example is al-Tusi's son, Sadr al-Din (sometimes known as "Pseudo-Tusi"), who wrote a book on the subject in 1298, based on al-Tusi's later thoughts, which presented another hypothesis equivalent to the parallel postulate. "He essentially revised both the Euclidean system of axioms and postulates and the proofs of many propositions from the *Elements*." His work was published in Rome in 1594 and was studied by European geometers, including Saccheri who criticised this work as well as that of Wallis.

Giordano Vitale, in his book *Euclide restituo* (1680, 1686), used the Saccheri quadrilateral to prove that if three points are equidistant on the base AB and the summit CD, then AB and CD are everywhere equidistant.

In a work titled *Euclides ab Omni Naevo Vindicatus* (*Euclid Freed from All Flaws*), published in 1733, Saccheri quickly discarded elliptic geometry as a possibility (some others of Euclid's axioms must be modified for elliptic geometry to work) and set to work proving a great number of results in hyperbolic geometry.

He finally reached a point where he believed that his results demonstrated the impossibility of hyperbolic geometry. His claim seems to have been based on Euclidean presuppositions, because no *logical* contradiction was present. In this attempt to prove Euclidean geometry he instead unintentionally discovered a new viable geometry, but did not realize it.

In 1766 Johann Lambert wrote, but did not publish, *Theorie der Parallellinien* in which he attempted, as Saccheri did, to prove the fifth postulate. He worked with a figure that today we call a *Lambert quadrilateral*, a quadrilateral with three right angles (can be considered half of a Saccheri quadrilateral). He quickly eliminated the possibility that the fourth angle is obtuse, as had Saccheri and Khayyam, and then proceeded to prove many theorems under the assumption of an acute angle. Unlike Saccheri, he never felt that he had reached a contradiction with this assumption. He had proved the non-Euclidean result that the sum of the angles in a triangle increases as the area of the triangle decreases, and this led him to speculate on the possibility of a model of the acute case on a sphere of imaginary radius. He did not carry this idea any further.

At this time it was widely believed that the universe worked according to the principles of Euclidean geometry.

Discovery of Non-Euclidean Geometry

The beginning of the 19th century would finally witness decisive steps in the creation of non-Euclidean geometry. Circa 1813, Carl Friedrich Gauss and independently around 1818, the German professor of law Ferdinand Karl Schweikart had the germinal ideas of non-Euclidean geometry worked out, but neither published any results. Then, around 1830, the Hungarian mathematician János Bolyai and the Russian mathematician Nikolai Ivanovich Lobachevsky separately published treatises on hyperbolic geometry. Consequently, hyperbolic geometry is called Bolyai-Lobachevskian geometry, as both mathematicians, independent of each other, are the basic authors of non-Euclidean geometry. Gauss mentioned to Bolyai's father, when shown the younger Bolyai's work, that he had developed such a geometry several years before, though he did not publish. While Lobachevsky created a non-Euclidean geometry by negating the parallel postulate, Bolyai worked out a geometry where both the Euclidean and the hyperbolic geometry are possible depending on a parameter k. Bolyai ends his work by mentioning that it is not possible to decide through mathematical reasoning alone if the geometry of the physical universe is Euclidean or non-Euclidean; this is a task for the physical sciences.

Bernhard Riemann, in a famous lecture in 1854, founded the field of Riemannian geometry, discussing in particular the ideas now called manifolds, Riemannian metric, and curvature. He constructed an infinite family of geometries which are not Euclidean by giving a formula for a family of Riemannian metrics on the unit ball in Euclidean space. The simplest of these is called elliptic geometry and it is considered to be a non-Euclidean geometry due to its lack of parallel lines.

By formulating the geometry in terms of a curvature tensor, Riemann allowed non-Euclidean geometry to be applied to higher dimensions.

Terminology

It was Gauss who coined the term "non-Euclidean geometry". He was referring to his own work which today we call *hyperbolic geometry*. Several modern authors still consider "non-Euclidean geometry" and "hyperbolic geometry" to be synonyms.

Arthur Cayley noted that distance between points inside a conic could be defined in terms of logarithm and the projective cross-ratio function. The method has become called the Cayley-Klein metric because Felix Klein exploited it to describe the non-euclidean geometries in articles in 1871 and 73 and later in book form. The Cayley-Klein metrics provided working models of hyperbolic and elliptic metric geometries, as well as Euclidean geometry.

Klein is responsible for the terms "hyperbolic" and "elliptic" (in his system he called Euclidean geometry "parabolic", a term which generally fell out of use). His influence has led to the current usage of the term "non-Euclidean geometry" to mean either "hyperbolic" or "elliptic" geometry.

There are some mathematicians who would extend the list of geometries that should be called "non-Euclidean" in various ways.

Axiomatic Basis of non-Euclidean Geometry

Euclidean geometry can be axiomatically described in several ways. Unfortunately, Euclid's original system of five postulates (axioms) is not one of these as his proofs relied on several unstated assumptions which should also have been taken as axioms. Hilbert's system consisting of 20 axioms most closely follows the approach of Euclid and provides the justification for all of Euclid's proofs. Other systems, using different sets of undefined terms obtain the same geometry by different paths. In all approaches, however, there is an axiom which is logically equivalent to Euclid's fifth postulate, the parallel postulate. Hilbert uses the Playfair axiom form, while Birkhoff, for instance, uses the axiom which says that "there exists a pair of similar but not congruent triangles." In any of these systems, removal of the one axiom which is equivalent to the parallel postulate, in whatever form it takes, and leaving all the other axioms intact, produces absolute geometry. As the first 28 propositions of Euclid (in *The Elements*) do not require the use of the parallel postulate or anything equivalent to it, they are all true statements in absolute geometry.

To obtain a non-Euclidean geometry, the parallel postulate (or its equivalent) *must* be replaced by its negation. Negating the Playfair's axiom form, since it is a compound statement (... there exists one and only one ...), can be done in two ways:

- Either there will exist more than one line through the point parallel to the given line or there will exist no lines through the point parallel to the given line. In the first case, replacing the parallel postulate (or its equivalent) with the statement "In a plane, given a point P and a line ℓ not passing through P, there exist two lines through P which do not meet ℓ" and keeping all the other axioms, yields hyperbolic geometry.

- The second case is not dealt with as easily. Simply replacing the parallel postulate with the statement, "In a plane, given a point P and a line ℓ not passing through P, all the lines through P meet ℓ", does not give a consistent set of axioms. This follows since parallel lines exist in absolute geometry, but this statement says that there are no parallel lines. This problem was known (in a different guise) to Khayyam, Saccheri and Lambert and was the basis for their rejecting what was known as the "obtuse angle case". In order to obtain a consistent set of axioms which includes this axiom about having no parallel lines, some of the other axioms must be tweaked. The adjustments to be made depend upon the axiom system being used. Among others these tweaks will have the effect of modifying Euclid's second postulate from the statement that line segments can be extended indefinitely to the statement that lines are unbounded. Riemann's elliptic geometry emerges as the most natural geometry satisfying this axiom.

Models of Non-Euclidean Geometry

Two dimensional Euclidean geometry is modelled by our notion of a "flat plane."

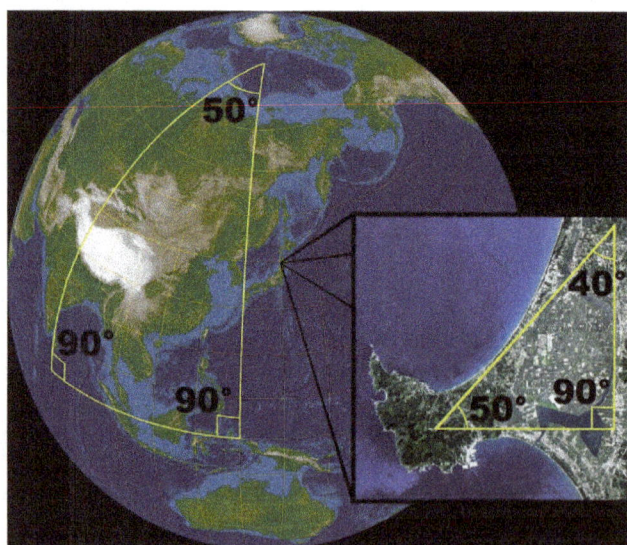

On a sphere, the sum of the angles of a triangle is not equal to 180°. The surface of a sphere is not a Euclidean space, but locally the laws of the Euclidean geometry are good approximations. In a small triangle on the face of the earth, the sum of the angles is very nearly 180°.

Elliptic Geometry

The simplest model for elliptic geometry is a sphere, where lines are "great circles" (such as the equator or the meridians on a globe), and points opposite each other (called antipodal points) are identified (considered to be the same). This is also one of the standard models of the real projective plane. The difference is that as a model of elliptic geometry a metric is introduced permitting the measurement of lengths and angles, while as a model of the projective plane there is no such metric.

In the elliptic model, for any given line ℓ and a point A, which is not on ℓ, all lines through A will intersect ℓ.

Hyperbolic Geometry

Even after the work of Lobachevsky, Gauss, and Bolyai, the question remained: "Does such a model exist for hyperbolic geometry?". The model for hyperbolic geometry was answered by Eugenio Beltrami, in 1868, who first showed that a surface called the pseudosphere has the appropriate curvature to model a portion of hyperbolic space and in a second paper in the same year, defined the Klein model which models the entirety of hyperbolic space, and used this to show that Euclidean geometry and hyperbolic geometry were equiconsistent so that hyperbolic geometry was logically consistent if and only if Euclidean geometry was. (The reverse implication follows from the horosphere model of Euclidean geometry.)

In the hyperbolic model, within a two-dimensional plane, for any given line ℓ and a point A, which is not on ℓ, there are infinitely many lines through A that do not intersect ℓ.

In these models the concepts of non-Euclidean geometries are being represented by Euclidean objects in a Euclidean setting. This introduces a perceptual distortion wherein the straight lines of the non-Euclidean geometry are being represented by Euclidean curves which visually bend. This "bending" is not a property of the non-Euclidean lines, only an artifice of the way they are being represented.

Three-dimensional Non-Euclidean Geometry

In three dimensions, there are eight models of geometries. There are Euclidean, elliptic, and hyperbolic geometries, as in the two-dimensional case; mixed geometries that are partially Euclidean and partially hyperbolic or spherical; twisted versions of the mixed geometries; and one unusual geometry that is completely anisotropic (i.e. every direction behaves differently).

Uncommon Properties

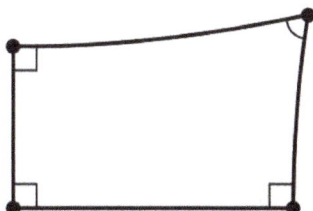

Lambert quadrilateral in hyperbolic geometry

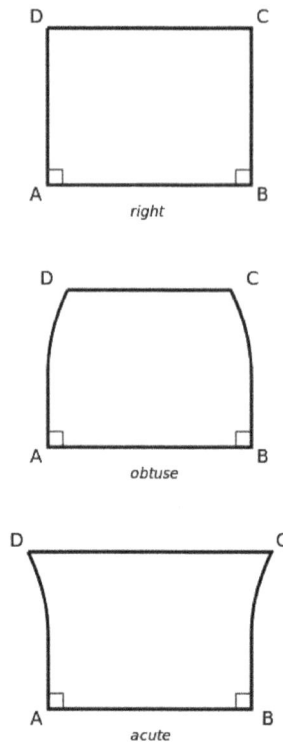

Saccheri quadrilaterals in the three geometries

Euclidean and non-Euclidean geometries naturally have many similar properties, namely those which do not depend upon the nature of parallelism. This commonality is the subject of absolute geometry (also called *neutral geometry*). However, the properties which distinguish one geometry from the others are the ones which have historically received the most attention.

Besides the behavior of lines with respect to a common perpendicular, mentioned in the introduction, we also have the following:

- A Lambert quadrilateral is a quadrilateral which has three right angles. The fourth angle of a Lambert quadrilateral is acute if the geometry is hyperbolic, a right angle if the geometry is Euclidean or obtuse if the geometry is elliptic. Consequently, rectangles exist (a statement equivalent to the parallel postulate) only in Euclidean geometry.

- A Saccheri quadrilateral is a quadrilateral which has two sides of equal length, both perpendicular to a side called the *base*. The other two angles of a Saccheri quadrilateral are called the *summit angles* and they have equal measure. The summit angles of a Saccheri quadrilateral are acute if the geometry is hyperbolic, right angles if the geometry is Euclidean and obtuse angles if the geometry is elliptic.

- The sum of the measures of the angles of any triangle is less than 180° if the geometry is hyperbolic, equal to 180° if the geometry is Euclidean, and greater than 180° if the geometry is elliptic. The *defect* of a triangle is the numerical value (180° - sum of the measures of the angles of the triangle). This result may also be stated as: the defect of triangles in hyperbolic geometry is positive, the defect of triangles in Euclidean geometry is zero, and the defect of triangles in elliptic geometry is negative.

Importance

Before the models of a non-Euclidean plane were presented by Beltrami, Klein, and Poincaré, Euclidean geometry stood unchallenged as the mathematical model of space. Furthermore, since the substance of the subject in synthetic geometry was a chief exhibit of rationality, the Euclidean point of view represented absolute authority.

The discovery of the non-Euclidean geometries had a ripple effect which went far beyond the boundaries of mathematics and science. The philosopher Immanuel Kant's treatment of human knowledge had a special role for geometry. It was his prime example of synthetic a priori knowledge; not derived from the senses nor deduced through logic — our knowledge of space was a truth that we were born with. Unfortunately for Kant, his concept of this unalterably true geometry was Euclidean. Theology was also affected by the change from absolute truth to relative truth in the way that mathematics is related to the world around it, that was a result of this paradigm shift.

Non-Euclidean geometry is an example of a scientific revolution in the history of science, in which mathematicians and scientists changed the way they viewed their subjects. Some geometers called Lobachevsky the "Copernicus of Geometry" due to the revolutionary character of his work.

The existence of non-Euclidean geometries impacted the intellectual life of Victorian England in many ways and in particular was one of the leading factors that caused a re-examination of the teaching of geometry based on Euclid's Elements. This curriculum issue was hotly debated at the time and was even the subject of a book, *Euclid and his Modern Rivals*, written by Charles Lutwidge Dodgson (1832–1898) better known as Lewis Carroll, the author of *Alice in Wonderland*.

Planar Algebras

In analytic geometry a plane is described with Cartesian coordinates : $C = \{ (x,y) : x, y \in R \}$. The points are sometimes identified with complex numbers $z = x + y \varepsilon$ where $\varepsilon^2 \in \{-1, 0, 1\}$.

The Euclidean plane corresponds to the case $\varepsilon^2 = -1$ since the modulus of z is given by

$$zz^* = (x + y\epsilon)(x - y\epsilon) = x^2 + y^2$$

and this quantity is the square of the Euclidean distance between z and the origin. For instance, $\{z \mid z z^* = 1\}$ is the unit circle.

For planar algebra, non-Euclidean geometry arises in the other cases. When $\varepsilon^2 = +1$, then z is a split-complex number and conventionally j replaces epsilon. Then

$$zz^* = (x + y\mathbf{j})(x - y\mathbf{j}) = x^2 - y^2$$

and $\{z \mid z z^* = 1\}$ is the unit hyperbola.

When $\varepsilon^2 = 0$, then z is a dual number.

This approach to non-Euclidean geometry explains the non-Euclidean angles: the parameters of slope in the dual number plane and hyperbolic angle in the split-complex plane correspond to angle in Euclidean geometry. Indeed, they each arise in polar decomposition of a complex number z.

Kinematic Geometries

Hyperbolic geometry found an application in kinematics with the cosmology introduced by Hermann Minkowski in 1908. Minkowski introduced terms like worldline and proper time into mathematical physics. He realized that the submanifold, of events one moment of proper time into the future, could be considered a hyperbolic space of three dimensions. Already in the 1890s Alexander Macfarlane was charting this submanifold through his Algebra of Physics and hyperbolic quaternions, though Macfarlane did not use cosmological language as Minkowski did in 1908. The relevant structure is now called the hyperboloid model of hyperbolic geometry.

The non-Euclidean planar algebras support kinematic geometries in the plane. For instance, the split-complex number $z = e^{aj}$ can represent a spacetime event one moment into the future of a frame of reference of rapidity a. Furthermore, multiplication by z amounts to a Lorentz boost mapping the frame with rapidity zero to that with rapidity a.

Kinematic study makes use of the dual numbers $z = x + y\epsilon$, $\epsilon^2 = 0$, to represent the classical description of motion in absolute time and space: The equations $x' = x + vt$, $t' = t$ are equivalent to a shear mapping in linear algebra:

$$\begin{pmatrix} x' \\ t' \end{pmatrix} = \begin{pmatrix} 1 & v \\ 0 & 1 \end{pmatrix}\begin{pmatrix} x \\ t \end{pmatrix}.$$

With dual numbers the mapping is $t' + x'\epsilon = (1 + v\epsilon)(t + x\epsilon) = t + (x + vt)\epsilon$.

Another view of special relativity as a non-Euclidean geometry was advanced by E. B. Wilson and Gilbert Lewis in *Proceedings of the American Academy of Arts and Sciences* in 1912. They revamped the analytic geometry implicit in the split-complex number algebra into synthetic geometry of premises and deductions.

Fiction

Non-Euclidean geometry often makes appearances in works of science fiction and fantasy.

- In 1895 H. G. Wells published the short story "The Remarkable Case of Davidson's Eyes". To appreciate this story one should know how antipodal points on a sphere are identified in a model of the elliptic plane. In the story, in the midst of a thunderstorm, Sidney Davidson sees "Waves and a remarkably neat schooner" while working in an electrical laboratory at Harlow Technical College. At the story's close Davidson proves to have witnessed H.M.S. *Fulmar* off Antipodes Island.

- Non-Euclidean geometry is sometimes connected with the influence of the 20th century horror fiction writer H. P. Lovecraft. In his works, many unnatural things follow their own unique laws of geometry: In Lovecraft's Cthulhu Mythos, the sunken city of R'lyeh is characterized by its non-Euclidean geometry. It is heavily implied this is achieved as a side effect of not following the natural laws of this universe rather than simply using an alternate geometric model, as the sheer innate wrongness of it is said to be capable of driving those who look upon it insane.

- The main character in Robert Pirsig's *Zen and the Art of Motorcycle Maintenance* mentioned Riemannian Geometry on multiple occasions.

- In *The Brothers Karamazov*, Dostoevsky discusses non-Euclidean geometry through his main character Ivan.

- Christopher Priest's novel *Inverted World* describes the struggle of living on a planet with the form of a rotating pseudosphere.

- Robert Heinlein's *The Number of the Beast* utilizes non-Euclidean geometry to explain instantaneous transport through space and time and between parallel and fictional universes.

- Alexander Bruce's *Antichamber* uses non-Euclidean geometry to create a minimal, Escher-like world, where geometry and space follow unfamiliar rules.

- Zeno Rogue's *HyperRogue* is a roguelike game set on the hyperbolic plane, allowing the player to experience many properties of this geometry. Many mechanics, quests, and locations are strongly dependent on the features of hyperbolic geometry.

- In the Renegade Legion science fiction setting for FASA's wargame, role-playing-game and fiction, faster-than-light travel and communications is possible through the use of Hsieh Ho's Polydimensional Non-Euclidean Geometry, published sometime in the middle of the 22nd century.

- In Ian Stewart's *Flatterland* the protagonist Victoria Line visit all kinds of non-Euclidean worlds.

- In Jean-Pierre Petit's *Here's looking at Euclid (and not looking at Euclid)* Archibald Higgins stumbles upon spherical geometry

References

- Greenberg, Marvin Jay (2007), Euclidean and Non-Euclidean Geometries: Development and History (4th ed.), New York: W. H. Freeman, ISBN 0-7167-9948-0

- Faber, Richard L. (1983), Foundations of Euclidean and Non-Euclidean Geometry, New York: Marcel Dekker, ISBN 0-8247-1748-1

- Falb, Peter (1990). Methods of Algebraic Geometry in Control Theory Part II Multivariable Linear Systems and Projective Algebraic Geometry. Springer. ISBN 978-0-8176-4113-9.

- Allen Tannenbaum (1982), Invariance and Systems Theory: Algebraic and Geometric Aspects, Lecture Notes in Mathematics, volume 845, Springer-Verlag, ISBN 9783540105657

- Cox, David A.; Sturmfels, Bernd. Manocha, Dinesh N., ed. Applications of Computational Algebraic Geometry. American Mathematical Soc. ISBN 978-0-8218-6758-7.

- Griffiths, Phillip; Harris, Joseph (1978). Principles of Algebraic Geometry. John Wiley & Sons. ISBN 0-471-32792-1. MR 0507725.

- Kollár, János; Mori, Shigefumi (1998), Birational Geometry of Algebraic Varieties, Cambridge University Press, ISBN 0-521-63277-3, MR 1658959

- Katz, Victor J. (1998), A History of Mathematics: An Introduction (2nd Ed.), Reading: Addison Wesley Longman, ISBN 0-321-01618-1

- Beutelspacher, Albrecht; Rosenbaum, Ute (1998). Projective Geometry: from foundations to applications. Cambridge: Cambridge University Press. ISBN 0-521-48277-1.

- Dembowski, Peter (1968), Finite geometries, Ergebnisse der Mathematik und ihrer Grenzgebiete, Band 44, Berlin, New York: Springer-Verlag, ISBN 3-540-61786-8, MR 0233275

- Richard Hartley and Andrew Zisserman, 2003. Multiple view geometry in computer vision, 2nd ed. Cambridge University Press. ISBN 0-521-54051-8

- Bezdek, András, (2003). Discrete geometry: in honor of W. Kuperberg's 60th birthday. New York, N.Y: Marcel Dekker. ISBN 0-8247-0968-3.

- Brass, Peter; Moser, William; Pach, János (2005). Research problems in discrete geometry. Berlin: Springer. ISBN 0-387-23815-8.

- Vladimir Boltyanski, Horst Martini, Petru S. Soltan, (1997). Excursions into Combinatorial Geometry. Springer. ISBN 3-540-61341-2.

- H. S. M. Coxeter (1942) Non-Euclidean Geometry, University of Toronto Press, reissued 1998 by Mathematical Association of America, ISBN 0-88385-522-4.

- Faber, Richard L. (1983), Foundations of Euclidean and Non-Euclidean Geometry, New York: Marcel Dekker, ISBN 0-8247-1748-1

- Greenberg, Marvin Jay Euclidean and Non-Euclidean Geometries: Development and History, 4th ed., New York: W. H. Freeman, 2007. ISBN 0-7167-9948-0

- Bernard H. Lavenda, (2012) " A New Perspective on Relativity : An Odyssey In Non-Euclidean Geometries", World Scientific, pp. 696, ISBN 9789814340489.

- Richards, Joan L. (1988), Mathematical Visions: The Pursuit of Geometry in Victorian England, Boston: Academic Press, ISBN 0-12-587445-6

Key Concepts of Geometry

Geometry has a number of important concepts. Some of these concepts are lines, line segment, plane, angle, similarity and congruence. A line is a straight path that joins points and has no endpoints whereas a plane is a two-dimensional surface that lengthens infinitely. This text elucidates the key theories of geometry.

Line (Geometry)

The notion of line or straight line was introduced by ancient mathematicians to represent straight objects (i.e., having no curvature) with negligible width and depth. Lines are an idealization of such objects. Until the 17th century, lines were defined in this manner: "The [straight or curved] line is the first species of quantity, which has only one dimension, namely length, without any width nor depth, and is nothing else than the flow or run of the point which [...] will leave from its imaginary moving some vestige in length, exempt of any width. [...] The straight line is that which is equally extended between its points."

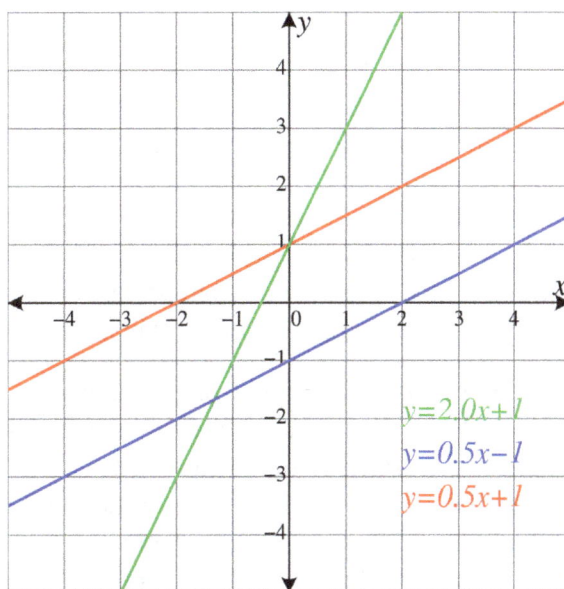

The red and blue lines on this graph have the same slope (gradient); the red and green lines have the same y-intercept (cross the y-axis at the same place).

A representation of one line segment.

Euclid described a line as "breadthless length" which "lies equally with respect to the points on itself"; he introduced several postulates as basic unprovable properties from which he constructed all of geometry, which is now called Euclidean geometry to avoid confusion with other geometries which have been introduced since the end of 19th century (such as non-Euclidean, projective and affine geometry).

In modern mathematics, given the multitude of geometries, the concept of a line is closely tied to the way the geometry is described. For instance, in analytic geometry, a line in the plane is often defined as the set of points whose coordinates satisfy a given linear equation, but in a more abstract setting, such as incidence geometry, a line may be an independent object, distinct from the set of points which lie on it.

When a geometry is described by a set of axioms, the notion of a line is usually left undefined (a so-called primitive object). The properties of lines are then determined by the axioms which refer to them. One advantage to this approach is the flexibility it gives to users of the geometry. Thus in differential geometry a line may be interpreted as a geodesic (shortest path between points), while in some projective geometries a line is a 2-dimensional vector space (all linear combinations of two independent vectors). This flexibility also extends beyond mathematics and, for example, permits physicists to think of the path of a light ray as being a line.

A line segment is a part of a line that is bounded by two distinct end points and contains every point on the line between its end points. Depending on how the line segment is defined, either of the two end points may or may not be part of the line segment. Two or more line segments may have some of the same relationships as lines, such as being parallel, intersecting, or skew, but unlike lines they may be none of these, if they are coplanar and either do not intersect or are collinear.

Definitions Versus Descriptions

All definitions are ultimately circular in nature since they depend on concepts which must themselves have definitions, a dependence which can not be continued indefinitely without returning to the starting point. To avoid this vicious circle certain concepts must be taken as primitive concepts; terms which are given no definition. In geometry, it is frequently the case that the concept of line is taken as a primitive. In those situations where a line is a defined concept, as in coordinate geometry, some other fundamental ideas are taken as primitives. When the line concept is a primitive, the behaviour and properties of lines are dictated by the axioms which they must satisfy.

In a non-axiomatic or simplified axiomatic treatment of geometry, the concept of a primitive notion may be too abstract to be dealt with. In this circumstance it is possible that a *description* or *mental image* of a primitive notion is provided to give a foundation to build the notion on which would formally be based on the (unstated) axioms. Descriptions of this type may be referred to, by some authors, as definitions in this informal style of presentation. These are not true definitions and could not be used in formal proofs of statements. The "definition" of line in Euclid's Elements falls into this category. Even in the case where a specific geometry is being considered (for example, Euclidean geometry), there is no generally accepted agreement among authors as to what an informal description of a line should be when the subject is not being treated formally.

Ray

Given a line and any point A on it, we may consider A as decomposing this line into two parts.

Each such part is called a ray (or half-line) and the point A is called its *initial point*. The point A is considered to be a member of the ray. Intuitively, a ray consists of those points on a line passing through A and proceeding indefinitely, starting at A, in one direction only along the line. However, in order to use this concept of a ray in proofs a more precise definition is required.

Given distinct points A and B, they determine a unique ray with initial point A. As two points define a unique line, this ray consists of all the points between A and B (including A and B) and all the points C on the line through A and B such that B is between A and C. This is, at times, also expressed as the set of all points C such that A is not between B and C. A point D, on the line determined by A and B but not in the ray with initial point A determined by B, will determine another ray with initial point A. With respect to the AB ray, the AD ray is called the *opposite ray*.

Thus, we would say that two different points, A and B, define a line and a decomposition of this line into the disjoint union of an open segment (A, B) and two rays, BC and AD (the point D is not drawn in the diagram, but is to the left of A on the line AB). These are not opposite rays since they have different initial points.

In Euclidean geometry two rays with a common endpoint form an angle.

The definition of a ray depends upon the notion of betweenness for points on a line. It follows that rays exist only for geometries for which this notion exists, typically Euclidean geometry or affine geometry over an ordered field. On the other hand, rays do not exist in projective geometry nor in a geometry over a non-ordered field, like the complex numbers or any finite field.

In topology, a ray in a space X is a continuous embedding $R^+ \to X$. It is used to define the important concept of end of the space.

Euclidean Geometry

When geometry was first formalised by Euclid in the *Elements*, he defined a general line (straight or curved) to be "breadthless length" with a straight line being a line "which lies evenly with the points on itself". These definitions serve little purpose since they use terms which are not, themselves, defined. In fact, Euclid did not use these definitions in this work and probably included them just to make it clear to the reader what was being discussed. In modern geometry, a line is simply taken as an undefined object with properties given by axioms, but is sometimes defined as a set of points obeying a linear relationship when some other fundamental concept is left undefined.

In an axiomatic formulation of Euclidean geometry, such as that of Hilbert (Euclid's original axioms contained various flaws which have been corrected by modern mathematicians), a line is stated to have certain properties which relate it to other lines and points. For example, for any two distinct points, there is a unique line containing them, and any two distinct lines intersect in at most one point. In two dimensions, i.e., the Euclidean plane, two lines which do not intersect are called parallel. In higher dimensions, two lines that do not intersect are parallel if they are contained in a plane, or skew if they are not.

Any collection of finitely many lines partitions the plane into convex polygons (possibly unbounded); this partition is known as an arrangement of lines.

Cartesian Plane

Lines in a Cartesian plane or, more generally, in affine coordinates, can be described algebraically by *linear* equations. In two dimensions, the equation for non-vertical lines is often given in the *slope-intercept form*:

$$y = mx + b$$

where:

m is the slope or gradient of the line.

b is the y-intercept of the line.

x is the independent variable of the function $y = f(x)$.

The slope of the line through points $A(x_a, y_a)$ and $B(x_b, y_b)$, when $x_a \neq x_b$, is given by $m = (y_b - y_a)/(x_b - x_a)$ and the equation of this line can be written $y = m(x - x_a) + y_a$.

In $\mathbb{R}^2$, every line L (including vertical lines) is described by a linear equation of the form

$$L = \{(x, y) \mid ax + by = c\}$$

with fixed real coefficients a, b and c such that a and b are not both zero. Using this form, vertical lines correspond to the equations with $b = 0$.

There are many variant ways to write the equation of a line which can all be converted from one to another by algebraic manipulation. These forms are generally named by the type of information (data) about the line that is needed to write down the form. Some of the important data of a line is its slope, x-intercept, known points on the line and y-intercept.

The equation of the line passing through two different points $P_0(x_0, y_0)$ and $P_1(x_1, y_1)$ may be written as

$$(y - y_0)(x_1 - x_0) = (y_1 - y_0)(x - x_0).$$

If $x_0 \neq x_1$, this equation may be rewritten as

$$y = (x - x_0)\frac{y_1 - y_0}{x_1 - x_0} + y_0$$

or

$$y = x\frac{y_1 - y_0}{x_1 - x_0} + \frac{x_1 y_0 - x_0 y_1}{x_1 - x_0}.$$

In three dimensions, lines can *not* be described by a single linear equation, so they are frequently described by parametric equations:

$$x = x_0 + at$$

$$y = y_0 + bt$$

$$z = z_0 + ct$$

where:

x, y, and z are all functions of the independent variable t which ranges over the real numbers.

(x_0, y_0, z_0) is any point on the line.

a, b, and c are related to the slope of the line, such that the vector (a, b, c) is parallel to the line.

They may also be described as the simultaneous solutions of two linear equations

$$a_1 x + b_1 y + c_1 z - d_1 = 0$$

$$a_2 x + b_2 y + c_2 z - d_2 = 0$$

such that (a_1, b_1, c_1) and (a_2, b_2, c_2) are not proportional (the relations $a_1 = ta_2, b_1 = tb_2, c_1 = tc_2$ imply $t = 0$). This follows since in three dimensions a single linear equation typically describes a plane and a line is what is common to two distinct intersecting planes.

Normal Form

The *normal segment* for a given line is defined to be the line segment drawn from the origin perpendicular to the line. This segment joins the origin with the closest point on the line to the origin. The *normal form* of the equation of a straight line on the plane is given by:

$$y \sin \theta + x \cos \theta - p = 0,$$

where θ is the angle of inclination of the normal segment (the oriented angle from the unit vector of the x axis to this segment), and p is the (positive) length of the normal segment. The normal form can be derived from the general form by dividing all of the coefficients by

$$\frac{|c|}{-c} \sqrt{a^2 + b^2}.$$

This form is also called the Hesse normal form, after the German mathematician Ludwig Otto Hesse.

Unlike the slope-intercept and intercept forms, this form can represent any line but also requires only two finite parameters, θ and p, to be specified. Note that if $p > 0$, then θ is uniquely defined modulo 2π. On the other hand, if the line is through the origin ($c = 0, p = 0$), one drops the $|c|/(-c)$ term to compute $\sin\theta$ and $\cos\theta$, and θ is only defined modulo π.

Polar Coordinates

In polar coordinates on the Euclidean plane the slope-intercept form of the equation of a line is expressed as:

$$r = \frac{mr\cos\theta + b}{\sin\theta},$$

where m is the slope of the line and b is the y-intercept. When $\theta = 0$ the graph will be undefined. The equation can be rewritten to eliminate discontinuities in this manner:

$$r\sin\theta = mr\cos\theta + b.$$

In polar coordinates on the Euclidean plane, the intercept form of the equation of a line that is non-horizontal, non-vertical, and does not pass through pole may be expressed as,

$$r = \frac{1}{\dfrac{\cos\theta}{x_o} + \dfrac{\sin\theta}{y_o}}$$

where x_o and y_o represent the x and y intercepts respectively. The above equation is not applicable for vertical and horizontal lines because in these cases one of the intercepts does not exist. Moreover, it is not applicable on lines passing through the pole since in this case, both x and y intercepts are zero (which is not allowed here since x_o and y_o are denominators). A vertical line that doesn't pass through the pole is given by the equation

$$r\cos\theta = x_o.$$

Similarly, a horizontal line that doesn't pass through the pole is given by the equation

$$r\sin\theta = y_o.$$

The equation of a line which passes through the pole is simply given as:

$$\theta = m$$

where m is the slope of the line.

Vector Equation

The vector equation of the line through points A and B is given by $\mathbf{r} = \mathbf{OA} + \lambda\mathbf{AB}$ (where λ is a scalar).

If a is vector OA and b is vector OB, then the equation of the line can be written: $\mathbf{r} = \mathbf{a} + \lambda(\mathbf{b} - \mathbf{a})$.

A ray starting at point A is described by limiting λ. One ray is obtained if $\lambda \geq 0$, and the opposite ray comes from $\lambda \leq 0$.

Euclidean Space

In three-dimensional space, a first degree equation in the variables x, y, and z defines a plane, so two such equations, provided the planes they give rise to are not parallel, define a line which is the intersection of the planes. More generally, in n-dimensional space n-1 first-degree equations in the n coordinate variables define a line under suitable conditions.

In more general Euclidean space, R^n (and analogously in every other affine space), the line L passing through two different points a and b (considered as vectors) is the subset

$$L = \{(1-t)a + tb \mid t \in \mathbb{R}\}$$

The direction of the line is from a ($t = 0$) to b ($t = 1$), or in other words, in the direction of the vector $b - a$. Different choices of a and b can yield the same line.

Collinear Points

Three points are said to be *collinear* if they lie on the same line. Three points *usually* determine a plane, but in the case of three collinear points this does *not* happen.

In affine coordinates, in n-dimensional space the points $X=(x_1, x_2, ..., x_n)$, $Y=(y_1, y_2, ..., y_n)$, and $Z=(z_1, z_2, ..., z_n)$ are collinear if the matrix

$$\begin{bmatrix} 1 & x_1 & x_2 & \cdots & x_n \\ 1 & y_1 & y_2 & \cdots & y_n \\ 1 & z_1 & z_2 & \cdots & z_n \end{bmatrix}$$

has a rank less than 3. In particular, for three points in the plane ($n = 2$), the above matrix is square and the points are collinear if and only if its determinant is zero.

Equivalently for three points in a plane, the points are collinear if and only if the slope between one pair of points equals the slope between any other pair of points (in which case the slope between the remaining pair of points will equal the other slopes). By extension, k points in a plane are collinear if and only if any ($k-1$) pairs of points have the same pairwise slopes.

In Euclidean geometry, the Euclidean distance $d(a,b)$ between two points a and b may be used to express the collinearity between three points by:

> The points a, b and c are collinear if and only if $d(x,a) = d(c,a)$ and $d(x,b) = d(c,b)$ implies $x=c$.

However, there are other notions of distance (such as the Manhattan distance) for which this property is not true.

In the geometries where the concept of a line is a primitive notion, as may be the case in some synthetic geometries, other methods of determining collinearity are needed.

Types of Lines

In a sense, all lines in Euclidean geometry are equal, in that, without coordinates, one can not tell them apart from one another. However, lines may play special roles with respect to other objects in the geometry and be divided into types according to that relationship. For instance, with respect to a conic (a circle, ellipse, parabola, or hyperbola), lines can be:

- tangent lines, which touch the conic at a single point;

- secant lines, which intersect the conic at two points and pass through its interior;

- exterior lines, which do not meet the conic at any point of the Euclidean plane; or

- a directrix, whose distance from a point helps to establish whether the point is on the conic.

In the context of determining parallelism in Euclidean geometry, a transversal is a line that intersects two other lines that may or not be parallel to each other.

For more general algebraic curves, lines could also be:

- i-secant lines, meeting the curve in i points counted without multiplicity, or

- asymptotes, which a curve approaches arbitrarily closely without touching it.

With respect to triangles we have:

- the Euler line,

- the Simson lines, and

- central lines.

For a convex quadrilateral with at most two parallel sides, the Newton line is the line that connects the midpoints of the two diagonals.

For a hexagon with vertices lying on a conic we have the Pascal line and, in the special case where the conic is a pair of lines, we have the Pappus line.

Parallel lines are lines in the same plane that never cross. Intersecting lines share a single point in common. Coincidental lines coincide with each other—every point that is on either one of them is also on the other.

Perpendicular lines are lines that intersect at right angles.

In three-dimensional space, skew lines are lines that are not in the same plane and thus do not intersect each other.

Projective Geometry

In many models of projective geometry, the representation of a line rarely conforms to the notion of the "straight curve" as it is visualised in Euclidean geometry. In elliptic geometry we see a typical example of this. In the spherical representation of elliptic geometry, lines are represented by great circles of a sphere with diametrically opposite points identified. In a different model of elliptic geometry, lines are represented by Euclidean planes passing through the origin. Even though these representations are visually distinct, they satisfy all the properties (such as, two points determining a unique line) that make them suitable representations for lines in this geometry.

Geodesics

The "shortness" and "straightness" of a line, interpreted as the property that the distance along the line between any two of its points is minimized, can be generalized and leads to the concept of geodesics in metric spaces.

Parallel (Geometry)

In geometry, parallel lines are lines in a plane which do not meet; that is, two lines in a plane that do not intersect or touch each other at any point are said to be parallel. By extension, a line and a plane, or two planes, in three-dimensional Euclidean space that do not share a point are said to be parallel. However, two lines in three-dimensional space which do not meet must be in a common plane to be considered parallel; otherwise they are called skew lines. Parallel planes are planes in the same three-dimensional space that never meet.

Line art drawing of parallel lines and curves.

Parallel lines are the subject of Euclid's parallel postulate. Parallelism is primarily a property of affine geometries and Euclidean space is a special instance of this type of geometry. Some other spaces, such as hyperbolic space, have analogous properties that are sometimes referred to as parallelism.

Symbol

The parallel symbol is $\parallel$. For example, $AB \parallel CD$ indicates that line AB is parallel to line CD.

In the Unicode character set, the "parallel" and "not parallel" signs have codepoints U+2225 ($\parallel$) and U+2226 ($\nparallel$), respectively. In addition, U+22D5 (⋕) represents the relation "equal and parallel to".

Euclidean Parallelism

Conditions for Parallelism

Given parallel straight lines l and m in Euclidean space, the following properties are equivalent:

1. Every point on line m is located at exactly the same (minimum) distance from line l (*equidistant lines*).

2. Line m is in the same plane as line l but does not intersect l (recall that lines extend to infinity in either direction).

3. When lines m and l are both intersected by a third straight line (a transversal) in the same plane, the corresponding angles of intersection with the transversal are congruent.

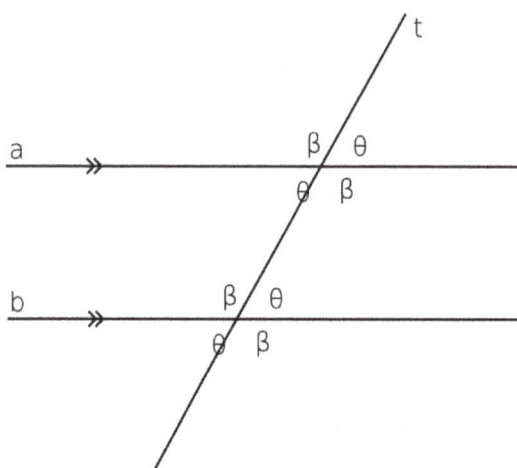

As shown by the tick marks, lines a and b are parallel. This can be proved because the transversal t produces congruent corresponding angles θ, shown here both to the right of the transversal, one above and adjacent to line a and the other above and adjacent to line b.

Since these are equivalent properties, any one of them could be taken as the definition of parallel lines in Euclidean space, but the first and third properties involve measurement, and so, are "more complicated" than the second. Thus, the second property is the one usually chosen as the defining property of parallel lines in Euclidean geometry. The other properties are then consequences of Euclid's Parallel Postulate. Another property that also involves measurement is that lines parallel to each other have the same gradient (slope).

History

The definition of parallel lines as a pair of straight lines in a plane which do not meet appears as Definition 23 in Book I of Euclid's Elements. Alternative definitions were discussed by other Greeks, often as part of an attempt to prove the parallel postulate. Proclus attributes a definition of parallel lines as equidistant lines to Posidonius and quotes Geminus in a similar vein. Simplicius also mentions Posidonius' definition as well as its modification by the philosopher Aganis.

At the end of the nineteenth century, in England, Euclid's Elements was still the standard textbook in secondary schools. The traditional treatment of geometry was being pressured to change by the new developments in projective geometry and non-Euclidean geometry, so several new textbooks for the teaching of geometry were written at this time. A major difference between these reform texts, both between themselves and between them and Euclid, is the treatment of parallel lines. These reform texts were not without their critics and one of them, Charles Dodgson (a.k.a. Lewis Carroll), wrote a play, *Euclid and His Modern Rivals*, in which these texts are lambasted.

One of the early reform textbooks was James Maurice Wilson's *Elementary Geometry* of 1868. Wilson based his definition of parallel lines on the primitive notion of *direction*. According to Wilhelm Killing the idea may be traced back to Leibniz. Wilson, without defining direction since it is a primitive, uses the term in other definitions such as his sixth definition, "Two straight lines that meet one another have different directions, and the difference of their directions is the *an-*

gle between them." Wilson (1868, p. 2) In definition 15 he introduces parallel lines in this way; "Straight lines which have the *same direction*, but are not parts of the same straight line, are called *parallel lines*." Wilson (1868, p. 12) Augustus De Morgan reviewed this text and declared it a failure, primarily on the basis of this definition and the way Wilson used it to prove things about parallel lines. Dodgson also devotes a large section of his play (Act II, Scene VI § 1) to denouncing Wilson's treatment of parallels. Wilson edited this concept out of the third and higher editions of his text.

Other properties, proposed by other reformers, used as replacements for the definition of parallel lines, did not fare much better. The main difficulty, as pointed out by Dodgson, was that to use them in this way required additional axioms to be added to the system. The equidistant line definition of Posidonius, expounded by Francis Cuthbertson in his 1874 text *Euclidean Geometry* suffers from the problem that the points that are found at a fixed given distance on one side of a straight line must be shown to form a straight line. This can not be proved and must be assumed to be true. The corresponding angles formed by a transversal property, used by W. D. Cooley in his 1860 text, *The Elements of Geometry, simplified and explained* requires a proof of the fact that if one transversal meets a pair of lines in congruent corresponding angles then all transversals must do so. Again, a new axiom is needed to justify this statement.

Construction

The three properties above lead to three different methods of construction of parallel lines.

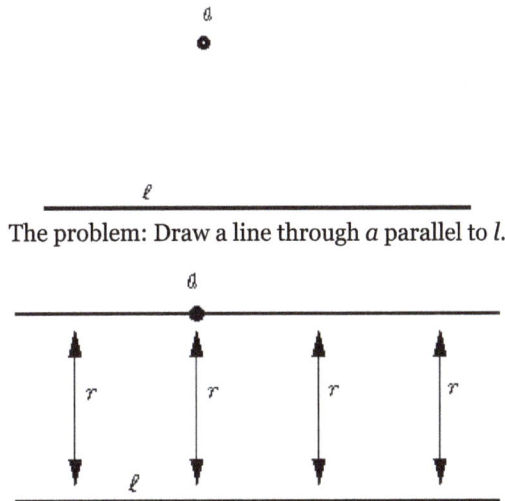

The problem: Draw a line through *a* parallel to *l*.

Property 1: Line *m* has everywhere the same distance to line *l*.

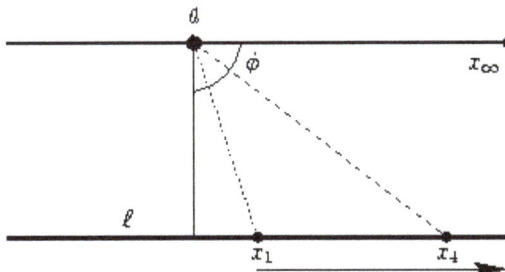

Property 2: Take a random line through *a* that intersects *l* in *x*. Move point *x* to infinity.

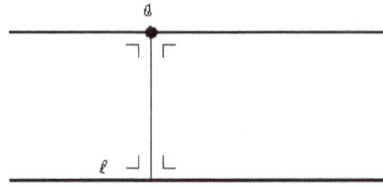

Property 3: Both l and m share a transversal line through a that intersect them at 90°.

Distance Between Two Parallel Lines

Because parallel lines in a Euclidean plane are equidistant there is a unique distance between the two parallel lines. Given the equations of two non-vertical, non-horizontal parallel lines,

$$y = mx + b_1$$

$$y = mx + b_2,$$

the distance between the two lines can be found by locating two points (one on each line) that lie on a common perpendicular to the parallel lines and calculating the distance between them. Since the lines have slope m, a common perpendicular would have slope $-1/m$ and we can take the line with equation $y = -x/m$ as a common perpendicular. Solve the linear systems

$$\begin{cases} y = mx + b_1 \\ y = -x/m \end{cases}$$

and

$$\begin{cases} y = mx + b_2 \\ y = -x/m \end{cases}$$

to get the coordinates of the points. The solutions to the linear systems are the points

$$\left(x_1, y_1\right) = \left(\frac{-b_1 m}{m^2 + 1}, \frac{b_1}{m^2 + 1}\right)$$

and

$$\left(x_2, y_2\right) = \left(\frac{-b_2 m}{m^2 + 1}, \frac{b_2}{m^2 + 1}\right).$$

These formulas still give the correct point coordinates even if the parallel lines are horizontal (i.e., $m = 0$). The distance between the points is

$$d = \sqrt{\left(\frac{b_1 m - b_2 m}{m^2 + 1}\right)^2 + \left(\frac{b_2 - b_1}{m^2 + 1}\right)^2},$$

which reduces to

$$d = \frac{|b_2 - b_1|}{\sqrt{m^2 + 1}}.$$

When the lines are given by the general form of the equation of a line (horizontal and vertical lines are included):

$$ax + by + c_1 = 0$$

$$ax + by + c_2 = 0,$$

their distance can be expressed as

$$d = \frac{|c_2 - c_1|}{\sqrt{a^2 + b^2}}.$$

Two Lines in Three-dimensional Space

Two lines in the same three-dimensional space that do not intersect need not be parallel. Only if they are in a common plane are they called parallel; otherwise they are called skew lines.

Two distinct lines l and m in three-dimensional space are parallel if and only if the distance from a point P on line m to the nearest point on line l is independent of the location of P on line m. This never holds for skew lines.

A Line and a Plane

A line m and a plane q in three-dimensional space, the line not lying in that plane, are parallel if and only if they do not intersect.

Equivalently, they are parallel if and only if the distance from a point P on line m to the nearest point in plane q is independent of the location of P on line m.

Two Planes

Similar to the fact that parallel lines must be located in the same plane, parallel planes must be situated in the same three-dimensional space and contain no point in common.

Two distinct planes q and r are parallel if and only if the distance from a point P in plane q to the nearest point in plane r is independent of the location of P in plane q. This will never hold if the two planes are not in the same three-dimensional space.

Extension to Non-Euclidean Geometry

In non-Euclidean geometry, it is more common to talk about geodesics than (straight) lines. A geodesic is the shortest path between two points in a given geometry. In physics this may be interpreted as the path that a particle follows if no force is applied to it. In non-Euclidean geometry (elliptic or hyperbolic geometry) the three Euclidean properties mentioned above are not equivalent and only the second one,(Line m is in the same plane as line l but does not intersect l) since it involves no measurements is useful in non-Euclidean geometries. In general geometry the three properties above give three different types of curves, equidistant curves, parallel geodesics and geodesics sharing a common perpendicular, respectively.

Hyperbolic Geometry

Intersecting, parallel and ultra parallel lines through a with respect to l in the hyperbolic plane. The parallel lines appear to intersect l just off the image. This is just an artifact of the visualisation. On a real hyperbolic plane the lines will get closer to each other and 'meet' in infinity.

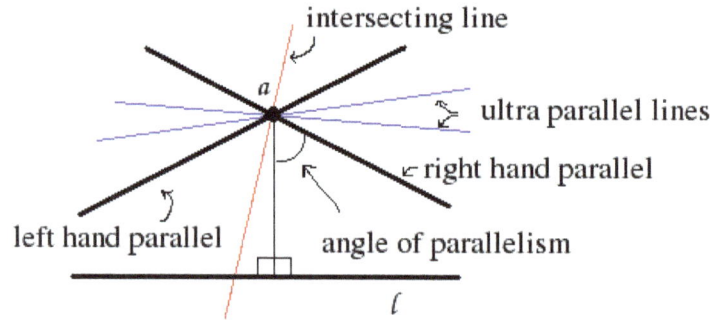

While in Euclidean geometry two geodesics can either intersect or be parallel, in hyperbolic geometry, there are three possibilities. Two geodesics belonging to the same plane can either be:

1. intersecting, if they intersect in a common point in the plane,

2. parallel, if they do not intersect in the plane, but converge to a common limit point at infinity (ideal point), or

3. ultra parallel, if they do not have a common limit point at infinity.

In the literature *ultra parallel* geodesics are often called *non-intersecting. Geodesics intersecting at infinity* are called *limiting parallel.*

As in the illustration through a point a not on line l there are two limiting parallel lines, one for each direction ideal point of line l. They separate the lines intersecting line l and those that are ultra parallel to line l.

Ultra parallel lines have single common perpendicular (ultraparallel theorem), and diverge on both sides of this common perpendicular.

Spherical or Elliptic Geometry

In spherical geometry, all geodesics are great circles. Great circles divide the sphere in two equal hemispheres and all great circles intersect each other. Thus, there are no parallel geodesics to a given geodesic, as all geodesics intersect. Equidistant curves on the sphere are called parallels of latitude analogous to the latitude lines on a globe. Parallels of latitude can be generated by the intersection of the sphere with a plane parallel to a plane through the center of the sphere.

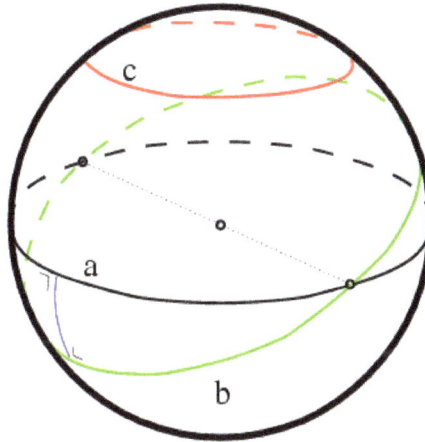

On the sphere there is no such thing as a parallel line. Line *a* is a great circle, the equivalent of a straight line in spherical geometry. Line *c* is equidistant to line *a* but is not a great circle. It is a parallel of latitude. Line *b* is another geodesic which intersects *a* in two antipodal points. They share two common perpendiculars (one shown in blue).

Reflexive Variant

In synthetic, affine geometry the relation of two parallel lines is a fundamental concept that is modified from the usage in Euclidean geometry. It is clear that the relation of parallelism is a symmetric relation and a transitive relation. These are two properties of an equivalence relation. In Euclidean geometry a line is *not* considered to be parallel to itself, but in affine geometry it is convenient to hold a line as parallel to itself, thus yielding parallelism as an equivalence relation.

Another way of describing this type of parallelism is the requirement that their intersection is *not* a singleton. Two lines are then parallel when they have all or none of their points in common. It has been noted that Playfair's axiom used in affine and Euclidean geometry is then equivalent to the statement that parallelism forms a transitive relation on the set of lines in the plane.

Perpendicular

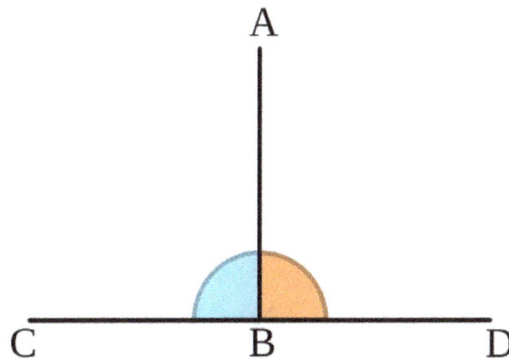

The segment AB is perpendicular to the segment CD because the two angles it creates (indicated in orange and blue) are each 90 degrees. The segment AB can be called *the perpendicular from A to the segment CD*, using "perpendicular" as a noun. The point *B* is called the *foot of the perpendicular from A to segment CD*, or simply, the *foot of A on CD*.

In elementary geometry, the property of being perpendicular (perpendicularity) is the relationship between two lines which meet at a right angle (90 degrees). The property extends to other related geometric objects.

A line is said to be perpendicular to another line if the two lines intersect at a right angle. Explicitly, a first line is perpendicular to a second line if (1) the two lines meet; and (2) at the point of intersection the straight angle on one side of the first line is cut by the second line into two congruent angles. Perpendicularity can be shown to be symmetric, meaning if a first line is perpendicular to a second line, then the second line is also perpendicular to the first. For this reason, we may speak of two lines as being perpendicular (to each other) without specifying an order.

Perpendicularity easily extends to segments and rays. For example, a line segment $\overline{AB}$ is perpendicular to a line segment $\overline{CD}$ if, when each is extended in both directions to form an infinite line, these two resulting lines are perpendicular in the sense above. In symbols, $\overline{AB} \perp \overline{CD}$ means line segment AB is perpendicular to line segment CD.

A line is said to be perpendicular to a plane if it is perpendicular to every line in the plane that it intersects. This definition depends on the definition of perpendicularity between lines.

Two planes in space are said to be perpendicular if the dihedral angle at which they meet is a right angle (90 degrees).

Perpendicularity is one particular instance of the more general mathematical concept of orthogonality; perpendicularity is the orthogonality of classical geometric objects. Thus, in advanced mathematics, the word "perpendicular" is sometimes used to describe much more complicated geometric orthogonality conditions, such as that between a surface and its normal.

Foot of a Perpendicular

The word "foot" is frequently used in connection with perpendiculars. This usage is exemplified in the top diagram, above, and its caption. The diagram can be in any orientation. The foot is not necessarily at the bottom.

Construction of the Perpendicular

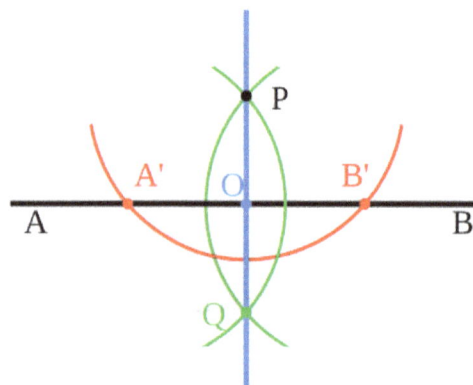

Construction of the perpendicular (blue) to the line AB through the point P.

A   g

Construction of the perpendicular to the line g at the point P (applicable not only at the end point A, M is freely selectable), animation

To make the perpendicular to the line AB through the point P using compass and straightedge, proceed as follows:

- Step 1 (red): construct a circle with center at P to create points A' and B' on the line AB, which are equidistant from P.

- Step 2 (green): construct circles centered at A' and B' having equal radius. Let Q and R be the points of intersection of these two circles.

- Step 3 (blue): connect Q and R to construct the desired perpendicular PQ.

To prove that the PQ is perpendicular to AB, use the SSS congruence theorem for ' and QPB' to conclude that angles OPA' and OPB' are equal. Then use the SAS congruence theorem for triangles OPA' and OPB' to conclude that angles POA and POB are equal.

The Pythagorean Theorem can be used as the basis of methods of constructing right angles. For example, by counting links, three pieces of chain can be made with lengths in the ratio 3:4:5. These can be laid out to form a triangle, which will have a right angle opposite its longest side. This method is useful for laying out gardens and fields, where the dimensions are large, and great accuracy is not needed. The chains can be used repeatedly whenever required.

In Relationship to Parallel Lines

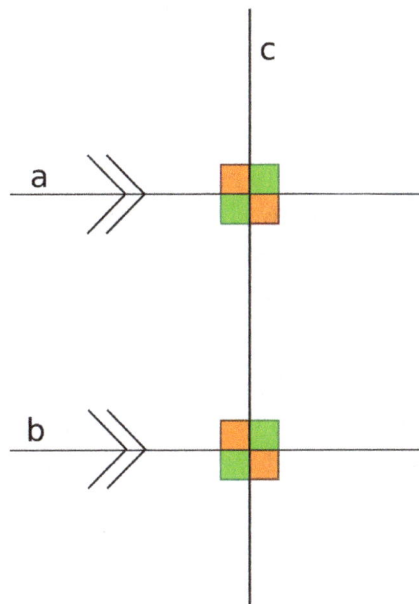

The arrowhead marks indicate that the lines *a* and *b*, cut by the transversal line *c*, are parallel.

If two lines (*a* and *b*) are both perpendicular to a third line (*c*), all of the angles formed along the third line are right angles. Therefore, in Euclidean geometry, any two lines that are both perpendicular to a third line are parallel to each other, because of the parallel postulate. Conversely, if one line is perpendicular to a second line, it is also perpendicular to any line parallel to that second line.

In the figure at the right, all of the orange-shaded angles are congruent to each other and all of the green-shaded angles are congruent to each other, because vertical angles are congruent and alternate interior angles formed by a transversal cutting parallel lines are congruent. Therefore, if lines a and b are parallel, any of the following conclusions leads to all of the others:

- One of the angles in the diagram is a right angle.

- One of the orange-shaded angles is congruent to one of the green-shaded angles.

- Line c is perpendicular to line a.

- Line c is perpendicular to line b.

In Computing Distances

The distance from a point to a line is the distance to the nearest point on that line. That is the point at which a segment from it to the given point is perpendicular to the line.

Likewise, the distance from a point to a curve is measured by a line segment that is perpendicular to a tangent line to the curve at the nearest point on the curve.

Perpendicular regression fits a line to data points by minimizing the sum of squared perpendicular distances from the data points to the line.

The distance from a point to a plane is measured as the length from the point along a segment that is perpendicular to the plane, meaning that it is perpendicular to all lines in the plane that pass through the nearest point in the plane to the given point.

Graph of Functions

In the two-dimensional plane, right angles can be formed by two intersected lines which the product of their slopes equals -1. Thus defining two linear functions: $y_1 = a_1 x + b_1$ and $y_2 = a_2 x + b_2$, the graphs of the functions will be perpendicular and will make four right angles where the lines intersect if and only if $a_1 a_2 = -1$. However, this method cannot be used if the slope is zero or undefined (the line is parallel to an axis).

For another method, let the two linear functions: $a_1 x + b_1 y + c_1 = 0$ and $a_2 x + b_2 y + c_2 = 0$. The lines will be perpendicular if and only if $a_1 a_2 + b_1 b_2 = 0$. This method is simplified from the dot product (or, more generally, the inner product) of vectors. In particular, two vectors are considered orthogonal if their inner product is zero.

In Circles and Other Conics

Circles

Each diameter of a circle is perpendicular to the tangent line to that circle at the point where the diameter intersects the circle.

A line segment through a circle's center bisecting a chord is perpendicular to the chord.

If the intersection of any two perpendicular chords divides one chord into lengths a and b and divides the other chord into lengths c and d, then $a^2 + b^2 + c^2 + d^2$ equals the square of the diameter.

The sum of the squared lengths of any two perpendicular chords intersecting at a given point is the same as that of any other two perpendicular chords intersecting at the same point, and is given by $8r^2 - 4p^2$ (where r is the circle's radius and p is the distance from the center point to the point of intersection).

Thales' theorem states that two lines both through the same point on a circle but going through opposite endpoints of a diameter are perpendicular. This is equivalent to saying that any diameter of a circle subtends a right angle at any point on the circle, except the two endpoints of the diameter.

Ellipses

The major and minor axes of an ellipse are perpendicular to each other and to the tangent lines to the ellipse at the points where the axes intersect the ellipse.

The major axis of an ellipse is perpendicular to the directrix and to each latus rectum.

Parabolas

In a parabola, the axis of symmetry is perpendicular to each of the latus rectum, the directrix, and the tangent line at the point where the axis intersects the parabola.

From a point on the tangent line to a parabola's vertex, the other tangent line to the parabola is perpendicular to the line from that point through the parabola's focus.

The orthoptic property of a parabola is that If two tangents to the parabola are perpendicular to each other, then they intersect on the directrix. Conversely, two tangents which intersect on the directrix are perpendicular. This implies that, seen from any point on its directrix, any parabola subtends a right angle.

Hyperbolas

The transverse axis of a hyperbola is perpendicular to the conjugate axis and to each directrix.

The product of the perpendicular distances from a point P on a hyperbola or on its conjugate hyperbola to the asymptotes is a constant independent of the location of P.

A rectangular hyperbola has asymptotes that are perpendicular to each other. It has an eccentricity equal to $\sqrt{2}$.

In Polygons

Triangles

The legs of a right triangle are perpendicular to each other.

The altitudes of a triangle are perpendicular to their respective bases. The perpendicular bisectors of the sides also play a prominent role in triangle geometry.

The Euler line of an isosceles triangle is perpendicular to the triangle's base.

The Droz-Farny line theorem concerns a property of two perpendicular lines intersecting at a triangle's orthocenter.

Harcourt's theorem concerns the relationship of line segments through a vertex and perpendicular to any line tangent to the triangle's incircle.

Quadrilaterals

In a square or other rectangle, all pairs of adjacent sides are perpendicular. A right trapezoid is a trapezoid that has two pairs of adjacent sides that are perpendicular.

Each of the four maltitudes of a quadrilateral is a perpendicular to a side through the midpoint of the opposite side.

An orthodiagonal quadrilateral is a quadrilateral whose diagonals are perpendicular. These include the square, the rhombus, and the kite. By Brahmagupta's theorem, in an orthodiagonal quadrilateral that is also cyclic, a line through the midpoint of one side and through the intersection point of the diagonals is perpendicular to the opposite side.

By van Aubel's theorem, if squares are constructed externally on the sides of a quadrilateral, the line segments connecting the centers of opposite squares are perpendicular and equal in length.

Lines in Three Dimensions

Up to three lines in three-dimensional space can be pairwise perpendicular, as exemplified by the x, y, and z axes of a three-dimensional Cartesian coordinate system.

Line Segment

In geometry, a line segment is a part of a line that is bounded by two distinct end points, and contains every point on the line between its endpoints. A closed line segment includes both endpoints, while an open line segment excludes both endpoints; a half-open line segment includes exactly one of the endpoints.

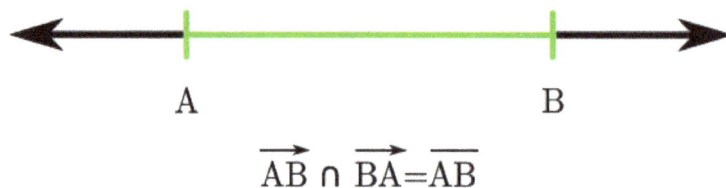

$$\overrightarrow{AB} \cap \overrightarrow{BA} = \overline{AB}$$

The geometric definition of a closed line segment: the intersection of all points at or to the right of A with all points at or to the left of B

historical image – create a line segment (1699)

Examples of line segments include the sides of a triangle or square. More generally, when both of the segment's end points are vertices of a polygon or polyhedron, the line segment is either an edge (of that polygon or polyhedron) if they are adjacent vertices, or otherwise a diagonal. When the end points both lie on a curve such as a circle, a line segment is called a chord (of that curve).

In Real or Complex Vector Spaces

If V is a vector space over $\mathbb{R}$ or $\mathbb{C}$, and L is a subset of V, then L is a line segment if L can be parameterized as

$$L = \{\mathbf{u} + t\mathbf{v} \mid t \in [0,1]\}$$

for some vectors $\mathbf{u}, \mathbf{v} \in V$, in which case the vectors u and u + v are called the end points of L.

Sometimes one needs to distinguish between "open" and "closed" line segments. Then one defines a closed line segment as above, and an open line segment as a subset L that can be parametrized as

$$L = \{\mathbf{u} + t\mathbf{v} \mid t \in (0,1)\}$$

for some vectors $\mathbf{u}, \mathbf{v} \in V$.

Equivalently, a line segment is the convex hull of two points. Thus, the line segment can be expressed as a convex combination of the segment's two end points.

In geometry, it is sometimes defined that a point B is between two other points A and C, if the distance AB added to the distance BC is equal to the distance AC. Thus in $\mathbb{R}^2$ the line segment with endpoints $A = (a_x, a_y)$ and $C = (c_x, c_y)$ is the following collection of points:

$$\{(x,y) \mid \sqrt{(x-c_x)^2 + (y-c_y)^2} + \sqrt{(x-a_x)^2 + (y-a_y)^2} = \sqrt{(c_x-a_x)^2 + (c_y-a_y)^2}\}.$$

Properties

- A line segment is a connected, non-empty set.

- If V is a topological vector space, then a closed line segment is a closed set in V. However, an open line segment is an open set in V if and only if V is one-dimensional.

- More generally than above, the concept of a line segment can be defined in an ordered geometry.

- A pair of line segments can be any one of the following: intersecting, parallel, skew, or none of these. The last possibility is a way that line segments differ from lines: if two nonparallel lines are in the same Euclidean plane they must cross each other, but that need not be true of segments.

In Proofs

In an axiomatic treatment of geometry, the notion of betweenness is either assumed to satisfy a certain number of axioms, or else be defined in terms of an isometry of a line (used as a coordinate system).

Segments play an important role in other theories. For example, a set is convex if the segment that joins any two points of the set is contained in the set. This is important because it transforms some of the analysis of convex sets to the analysis of a line segment. The Segment Addition Postulate can be used to add congruent segment or segments with equal lengths and consequently substitute other segments into another statement to make segments congruent.

As a Degenerate Ellipse

A line segment can be viewed as a degenerate case of an ellipse in which the semiminor axis goes to zero, the foci go to the endpoints, and the eccentricity goes to one. As a degenerate orbit this is a radial elliptic trajectory.

In Other Geometric Shapes

In addition to appearing as the edges and diagonals of polygons and polyhedra, line segments appear in numerous other locations relative to other geometric shapes.

Triangles

Some very frequently considered segments in a triangle include the three altitudes (each perpendicularly connecting a side or its extension to the opposite vertex), the three medians (each connecting a side's midpoint to the opposite vertex), the perpendicular bisectors of the sides (perpendicularly connecting the midpoint of a side to one of the other sides), and the internal angle bisectors (each connecting a vertex to the opposite side). In each case there are various equalities relating these segment lengths to others as well as various inequalities.

Other segments of interest in a triangle include those connecting various triangle centers to each other, most notably the incenter, the circumcenter, the nine-point center, the centroid, and the orthocenter.

Quadrilaterals

In addition to the sides and diagonals of a quadrilateral, some important segments are the two bimedians (connecting the midpoints of opposite sides) and the four maltitudes (each perpendicularly connecting one side to the midpoint of the opposite side).

Circles and Ellipses

Any straight line segment connecting two points on a circle or ellipse is called a chord. Any chord in a circle which has no longer chord is called a diameter, and any segment connecting the circle's center (the midpoint of a diameter) to a point on the circle is called a radius.

In an ellipse, the longest chord is called the *major axis*, and a segment from the midpoint of the major axis (the ellipse's center) to either endpoint of the major axis is called a *semi-major axis*. Similarly, the shortest chord of an ellipse is called the *minor axis*, and the segment from its midpoint (the ellipse's center) to either of its endpoints is called a *semi-minor axis*. The chords of an ellipse which are perpendicular to the major axis and pass through one of its foci are called the latera recta of the ellipse. The *interfocal segment* connects the two foci.

Directed Line Segment

When a line segment is given an orientation (direction) it suggests a translation or perhaps a force tending to make a translation. The magnitude and direction are indicative of a potential change. This suggestion has been absorbed into mathematical physics through the concept of a Euclidean vector. The collection of all directed line segments is usually reduced by making "equivalent" any pair having the same length and orientation. This application of an equivalence relation dates from Giusto Bellavitis's introduction of the concept of equipollence of directed line segments in 1835.

Generalizations

Analogous to straight line segments above, one can define arcs as segments of a curve.

Diagonal

In geometry, a diagonal is a line segment joining two vertices of a polygon or polyhedron, when those vertices are not on the same edge. Informally, any sloping line is called diagonal. It was used by both Strabo and Euclid to refer to a line connecting two vertices of a rhombus or cuboid, and later adopted into Latin as *diagonus* ("slanting line").

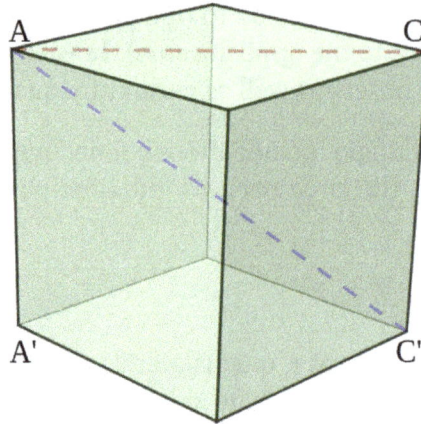

The diagonals of a cube with side length 1. AC' (shown in blue) is a space diagonal with length $\sqrt{3}$, while AC (shown in red) is a face diagonal and has length $\sqrt{2}$.

In matrix algebra, a diagonal of a square matrix is a set of entries extending from one corner to the farthest corner.

There are also other, non-mathematical uses.

Non-mathematical Uses

In engineering, a diagonal brace is a beam used to brace a rectangular structure (such as scaffolding) to withstand strong forces pushing into it; although called a diagonal, due to practical considerations diagonal braces are often not connected to the corners of the rectangle.

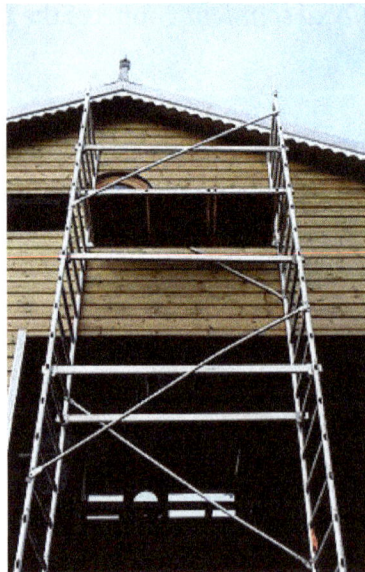

A stand of basic scaffolding on a house construction site, with diagonal braces to maintain its structure

Diagonal pliers are wire-cutting pliers defined by the cutting edges of the jaws intersects the joint rivet at an angle or "on a diagonal", hence the name.

A diagonal lashing is a type of lashing used to bind spars or poles together applied so that the lashings cross over the poles at an angle.

In association football, the diagonal system of control is the method referees and assistant referees use to position themselves in one of the four quadrants of the pitch.

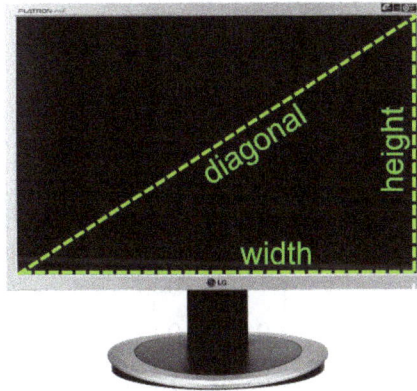

The diagonal is a common measurement of display size.

Polygons

As applied to a polygon, a diagonal is a line segment joining any two non-consecutive vertices. Therefore, a quadrilateral has two diagonals, joining opposite pairs of vertices. For any convex polygon, all the diagonals are inside the polygon, but for re-entrant polygons, some diagonals are outside of the polygon.

Any n-sided polygon ($n \geq 3$), convex or concave, has $\frac{n(n-3)}{2}$ diagonals, as each vertex has diagonals to all other vertices except itself and the two adjacent vertices, or $n - 3$ diagonals, and each diagonal is shared by two vertices.

Sides	Diagonals	Sides	Diagonals	Sides	Diagonals	Sides	Diagonals	Sides	Diagonals
3	0	11	44	19	152	27	324	35	560
4	2	12	54	20	170	28	350	36	594
5	5	13	65	21	189	29	377	37	629
6	9	14	77	22	209	30	405	38	665
7	14	15	90	23	230	31	434	39	702
8	20	16	104	24	252	32	464	40	740
9	27	17	119	25	275	33	495	41	779
10	35	18	135	26	299	34	527	42	819

Regions Formed by Diagonals

In a convex polygon, if no three diagonals are concurrent at a single point, the number of regions that the diagonals divide the interior into is given by

$$\binom{n}{4} + \binom{n-1}{2} = \frac{(n-1)(n-2)(n^2-3n+12)}{24}.$$

For n-gons with $n=3, 4, \ldots$ the number of regions is

1, 4, 11, 25, 50, 91, 154, 246...

This is OEIS sequence A006522.

Matrices

In the case of a square matrix, the *main* or *principal diagonal* is the diagonal line of entries running from the top-left corner to the bottom-right corner. For a matrix A with row index specified by i and column index specified by j, these would be entries A_{ij} with $i = j$ For example, the identity matrix can be defined as having entries of 1 on the main diagonal and zeroes elsewhere:

$$\begin{pmatrix} 1 & 0 & 0 \\ 0 & 1 & 0 \\ 0 & 0 & 1 \end{pmatrix}$$

The top-right to bottom-left diagonal is sometimes described as the *minor* diagonal or *antidiagonal*. The *off-diagonal* entries are those not on the main diagonal. A *diagonal matrix* is one whose off-diagonal entries are all zero.

A *superdiagonal* entry is one that is directly above and to the right of the main diagonal. Just as diagonal entries are those A_{ij} with $j = i$, the superdiagonal entries are those with $j = i+1$. For example, the non-zero entries of the following matrix all lie in the superdiagonal:

$$\begin{pmatrix} 0 & 2 & 0 \\ 0 & 0 & 3 \\ 0 & 0 & 0 \end{pmatrix}$$

Likewise, a *subdiagonal* entry is one that is directly below and to the left of the main diagonal, that is, an entry A_{ij} with $j = i-1$. General matrix diagonals can be specified by an index k measured relative to the main diagonal: the main diagonal has $k = 0$; the superdiagonal has $k = 1$; the subdiagonal has $k = -1$; and in general, the k – diagonal consists of the entries A_{ij} with $j = i+k$.

Geometry

By analogy, the subset of the Cartesian product $X \times X$ of any set X with itself, consisting of all pairs (x,x), is called the diagonal, and is the graph of the equality relation on X or equivalently the graph of the identity function from X to x. This plays an important part in geometry; for example, the fixed points of a mapping F from X to itself may be obtained by intersecting the graph of F with the diagonal.

In geometric studies, the idea of intersecting the diagonal *with itself* is common, not directly, but by perturbing it within an equivalence class. This is related at a deep level with the Euler characteristic and the zeros of vector fields. For example, the circle S^1 has Betti numbers 1, 1, 0, 0, 0, and therefore Euler characteristic 0. A geometric way of expressing this is to look at the diagonal on the two-torus $S^1 \times S^1$ and observe that it can move *off itself* by the small motion (θ, θ) to $(\theta, \theta + \varepsilon)$. In general, the intersection number of the graph of a function with the diagonal may be computed using homology via the Lefschetz fixed point theorem; the self-intersection of the diagonal is the special case of the identity function.

Point (Geometry)

In modern mathematics, a point refers usually to an element of some set called a space.

More specifically, in Euclidean geometry, a point is a primitive notion upon which the geometry is built. Being a primitive notion means that a point cannot be defined in terms of previously defined objects. That is, a point is defined only by some properties, called axioms, that it must satisfy. In particular, the geometric points do not have any length, area, volume, or any other dimensional attribute. A common interpretation is that the concept of a point is meant to capture the notion of a unique location in Euclidean space.

Points in Euclidean Geometry

Points, considered within the framework of Euclidean geometry, are one of the most fundamental objects. Euclid originally defined the point as "that which has no part". In two-dimensional Euclidean space, a point is represented by an ordered pair (x, y) of numbers, where the first number conventionally represents the horizontal and is often denoted by x, and the second number conventionally represents the vertical and is often denoted by y. This idea is easily generalized to three-dimensional Euclidean space, where a point is represented by an ordered triplet (x, y, z) with the additional third number representing depth and often denoted by z. Further generalizations are represented by an ordered tuplet of n terms, $(a_1, a_2, \ldots, a_n)$ where n is the dimension of the space in which the point is located.

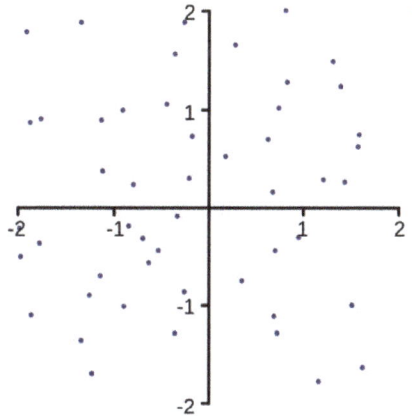

A finite set of points (blue) in two-dimensional Euclidean space.

Many constructs within Euclidean geometry consist of an infinite collection of points that conform to certain axioms. This is usually represented by a set of points; As an example, a line is an infinite set of points of the form $L = \{(a_1, a_2, \ldots a_n) \mid a_1 c_1 + a_2 c_2 + \ldots a_n c_n = d\}$, where c_1 through c_n and d are constants and n is the dimension of the space. Similar constructions exist that define the plane, line segment and other related concepts. By the way, a degenerate line segment consists of only one point.

In addition to defining points and constructs related to points, Euclid also postulated a key idea about points; he claimed that any two points can be connected by a straight line. This is easily confirmed under modern expansions of Euclidean geometry, and had lasting consequences at its introduction, allowing the construction of almost all the geometric concepts of the time. However, Euclid's postulation of points was neither complete nor definitive, as he occasionally assumed facts about points that did not follow directly from his axioms, such as the ordering of points on the line or the existence of specific points. In spite of this, modern expansions of the system serve to remove these assumptions.

Dimension of a Point

There are several inequivalent definitions of dimension in mathematics. In all of the common definitions, a point is 0-dimensional.

Vector Space Dimension

The dimension of a vector space is the maximum size of a linearly independent subset. In a vector space consisting of a single point (which must be the zero vector 0), there is no linearly independent subset. The zero vector is not itself linearly independent, because there is a non trivial linear combination making it zero: $1 \cdot \mathbf{0} = \mathbf{0}$.

Topological Dimension

The topological dimension of a topological space X is defined to be the minimum value of n, such that every finite open cover $\mathcal{A}$ of X admits a finite open cover $\mathcal{B}$ of X which refines $\mathcal{A}$ in which no point is included in more than $n+1$ elements. If no such minimal n exists, the space is said to be of infinite covering dimension.

A point is zero-dimensional with respect to the covering dimension because every open cover of the space has a refinement consisting of a single open set.

Hausdorff Dimension

Let X be a metric space. If $S \subset X$ and $d \in [0, \infty)$, the d-dimensional Hausdorff content of S is the infimum of the set of numbers $\delta \geq 0$ such that there is some (indexed) collection of balls $\{B(x_i, r_i) : i \in I\}$ covering S with $r_i > 0$ for each $i \in I$ that satisfies $\sum_{i \in I} r_i^d < \delta$.

The Hausdorff dimension of X is defined by

$$\dim_{\mathrm{H}}(X) := \inf\{d \geq 0 : C_H^d(X) = 0\}.$$

A point has Hausdorff dimension 0 because it can be covered by a single ball of arbitrarily small radius.

Geometry Without Points

Although the notion of a point is generally considered fundamental in mainstream geometry and topology, there are some systems that forgo it, e.g. noncommutative geometry and pointless topology. A "pointless" or "pointfree" space is defined not as a set, but via some structure (algebraic or logical respectively) which looks like a well-known function space on the set: an algebra of continuous functions or an algebra of sets respectively. More precisely, such structures generalize well-known spaces of functions in a way that the operation "take a value at this point" may not be defined. A further tradition starts from some books of A. N. Whitehead in which the notion of region is assumed as a primitive together with the one of *inclusion* or *connection*.

Point Masses and the Dirac Delta Function

Often in physics and mathematics, it is useful to think of a point as having non-zero mass or charge (this is especially common in classical electromagnetism, where electrons are idealized as points with non-zero charge). The Dirac delta function, or δ function, is (informally) a generalized function on the real number line that is zero everywhere except at zero, with an integral of one over the entire real line. The delta function is sometimes thought of as an infinitely high, infinitely thin spike at the origin, with total area one under the spike, and physically represents an idealized point mass or point charge. It was introduced by theoretical physicist Paul Dirac. In the context of signal processing it is often referred to as the unit impulse symbol (or function). Its discrete analog is the Kronecker delta function which is usually defined on a finite domain and takes values 0 and 1.

Vertex

Vertex means the "top", or the highest point of something. It may refer to:

Mathematics and Computer Science

- Vertex (geometry), a point where higher-dimensional geometric objects meet

- Vertex (graph theory), a node in a graph

- Vertex (curve), a local extreme point of curvature

- Vertex of a representation, a certain type of subgroup in finite group theory

- Vertex (computer graphics), a point in space with additional attributes

Physics

- Vertex (physics), a point where particles collide and interact

- Vertex (optics), a point where the optical axis crosses a lens surface

- Vertex operator algebra in conformal field theory

- Vertex function describing the interaction between a photon and an electron

- Vertex model in statistical mechanics, a discrete model of a physical system in which weights are associated with vertices of a grid graph.

Biology and Anatomy

- Vertex (anatomy), the uppermost surface of the head of an arthropod or vertebrate

- Vertex (urinary bladder), alternative name of the apex of urinary bladder

- Vertex presentation, head-first birth of a baby

Companies

- Vertex (company), a business services provider

- Vertex Pharmaceuticals, a biotech company

- Vertex Inc, an annuity software and business consulting company

Music

- A song in In the Groove (video game)

- *Vertex* (album), an album by Buck 65

- *Vertex* (band), a band formed in 1996

Collinearity

In geometry, collinearity of a set of points is the property of their lying on a single line. A set of points with this property is said to be collinear (sometimes spelled as colinear). In greater generality, the term has been used for aligned objects, that is, things being "in a line" or "in a row".

Points on a Line

In any geometry, the set of points on a line are said to be collinear. In Euclidean geometry this relation is intuitively visualized by points lying in a row on a "straight line". However, in most geometries (including Euclidean) a line is typically a primitive (undefined) object type, so such visualizations will not necessarily be appropriate. A model for the geometry offers an interpretation of how the points, lines and other object types relate to one another and a notion such as collinearity must be interpreted within the context of that model. For instance, in spherical geometry, where lines are represented in the standard model by great circles of a sphere, sets of collinear points lie on the same great circle. Such points do not lie on a "straight line" in the Euclidean sense, and are not thought of as being *in a row*.

A mapping of a geometry to itself which sends lines to lines is called a collineation; it preserves the collinearity property. The linear maps (or linear functions) of vector spaces, viewed as geometric maps, map lines to lines; that is, they map collinear point sets to collinear point sets and so, are collineations. In projective geometry these linear mappings are called *homographies* and are just one type of collineation.

Examples in Euclidean Geometry

Triangles

In any triangle the following sets of points are collinear:

- The orthocenter, the circumcenter, the centroid, the Exeter point, the de Longchamps point, and the center of the nine-point circle are collinear, all falling on a line called the Euler line.

- The de Longchamps point also has other collinearities.

- Any vertex, the tangency of the opposite side with an excircle, and the Nagel point are collinear in a line called a splitter of the triangle.

- The midpoint of any side, the point that is equidistant from it along the triangle's boundary in either direction (so these two points bisect the perimeter), and the center of the Spieker circle are collinear in a line called a cleaver of the triangle. (The Spieker circle is the incircle of the medial triangle, and its center is the center of mass of the perimeter of the triangle.)

- Any vertex, the tangency of the opposite side with the incircle, and the Gergonne point are collinear.

- From any point on the circumcircle of a triangle, the nearest points on each of the three extended sides of the triangle are collinear in the Simson line of the point on the circumcircle.

- The lines connecting the feet of the altitudes intersect the opposite sides at collinear points.

- A triangle's incenter, the midpoint of an altitude, and the point of contact of the corresponding side with the excircle relative to that side are collinear.

- Menelaus' theorem states that three points P_1, P_2, P_3 on the sides (some extended) of a triangle opposite vertices A_1, A_2, A_3 respectively are collinear if and only if the following products of segment lengths are equal:

$$P_1 A_2 \cdot P_2 A_3 \cdot P_3 A_1 = P_1 A_3 \cdot P_2 A_1 \cdot P_3 A_2.$$

- The incenter, the centroid, and the Spieker circle's center are collinear.

- The circumcenter, the Brocard midpoint, and the Lemoine point of a triangle are collinear.

Quadrilaterals

- In a convex quadrilateral $ABCD$ whose opposite sides intersect at E and F, the midpoints of AC, BD, and EF are collinear and the line through them is called the Newton line (sometimes known as the Newton-Gauss line). If the quadrilateral is a tangential quadrilateral, then its incenter also lies on this line.

- In a convex quadrilateral, the quasiorthocenter H, the "area centroid" G, and the quasicircumcenter O are collinear in this order, and $HG = 2GO$.

- Other collinearities of a tangential quadrilateral are given in Tangential quadrilateral#Collinear points.

- In a cyclic quadrilateral, the circumcenter, the vertex centroid (the intersection of the two bimedians), and the anticenter are collinear.

- In a cyclic quadrilateral, the area centroid, the vertex centroid, and the intersection of the diagonals are collinear.

- In a tangential trapezoid, the tangencies of the incircle with the two bases are collinear with the incenter.

- In a tangential trapezoid, the midpoints of the legs are collinear with the incenter.

Hexagons

- Pascal's theorem (also known as the Hexagrammum Mysticum Theorem) states that if an arbitrary six points are chosen on a conic section (i.e., ellipse, parabola or hyperbola) and joined by line segments in any order to form a hexagon, then the three pairs of opposite sides of the hexagon (extended if necessary) meet in three points which lie on a straight line, called the Pascal line of the hexagon. The converse is also true: the Braikenridge–Maclaurin theorem states that if the three intersection points of the three pairs of lines through opposite sides of a hexagon lie on a line, then the six vertices of the hexagon lie on a conic, which may be degenerate as in Pappus's hexagon theorem.

Conic Sections

- By Monge's theorem, for any three circles in a plane, none of which is inside one of the others, the three intersection points of the three pairs of lines, each externally tangent to two of the circles, are collinear.

- In an ellipse, the center, the two foci, and the two vertices with the smallest radius of curvature are collinear, and the center and the two vertices with the greatest radius of curvature are collinear.

- In a hyperbola, the center, the two foci, and the two vertices are collinear.

- Let a conic S and a point P lie in a plane. Construct three lines d_a, d_b, d_c through P such that they meet the conic at A, A'; B, B' ; C, C' respectively. Let D be a point on the polar of point P with respect to (S) or D lies on the conic (S). Let DA' ∩ BC =A_o; DB' ∩ AC = B_o; DC' ∩ AB= C_o. Then A_o, B_o, C_o are collinear.

Cones

- The center of mass of a conic solid of uniform density lies one-quarter of the way from the center of the base to the vertex, on the straight line joining the two.

Tetrahedrons

- The centroid of a tetrahedron is the midpoint between its Monge point and circumcenter.

These points define the *Euler line* of the tetrahedron that is analogous to the Euler line of a triangle. The center of the tetrahedron's twelve-point sphere also lies on the Euler line.

Algebra

Collinearity of Points Whose Coordinates are Given

In coordinate geometry, in n-dimensional space, a set of three or more distinct points are collinear if and only if, the matrix of the coordinates of these vectors is of rank 1 or less. For example, given three points $X = (x_1, x_2, \dots, x_n)$, $Y = (y_1, y_2, \dots, y_n)$, and $Z = (z_1, z_2, \dots, z_n)$, if the matrix

$$\begin{bmatrix} x_1 & x_2 & \dots & x_n \\ y_1 & y_2 & \dots & y_n \\ z_1 & z_2 & \dots & z_n \end{bmatrix}$$

is of rank 1 or less, the points are collinear.

Equivalently, for every subset of three points $X = (x_1, x_2, \dots, x_n)$, $Y = (y_1, y_2, \dots, y_n)$, and $Z = (z_1, z_2, \dots, z_n)$, if the matrix

$$\begin{bmatrix} 1 & x_1 & x_2 & \dots & x_n \\ 1 & y_1 & y_2 & \dots & y_n \\ 1 & z_1 & z_2 & \dots & z_n \end{bmatrix}$$

is of rank 2 or less, the points are collinear. In particular, for three points in the plane ($n = 2$), the above matrix is square and the points are collinear if and only if its determinant is zero; since that 3×3 determinant is plus or minus twice the area of a triangle with those three points as vertices, this is equivalent to the statement that the three points are collinear if and only if the triangle with those points as vertices has zero area.

Collinearity of Points Whose Pairwise Distances are Given

A set of at least three distinct points is called straight, meaning all the points are collinear, if and only if, for every three of those points A, B, and C, the following determinant of a Cayley–Menger determinant is zero (with $d(AB)$ meaning the distance between A and B, etc.):

$$\det \begin{bmatrix} 0 & d(AB)^2 & d(AC)^2 & 1 \\ d(AB)^2 & 0 & d(BC)^2 & 1 \\ d(AC)^2 & d(BC)^2 & 0 & 1 \\ 1 & 1 & 1 & 0 \end{bmatrix} = 0.$$

This determinant is, by Heron's formula, equal to −16 times the square of the area of a triangle with side lengths $d(AB)$, $d(BC)$, and $d(AC)$; so checking if this determinant equals zero is equivalent to checking whether the triangle with vertices A, B, and C has zero area (so the vertices are collinear).

Equivalently, a set of at least three distinct points are collinear if and only if, for every three of those points A, B, and C with $d(AC)$ greater than or equal to each of $d(AB)$ and $d(BC)$, the triangle inequality $d(AC) \leq d(AB) + d(BC)$ holds with equality.

Number Theory

Two numbers m and n are not coprime—that is, they share a common factor other than 1—if and only if for a rectangle plotted on a square lattice with vertices at $(0, 0)$, $(m, 0)$, (m, n), and $(0, n)$, at least one interior point is collinear with $(0, 0)$ and (m, n).

Concurrency (Plane Dual)

In various plane geometries the notion of interchanging the roles of "points" and "lines" while preserving the relationship between them is called plane duality. Given a set of collinear points, by plane duality we obtain a set of lines all of which meet at a common point. The property that this set of lines has (meeting at a common point) is called concurrency, and the lines are said to be concurrent lines. Thus, concurrency is the plane dual notion to collinearity.

Collinearity Graph

Given a partial geometry P, where two points determine at most one line, a collinearity graph of P is a graph whose vertices are the points of P, where two vertices are adjacent if and only if they determine a line in P.

Usage in Statistics and Econometrics

In statistics, collinearity refers to a linear relationship between two explanatory variables. Two variables are *perfectly collinear* if there is an exact linear relationship between the two, so the correlation between them is equal to 1 or −1. That is, X_1 and X_2 are perfectly collinear if there exist parameters λ_0 and λ_1 such that, for all observations i, we have

$$X_{2i} = \lambda_0 + \lambda_1 X_{1i}.$$

This means that if the various observations (X_{1i}, X_{2i}) are plotted in the (X_1, X_2) plane, these points are collinear in the sense defined earlier in this article.

Perfect multicollinearity refers to a situation in which k ($k \geq 2$) explanatory variables in a multiple regression model are perfectly linearly related, according to

$$X_{ki} = \lambda_0 + \lambda_1 X_{1i} + \lambda_2 X_{2i} + \ldots + \lambda_{k-1} X_{(k-1),i}$$

for all observations i. In practice, we rarely face perfect multicollinearity in a data set. More commonly, the issue of multicollinearity arises when there is a "strong linear relationship" among two or more independent variables, meaning that

$$X_{ki} = \lambda_0 + \lambda_1 X_{1i} + \lambda_2 X_{2i} + \ldots + \lambda_{k-1} X_{(k-1),i} + \varepsilon_i$$

where the variance of ε_i is relatively small.

The concept of *lateral collinearity* expands on this traditional view, and refers to collinearity between explanatory and criteria (i.e., explained) variables.

Usage in Other Areas

Antenna Arrays

In telecommunications, a collinear (or co-linear) antenna array is an array of dipole antennas mounted in such a manner that the corresponding elements of each antenna are parallel and aligned, that is they are located along a common line or axis.

An antenna mast with four collinear directional arrays.

Photography

The collinearity equations are a set of two equations, used in photogrammetry and remote sensing to relate coordinates in an image (sensor) plane (in two dimensions) to object coordinates (in three dimensions). In the photography setting, the equations are derived by considering the central projection of a point of the object through the optical centre of the camera to the image in the image (sensor) plane. The three points, object point, image point and optical centre, are always collinear. Another way to say this is that the line segments joining the object points with their image points are all concurrent at the optical centre.

Plane (Geometry)

In mathematics, a plane is a flat, two-dimensional surface that extends infinitely far. A plane is the two-dimensional analogue of a point (zero dimensions), a line (one dimension) and three-dimensional space. Planes can arise as subspaces of some higher-dimensional space, as with a room's walls extended infinitely far, or they may enjoy an independent existence in their own right, as in the setting of Euclidean geometry.

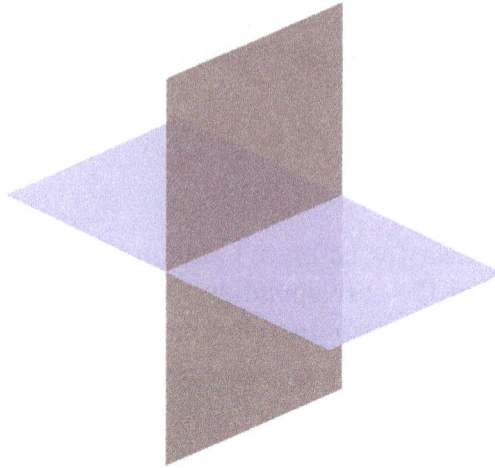

Two intersecting planes in three-dimensional space

When working exclusively in two-dimensional Euclidean space, the definite article is used, so, *the* plane refers to the whole space. Many fundamental tasks in mathematics, geometry, trigonometry, graph theory and graphing are performed in a two-dimensional space, or in other words, in the plane.

Euclidean Geometry

Euclid set forth the first great landmark of mathematical thought, an axiomatic treatment of geometry. He selected a small core of undefined terms (called *common notions*) and postulates (or axioms) which he then used to prove various geometrical statements. Although the plane in its modern sense is not directly given a definition anywhere in the *Elements*, it may be thought of as part of the common notions. In his work Euclid never makes use of numbers to measure length, angle, or area. In this way the Euclidean plane is not quite the same as the Cartesian plane.

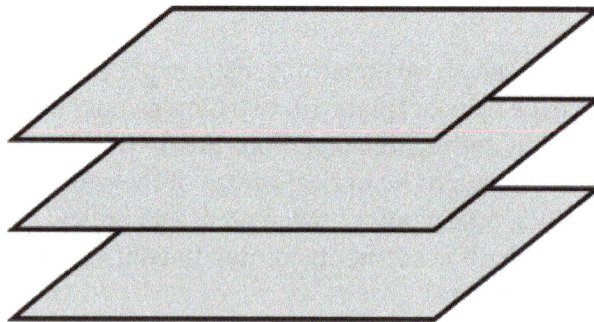

Three parallel planes.

Planes Embedded in 3-dimensional Euclidean Space

This section is solely concerned with planes embedded in three dimensions: specifically, in R³.

Determination by Contained Points and Lines

In a Euclidean space of any number of dimensions, a plane is uniquely determined by any of the following:

- Three non-collinear points (points not on a single line).

- A line and a point not on that line.

- Two distinct but intersecting lines.

- Two parallel lines.

Properties

The following statements hold in three-dimensional Euclidean space but not in higher dimensions, though they have higher-dimensional analogues:

- Two distinct planes are either parallel or they intersect in a line.

- A line is either parallel to a plane, intersects it at a single point, or is contained in the plane.

- Two distinct lines perpendicular to the same plane must be parallel to each other.

- Two distinct planes perpendicular to the same line must be parallel to each other.

Point-normal form and General form of the Equation of a Plane

In a manner analogous to the way lines in a two-dimensional space are described using a point-slope form for their equations, planes in a three dimensional space have a natural description using a point in the plane and a vector orthogonal to it (the normal vector) to indicate its "inclination".

Specifically, let r_0 be the position vector of some point $P_0 = (x_0, y_0, z_0)$, and let $n = (a, b, c)$ be a non-zero vector. The plane determined by the point P_0 and the vector n consists of those points P, with position vector r, such that the vector drawn from P_0 to P is perpendicular to n. Recalling that two vectors are perpendicular if and only if their dot product is zero, it follows that the desired plane can be described as the set of all points r such that

$$\mathbf{n} \cdot (\mathbf{r} - \mathbf{r}_0) = 0.$$

(The dot here means a dot product, not scalar multiplication.) Expanded this becomes

$$a(x - x_0) + b(y - y_0) + c(z - z_0) = 0,$$

which is the *point-normal* form of the equation of a plane. This is just a linear equation

$$ax + by + cz + d = 0,$$

where

$$d = -(ax_0 + by_0 + cz_0).$$

Conversely, it is easily shown that if a, b, c and d are constants and a, b, and c are not all zero, then the graph of the equation

$$ax + by + cz + d = 0,$$

is a plane having the vector $n = (a, b, c)$ as a normal. This familiar equation for a plane is called the *general form* of the equation of the plane.

Thus for example a regression equation of the form $y = d + ax + cz$ (with $b = -1$) establishes a best-fit plane in three-dimensional space when there are two explanatory variables.

Describing a plane with a point and two vectors lying on it

Alternatively, a plane may be described parametrically as the set of all points of the form

$$\mathbf{r} = \mathbf{r_0} + s\mathbf{v} + t\mathbf{w},$$

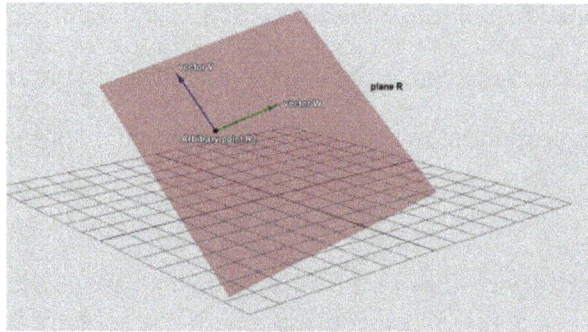

Vector description of a plane

where s and t range over all real numbers, v and w are given linearly independent vectors defining the plane, and r_o is the vector representing the position of an arbitrary (but fixed) point on the plane. The vectors v and w can be visualized as vectors starting at r_o and pointing in different directions along the plane. Note that v and w can be perpendicular, but cannot be parallel.

Describing a Plane through Three Points

Let $p_1 = (x_1, y_1, z_1)$, $p_2 = (x_2, y_2, z_2)$, and $p_3 = (x_3, y_3, z_3)$ be non-collinear points.

Method 1

The plane passing through p_1, p_2, and p_3 can be described as the set of all points (x,y,z) that satisfy the following determinant equations:

$$\begin{vmatrix} x - x_1 & y - y_1 & z - z_1 \\ x_2 - x_1 & y_2 - y_1 & z_2 - z_1 \\ x_3 - x_1 & y_3 - y_1 & z_3 - z_1 \end{vmatrix} = \begin{vmatrix} x - x_1 & y - y_1 & z - z_1 \\ x - x_2 & y - y_2 & z - z_2 \\ x - x_3 & y - y_3 & z - z_3 \end{vmatrix} = 0.$$

Method 2

To describe the plane by an equation of the form $ax + by + cz + d = 0$, solve the following system of equations:

$$ax_1 + by_1 + cz_1 + d = 0$$

$$ax_2 + by_2 + cz_2 + d = 0$$

$$ax_3 + by_3 + cz_3 + d = 0.$$

This system can be solved using Cramer's Rule and basic matrix manipulations. Let

$$D = \begin{vmatrix} x_1 & y_1 & z_1 \\ x_2 & y_2 & z_2 \\ x_3 & y_3 & z_3 \end{vmatrix}.$$

If D is non-zero (so for planes not through the origin) the values for a, b and c can be calculated as follows:

$$a = \frac{-d}{D} \begin{vmatrix} 1 & y_1 & z_1 \\ 1 & y_2 & z_2 \\ 1 & y_3 & z_3 \end{vmatrix}$$

$$b = \frac{-d}{D} \begin{vmatrix} x_1 & 1 & z_1 \\ x_2 & 1 & z_2 \\ x_3 & 1 & z_3 \end{vmatrix}$$

$$c = \frac{-d}{D} \begin{vmatrix} x_1 & y_1 & 1 \\ x_2 & y_2 & 1 \\ x_3 & y_3 & 1 \end{vmatrix}.$$

These equations are parametric in d. Setting d equal to any non-zero number and substituting it into these equations will yield one solution set.

Method 3

This plane can also be described by the "point and a normal vector" prescription above. A suitable normal vector is given by the cross product

$$\mathbf{n} = (\mathbf{p}_2 - \mathbf{p}_1) \times (\mathbf{p}_3 - \mathbf{p}_1),$$

and the point r_0 can be taken to be any of the given points p_1, p_2 or p_3.

Distance from a Point to a Plane

For a plane $\Pi : ax + by + cz + d = 0$ and a point $\mathbf{p}_1 = (x_1, y_1, z_1)$ not necessarily lying on the plane, the shortest distance from $\mathbf{p}_1$ to the plane is

$$D = \frac{|ax_1 + by_1 + cz_1 + d|}{\sqrt{a^2 + b^2 + c^2}}.$$

It follows that $\mathbf{p}_1$ lies in the plane if and only if $D=0$.

If $\sqrt{a^2 + b^2 + c^2} = 1$ meaning that a, b, and c are normalized then the equation becomes

$$D = |ax_1 + by_1 + cz_1 + d|.$$

Another vector form for the equation of a plane, known as the Hesse normal form relies on the parameter D. This form is:

$$\mathbf{n} \cdot \mathbf{r} - D_0 = 0,$$

where $\mathbf{n}$ is a unit normal vector to the plane, $\mathbf{r}$ a position vector of a point of the plane and D_0 the distance of the plane from the origin.

The general formula for higher dimensions can be quickly arrived at using vector notation. Let the hyperplane have equation $\mathbf{n} \cdot (\mathbf{r} - \mathbf{r}_0) = 0$, where the $\mathbf{n}$ is a normal vector and $\mathbf{r}_0 = (x_{10}, x_{20}, \ldots, x_{N0})$ is a position vector to a point in the hyperplane. We desire the perpendicular distance to the point $\mathbf{r}_1 = (x_{11}, x_{21}, \ldots, x_{N1})$. The hyperplane may also be represented by the scalar equation $\sum_{i=1}^{N} a_i x_i = -a_0$, for constants $\{a_i\}$. Likewise, a corresponding $\mathbf{n}$ may be represented as $(a_1, a_2, \ldots, a_N)$. We desire the scalar projection of the vector $\mathbf{r}_1 - \mathbf{r}_0$ in the direction of $\mathbf{n}$. Noting that $\mathbf{n} \cdot \mathbf{r}_0 = \mathbf{r}_0 \cdot \mathbf{n} = -a_0$ (as $\mathbf{r}_0$ satisfies the equation of the hyperplane) we have

$$D = \frac{|(\mathbf{r}_1 - \mathbf{r}_0) \cdot \mathbf{n}|}{|\mathbf{n}|}$$

$$= \frac{|\mathbf{r}_1 \cdot \mathbf{n} - \mathbf{r}_0 \cdot \mathbf{n}|}{|\mathbf{n}|}$$

$$= \frac{|\mathbf{r}_1 \cdot \mathbf{n} + a_0|}{|\mathbf{n}|}$$

$$= \frac{|a_1 x_{11} + a_2 x_{21} + \ldots + a_N x_{N1} + a_0|}{\sqrt{a_1^2 + a_2^2 + \ldots + a_N^2}}.$$

Line of Intersection between Two Planes

The line of intersection between two planes $\Pi_1 : \mathbf{n}_1 \cdot \mathbf{r} = h_1$ and $\Pi_2 : \mathbf{n}_2 \cdot \mathbf{r} = h_2$ where $\mathbf{n}_i$ are normalized is given by

$$\mathbf{r} = (c_1 \mathbf{n}_1 + c_2 \mathbf{n}_2) + \lambda (\mathbf{n}_1 \times \mathbf{n}_2)$$

where

$$c_1 = \frac{h_1 - h_2 (\mathbf{n}_1 \cdot \mathbf{n}_2)}{1 - (\mathbf{n}_1 \cdot \mathbf{n}_2)^2}$$

$$c_2 = \frac{h_2 - h_1(\mathbf{n}_1 \cdot \mathbf{n}_2)}{1 - (\mathbf{n}_1 \cdot \mathbf{n}_2)^2}.$$

This is found by noticing that the line must be perpendicular to both plane normals, and so parallel to their cross product $\mathbf{n}_1 \times \mathbf{n}_2$ (this cross product is zero if and only if the planes are parallel, and are therefore non-intersecting or entirely coincident).

The remainder of the expression is arrived at by finding an arbitrary point on the line. To do so, consider that any point in space may be written as $\mathbf{r} = c_1\mathbf{n}_1 + c_2\mathbf{n}_2 + \lambda(\mathbf{n}_1 \times \mathbf{n}_2)$, since $\{\mathbf{n}_1, \mathbf{n}_2, (\mathbf{n}_1 \times \mathbf{n}_2)\}$ is a basis. We wish to find a point which is on both planes (i.e. on their intersection), so insert this equation into each of the equations of the planes to get two simultaneous equations which can be solved for c_1 and .

If we further assume that $\mathbf{n}_1$ and $\mathbf{n}_2$ are orthonormal then the closest point on the line of intersection to the origin is $\mathbf{r}_0 = h_1\mathbf{n}_1 + h_2\mathbf{n}_2$. If that is not the case, then a more complex procedure must be used.

Dihedral Angle

Given two intersecting planes described by $\Pi_1 : a_1x + b_1y + c_1z + d_1 = 0$ and $\Pi_2 : a_2x + b_2y + c_2z + d_2 = 0,$, the dihedral angle between them is defined to be the angle α between their normal directions:

$$\cos \alpha = \frac{\hat{n}_1 \cdot \hat{n}_2}{|\hat{n}_1||\hat{n}_2|} = \frac{a_1a_2 + b_1b_2 + c_1c_2}{\sqrt{a_1^2 + b_1^2 + c_1^2}\sqrt{a_2^2 + b_2^2 + c_2^2}}.$$

Planes in Various Areas of Mathematics

In addition to its familiar geometric structure, with isomorphisms that are isometries with respect to the usual inner product, the plane may be viewed at various other levels of abstraction. Each level of abstraction corresponds to a specific category.

At one extreme, all geometrical and metric concepts may be dropped to leave the topological plane, which may be thought of as an idealized homotopically trivial infinite rubber sheet, which retains a notion of proximity, but has no distances. The topological plane has a concept of a linear path, but no concept of a straight line. The topological plane, or its equivalent the open disc, is the basic topological neighborhood used to construct surfaces (or 2-manifolds) classified in low-dimensional topology. Isomorphisms of the topological plane are all continuous bijections. The topological plane is the natural context for the branch of graph theory that deals with planar graphs, and results such as the four color theorem.

The plane may also be viewed as an affine space, whose isomorphisms are combinations of translations and non-singular linear maps. From this viewpoint there are no distances, but collinearity and ratios of distances on any line are preserved.

Differential geometry views a plane as a 2-dimensional real manifold, a topological plane which is provided with a differential structure. Again in this case, there is no notion of distance, but there is now a concept of smoothness of maps, for example a differentiable or smooth path (depending on the type of differential structure applied). The isomorphisms in this case are bijections with the chosen degree of differentiability.

In the opposite direction of abstraction, we may apply a compatible field structure to the geometric plane, giving rise to the complex plane and the major area of complex analysis. The complex field has only two isomorphisms that leave the real line fixed, the identity and conjugation.

In the same way as in the real case, the plane may also be viewed as the simplest, one-dimensional (over the complex numbers) complex manifold, sometimes called the complex line. However, this viewpoint contrasts sharply with the case of the plane as a 2-dimensional real manifold. The isomorphisms are all conformal bijections of the complex plane, but the only possibilities are maps that correspond to the composition of a multiplication by a complex number and a translation.

In addition, the Euclidean geometry (which has zero curvature everywhere) is not the only geometry that the plane may have. The plane may be given a spherical geometry by using the stereographic projection. This can be thought of as placing a sphere on the plane (just like a ball on the floor), removing the top point, and projecting the sphere onto the plane from this point). This is one of the projections that may be used in making a flat map of part of the Earth's surface. The resulting geometry has constant positive curvature.

Alternatively, the plane can also be given a metric which gives it constant negative curvature giving the hyperbolic plane. The latter possibility finds an application in the theory of special relativity in the simplified case where there are two spatial dimensions and one time dimension. (The hyperbolic plane is a timelike hypersurface in three-dimensional Minkowski space.)

Topological and Differential Geometric Notions

The one-point compactification of the plane is homeomorphic to a sphere; the open disk is homeomorphic to a sphere with the "north pole" missing; adding that point completes the (compact) sphere. The result of this compactification is a manifold referred to as the Riemann sphere or the complex projective line. The projection from the Euclidean plane to a sphere without a point is a diffeomorphism and even a conformal map.

The plane itself is homeomorphic (and diffeomorphic) to an open disk. For the hyperbolic plane such diffeomorphism is conformal, but for the Euclidean plane it is not.

Similarity (Geometry)

Two geometrical objects are called similar if they both have the same shape, or one has the same shape as the mirror image of the other. More precisely, one can be obtained from the other by uniformly scaling (enlarging or reducing), possibly with additional translation, rotation and reflection. This means that either object can be rescaled, repositioned, and reflected, so as to coincide precisely with the other object. If two objects are similar, each is congruent to the result of a particular uniform scaling of the other. A modern and novel perspective of similarity is to consider geometrical objects similar if one appears congruent to the other when zoomed in or out at some level.

Figures shown in the same color are similar

For example, all circles are similar to each other, all squares are similar to each other, and all equilateral triangles are similar to each other. On the other hand, ellipses are *not* all similar to each other, rectangles are not all similar to each other, and isosceles triangles are not all similar to each other.

If two angles of a triangle have measures equal to the measures of two angles of another triangle, then the triangles are similar. Corresponding sides of similar polygons are in proportion, and corresponding angles of similar polygons have the same measure.

This article assumes that a scaling can have a scale factor of 1, so that all congruent shapes are also similar, but some school textbooks specifically exclude congruent triangles from their definition of similar triangles by insisting that the sizes must be different if the triangles are to qualify as similar.

Similar Triangles

In geometry two triangles, $\triangle ABC$ and $\triangle A'B'C'$, are similar if and only if corresponding angles have the same measure: this implies that they are similar if and only if the lengths of corresponding sides are proportional. It can be shown that two triangles having congruent angles (*equiangular triangles*) are similar, that is, the corresponding sides can be proved to be proportional. This is known as the *AAA similarity theorem*. Due to this theorem, several authors simplify the definition of similar triangles to only require that the corresponding three angles are congruent.

There are several statements each of which is necessary and sufficient for two triangles to be similar:

1. The triangles have two congruent angles, which in Euclidean geometry implies that all their angles are congruent. That is:

> If $\angle BAC$ is equal in measure to $\angle B'A'C'$, and $\angle ABC$ is equal in measure to $\angle A'B'C'$, then this implies that $\angle ACB$ is equal in measure to $\angle A'C'B'$ and the triangles are similar.

2. All the corresponding sides have lengths in the same ratio:

> $AB/A'B' = BC/B'C' = AC/A'C'$. This is equivalent to saying that one triangle (or its mirror image) is an enlargement of the other.

3. Two sides have lengths in the same ratio, and the angles included between these sides have the same measure. For instance:

$AB/A'B' = BC/B'C'$ and $\angle ABC$ is equal in measure to $\angle A'B'C'$.

This is known as the *SAS Similarity Criterion*.

When two triangles $\triangle ABC$ and $\triangle A'B'C'$ are similar, one writes[:p. 22]

$\triangle ABC \sim \triangle A'B'C'$.

There are several elementary results concerning similar triangles in Euclidean geometry:

- Any two equilateral triangles are similar.

- Two triangles, both similar to a third triangle, are similar to each other (transitivity of similarity of triangles).

- Corresponding altitudes of similar triangles have the same ratio as the corresponding sides.

- Two right triangles are similar if the hypotenuse and one other side have lengths in the same ratio.

Given a triangle $\triangle ABC$ and a line segment DE one can, with straightedge and compass, find a point F such that $\triangle ABC \sim \triangle DEF$. The statement that the point F satisfying this condition exists is *Wallis's Postulate* and is logically equivalent to Euclid's Parallel Postulate. In hyperbolic geometry (where Wallis's Postulate is false) similar triangles are congruent.

In the axiomatic treatment of Euclidean geometry given by G.D. Birkhoff the SAS Similarity Criterion given above was used to replace both Euclid's Parallel Postulate and the SAS axiom which enabled the dramatic shortening of Hilbert's axioms.

Other Similar Polygons

The concept of similarity extends to polygons with more than three sides. Given any two similar polygons, corresponding sides taken in the same sequence (even if clockwise for one polygon and counterclockwise for the other) are proportional and corresponding angles taken in the same sequence are equal in measure. However, proportionality of corresponding sides is not by itself sufficient to prove similarity for polygons beyond triangles (otherwise, for example, all rhombi would be similar). Likewise, equality of all angles in sequence is not sufficient to guarantee similarity (otherwise all rectangles would be similar). A sufficient condition for similarity of polygons is that corresponding sides and diagonals are proportional.

For given n, all regular n-gons are similar.

Similar Curves

Several types of curves have the property that all examples of that type are similar to each other. These include:

- Circles

- Parabolas

- Hyperbolas of a specific eccentricity

- Ellipses of a specific eccentricity

- Catenaries

- Graphs of the logarithm function for different bases

- Graphs of the exponential function for different bases

- Logarithmic spirals

Similarity in Euclidean Space

A similarity (also called a similarity transformation or similitude) of a Euclidean space is a bijection f from the space onto itself that multiplies all distances by the same positive real number r, so that for any two points x and y we have

$$d(f(x), f(y)) = rd(x, y),$$

where "$d(x,y)$" is the Euclidean distance from x to y. The scalar r has many names in the literature including; the *ratio of similarity*, the *stretching factor* and the *similarity coefficient*. When $r = 1$ a similarity is called an isometry (rigid motion). Two sets are called similar if one is the image of the other under a similarity.

As a map $f : \mathbb{R}^n \to \mathbb{R}^n$, a similarity of ratio r takes the form

$$f(x) = rAx + t,$$

where $A \in O_n(\mathbb{R})$ is an $n \times n$ orthogonal matrix and $t \in \mathbb{R}^n$ is a translation vector.

Similarities preserve planes, lines, perpendicularity, parallelism, midpoints, inequalities between distances and line segments. Similarities preserve angles but do not necessarily preserve orientation, *direct similitudes* preserve orientation and *opposite similitudes* change it.

The similarities of Euclidean space form a group under the operation of composition called the *similarities group S*. The direct similitudes form a normal subgroup of S and the Euclidean group $E(n)$ of isometries also forms a normal subgroup. The similarities group S is itself a subgroup of the affine group, so every similarity is an affine transformation.

One can view the Euclidean plane as the complex plane, that is, as a 2-dimensional space over the reals. The 2D similarity transformations can then be expressed in terms of complex arithmetic and are given by $f(z) = az + b$ (direct similitudes) and $f(z) = a\overline{z} + b$ (opposite similitudes), where a and b are complex numbers, $a \neq 0$. When $|a| = 1$, these similarities are isometries.

Ratios of Sides, of Areas, and of Volumes

The ratio between the areas of similar figures is equal to the square of the ratio of corresponding lengths of those figures (for example, when the side of a square or the radius of a circle is multiplied by three, its area is multiplied by nine — i.e. by three squared). The altitudes of similar triangles are in the same ratio as corresponding sides. If a triangle has a side of length b and an altitude drawn to that side of length h then a similar triangle with corresponding side of length kb will have an altitude drawn to that side of length kh. The area of the first triangle is, $A = 1/2bh$, while the area of the similar triangle will be $A' = 1/2(kb)(kh) = k^2A$. Similar figures which can be decomposed into similar triangles will have areas related in the same way. The relationship holds for figures that are not rectifiable as well.

The ratio between the volumes of similar figures is equal to the cube of the ratio of corresponding lengths of those figures (for example, when the edge of a cube or the radius of a sphere is multiplied by three, its volume is multiplied by 27 — i.e. by three cubed).

Galileo's square–cube law concerns similar solids. If the ratio of similitude (ratio of corresponding sides) between the solids is k, then the ratio of surface areas of the solids will be k^2, while the ratio of volumes will be k^3.

Similarity in General Metric Spaces

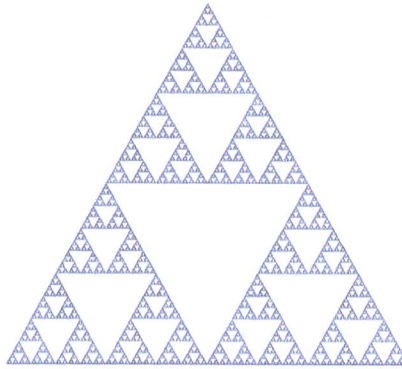

Sierpiński triangle. A space having self-similarity dimension log 3/log 2 = $\log_2 3$, which is approximately 1.58. (From Hausdorff dimension.)

In a general metric space (X, d), an exact similitude is a function f from the metric space X into itself that multiplies all distances by the same positive scalar r, called f's contraction factor, so that for any two points x and y we have

$$d(f(x), f(y)) = rd(x, y).$$

Weaker versions of similarity would for instance have f be a bi-Lipschitz function and the scalar r a limit

$$\lim \frac{d(f(x), f(y))}{d(x, y)} = r.$$

This weaker version applies when the metric is an effective resistance on a topologically self-similar set.

A self-similar subset of a metric space (X, d) is a set K for which there exists a finite set of similitudes $\{f_s\}_{s \in S}$ with contraction factors $0 \le r_s < 1$ such that K is the unique compact subset of X for which

$$\bigcup_{s \in S} f_s(K) = K.$$

These self-similar sets have a self-similar measure μ^D with dimension D given by the formula

$$\sum_{s \in S} (r_s)^D = 1$$

which is often (but not always) equal to the set's Hausdorff dimension and packing dimension. If the overlaps between the $f_s(K)$ are "small", we have the following simple formula for the measure:

$$\mu^D(f_{s_1} \circ f_{s_2} \circ \cdots \circ f_{s_n}(K)) = (r_{s_1} \cdot r_{s_2} \cdots r_{s_n})^D.$$

Topology

In topology, a metric space can be constructed by defining a similarity instead of a distance. The similarity is a function such that its value is greater when two points are closer (contrary to the distance, which is a measure of dissimilarity: the closer the points, the lesser the distance).

The definition of the similarity can vary among authors, depending on which properties are desired. The basic common properties are

1. Positive defined:

$$\forall(a,b), S(a,b) \geq 0$$

2. Majored by the similarity of one element on itself (auto-similarity):

$$S(a,b) \leq S(a,a) \quad \text{and} \quad \forall(a,b), S(a,b) = S(a,a) \Leftrightarrow a = b$$

More properties can be invoked, such as reflectivity ($\forall(a,b)S(a,b) = S(b,a)$) or finiteness ($\forall(a,b) \, S(a,b) < \infty$). The upper value is often set at 1 (creating a possibility for a probabilistic interpretation of the similitude).

Self-similarity

Self-similarity means that a pattern is non-trivially similar to itself, e.g., the set $\{..., 0.5, 0.75, 1, 1.5, 2, 3, 4, 6, 8, 12, ...\}$ of numbers of the form $\{2^i, 3 \cdot 2^i\}$ where i ranges over all integers. When this set is plotted on a logarithmic scale it has one-dimensional translational symmetry: adding or subtracting the logarithm of two to the logarithm of one of these numbers produces the logarithm of another of these numbers. In the given set of numbers themselves, this corresponds to a similarity transformation in which the numbers are multiplied or divided by two.

Psychology

The intuition for the notion of geometric similarity already appears in human children, as can be seen in their drawings.

Congruence (Geometry)

In geometry, two figures or objects are congruent if they have the same shape and size, or if one has the same shape and size as the mirror image of the other.

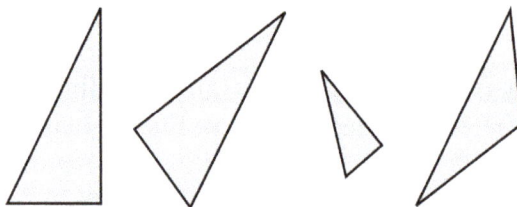

An example of congruence. The two triangles on the left are congruent, while the third is similar to them. The last triangle is neither similar nor congruent to any of the others. Note that congruence permits alteration of some properties, such as location and orientation, but leaves others unchanged, like distance and angles. The unchanged properties are called invariants.

More formally, two sets of points are called congruent if, and only if, one can be transformed into the other by an isometry, i.e., a combination of rigid motions, namely a translation, a rotation, and a reflection. This means that either object can be repositioned and reflected (but not resized) so as to coincide precisely with the other object. So two distinct plane figures on a piece of paper are congruent if we can cut them out and then match them up completely. Turning the paper over is permitted.

In elementary geometry the word *congruent* is often used as follows. The word *equal* is often used in place of *congruent* for these objects.

- Two line segments are congruent if they have the same length.

- Two angles are congruent if they have the same measure.

- Two circles are congruent if they have the same diameter.

In this sense, *two plane figures are congruent* implies that their corresponding characteristics are "congruent" or "equal" including not just their corresponding sides and angles, but also their corresponding diagonals, perimeters and areas.

The related concept of similarity applies if the objects differ in size but not in shape.

Determining Congruence of Polygons

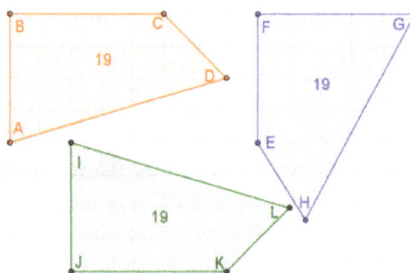

The orange and green quadrilaterals are congruent; the blue is not congruent to them. All three have the same perimeter and area. (The ordering of the sides of the blue quadrilateral is "mixed" which results in two of the interior angles and one of the diagonals not being congruent.)

For two polygons to be congruent, they must have an equal number of sides (and hence an equal number—the same number—of vertices). Two polygons with n sides are congruent if and only if they each have numerically identical sequences (even if clockwise for one polygon and counterclockwise for the other) side-angle-side-angle-... for n sides and n angles.

Congruence of polygons can be established graphically as follows:

- First, match and label the corresponding vertices of the two figures.

- Second, draw a vector from one of the vertices of the one of the figures to the corresponding vertex of the other figure. *Translate* the first figure by this vector so that these two vertices match.

- Third, *rotate* the translated figure about the matched vertex until one pair of corresponding sides matches.

- Fourth, *reflect* the rotated figure about this matched side until the figures match.

If at any time the step cannot be completed, the polygons are not congruent.

Congruence of Triangles

Two triangles are congruent if their corresponding sides are equal in length, in which case their corresponding angles are equal in size.

If triangle ABC is congruent to triangle DEF, the relationship can be written mathematically as:

$$\triangle ABC \cong \triangle DEF$$

In many cases it is sufficient to establish the equality of three corresponding parts and use one of the following results to deduce the congruence of the two triangles.

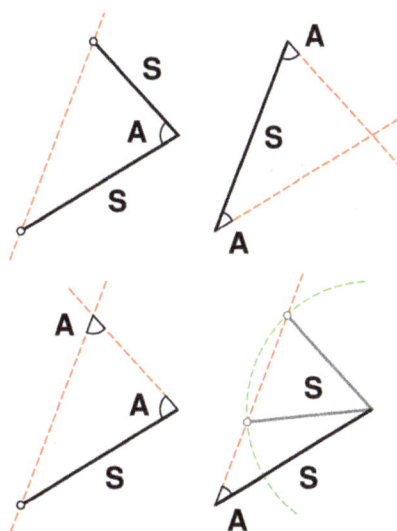

The shape of a triangle is determined up to congruence by specifying two sides and the angle between them (SAS), two angles and the side between them (ASA) or two angles and a corresponding adjacent side (AAS). Specifying two sides and an adjacent angle (SSA), however, can yield two distinct possible triangles.

Determining Congruence

Sufficient evidence for congruence between two triangles in Euclidean space can be shown through the following comparisons:

- SAS (Side-Angle-Side): If two pairs of sides of two triangles are equal in length, and the included angles are equal in measurement, then the triangles are congruent.

- SSS (Side-Side-Side): If three pairs of sides of two triangles are equal in length, then the triangles are congruent.

- ASA (Angle-Side-Angle): If two pairs of angles of two triangles are equal in measurement, and the included sides are equal in length, then the triangles are congruent. The ASA Postulate was contributed by Thales of Miletus (Greek). In most systems of axioms, the three criteria—SAS, SSS and ASA—are established as theorems. In the School Mathematics Study Group system SAS is taken as one (#15) of 22 postulates.

- AAS (Angle-Angle-Side): If two pairs of angles of two triangles are equal in measurement, and a pair of **corresponding** non-included sides are equal in length, then the triangles are congruent. *(In British usage, **ASA** and **AAS** are usually combined into a single condition **AAcorrS** - any two angles and a corresponding side.)* For American usage, AAS is equivalent to an ASA condition, by the fact that if any two angles are given, so is the third angle, since their sum should be 180°.

- RHS (Right-angle-Hypotenuse-Side), also known as HL (Hypotenuse-Leg): If two right-angled triangles have their hypotenuses equal in length, and a pair of shorter sides are equal in length, then the triangles are congruent.

Side-side-angle

The SSA condition (Side-Side-Angle) which specifies two sides and a non-included angle (also known as ASS, or Angle-Side-Side) does not by itself prove congruence. In order to show congruence, additional information is required such as the measure of the corresponding angles and in some cases the lengths of the two pairs of corresponding sides. There are a few possible cases:

If two triangles satisfy the SSA condition and the length of the side opposite the angle is greater than or equal to the length of the adjacent side (SsA, or long side-short side-angle), then the two triangles are congruent. The opposite side is sometimes longer when the corresponding angles are acute, but it is *always* longer when the corresponding angles are right or obtuse. Where the angle is a right angle, also known as the Hypotenuse-Leg (HL) postulate or the Right-angle-Hypotenuse-Side (RHS) condition, the third side can be calculated using the Pythagorean Theorem thus allowing the SSS postulate to be applied.

If two triangles satisfy the SSA condition and the corresponding angles are acute and the length of the side opposite the angle is equal to the length of the adjacent side multiplied by the sine of the angle, then the two triangles are congruent.

If two triangles satisfy the SSA condition and the corresponding angles are acute and the length of the side opposite the angle is greater than the length of the adjacent side multiplied by the sine of the angle (but less than the length of the adjacent side), then the two triangles cannot be shown to be congruent. This is the ambiguous case and two different triangles can be formed from the given information, but further information distinguishing them can lead to a proof of congruence.

Angle-angle-angle

In Euclidean geometry, AAA (Angle-Angle-Angle) (or just AA, since in Euclidean geometry the angles of a triangle add up to 180°) does not provide information regarding the size of the two triangles and hence proves only similarity and not congruence in Euclidean space.

However, in spherical geometry and hyperbolic geometry (where the sum of the angles of a triangle varies with size) AAA is sufficient for congruence on a given curvature of surface.

Definition of congruence in Analytic Geometry

In a Euclidean system, congruence is fundamental; it is the counterpart of equality for numbers. In analytic geometry, congruence may be defined intuitively thus: two mappings of figures onto one Cartesian coordinate system are congruent if and only if, for *any* two points in the first mapping, the Euclidean distance between them is equal to the Euclidean distance between the corresponding points in the second mapping.

A more formal definition states that two subsets A and B of Euclidean space R^n are called congruent if there exists an isometry $f: R^n \rightarrow R^n$ (an element of the Euclidean group $E(n)$) with $f(A) = B$. Congruence is an equivalence relation.

Congruent Conic Sections

Two conic sections are congruent if their eccentricities and one other distinct parameter characterizing them are equal. Their eccentricities establish their shapes, equality of which is sufficient to establish similarity, and the second parameter then establishes size. Since two circles, parabolas, or rectangular hyperbolas always have the same eccentricity (specifically 0 in the case of circles, 1 in the case of parabolas, and $\sqrt{2}$ in the case of rectangular hyperbolas), two circles, parabolas, or rectangular hyperbolas need to have only one other common parameter value, establishing their size, for them to be congruent.

Congruent Polyhedra

For two polyhedra with the same number E of edges, the same number of faces, and the same number of sides on corresponding faces, there exists a set of at most E measurements that can establish whether or not the polyhedra are congruent. For cubes, which have 12 edges, only 9 measurements are necessary.

Congruent Triangles on a Sphere

As with plane triangles, on a sphere two triangles sharing the same sequence of angle-side-angle (ASA) are necessarily congruent (that is, they have three identical sides and three identical angles). This can be seen as follows: One can situate one of the vertices with a given angle at the south pole and run the side with given length up the prime meridian. Knowing both angles at either end of the segment of fixed length ensures that the other two sides emanate with a uniquely determined trajectory, and thus will meet each other at a uniquely determined point; thus ASA is valid.

The congruence theorems side-angle-side (SAS) and side-side-side (SSS) also hold on a sphere; in addition, if two spherical triangles have an identical angle-angle-angle (AAA) sequence, they are congruent (unlike for plane triangles).

The plane-triangle congruence theorem angle-angle-side (AAS) does not hold for spherical triangles. As in plane geometry, side-side-angle (SSA) does not imply congruence.

Angle

In planar geometry, an angle is the figure formed by two rays, called the *sides* of the angle, sharing a common endpoint, called the *vertex* of the angle. Angles formed by two rays lie in a plane, but this plane does not have to be a Euclidean plane. Angles are also formed by the intersection of two planes in Euclidean and other spaces. These are called dihedral angles. Angles formed by the intersection of two curves in a plane are defined as the angle determined by the tangent rays at the point of intersection. Similar statements hold in space, for example, the spherical angle formed by two great circles on a sphere is the dihedral angle between the planes determined by the great circles.

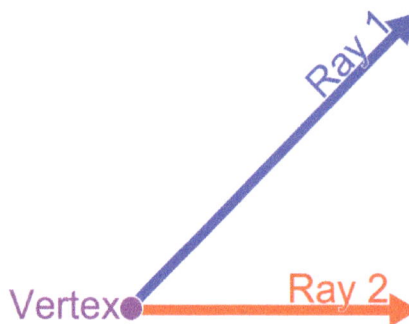

An angle enclosed by rays emanating from a vertex.

∠, the angle symbol in Unicode is U+2220.

Angle is also used to designate the measure of an angle or of a rotation. This measure is the ratio of the length of a circular arc to its radius. In the case of a geometric angle, the arc is centered at the vertex and delimited by the sides. In the case of a rotation, the arc is centered at the center of the rotation and delimited by any other point and its image by the rotation.

The word *angle* comes from the Latin word *angulus*, meaning "corner". Both are connected with the Proto-Indo-European root *ank-*, meaning "to bend" or "bow".

Euclid defines a plane angle as the inclination to each other, in a plane, of two lines which meet each other, and do not lie straight with respect to each other. According to Proclus an angle must be either a quality or a quantity, or a relationship. The first concept was used by Eudemus, who regarded an angle as a deviation from a straight line; the second by Carpus of Antioch, who regarded it as the interval or space between the intersecting lines; Euclid adopted the third concept, although his definitions of right, acute, and obtuse angles are certainly quantitative.

Identifying Angles

In mathematical expressions, it is common to use Greek letters ($\alpha, \beta, \gamma, \theta, \varphi, \dots$) to serve as variables standing for the size of some angle. (To avoid confusion with its other meaning, the symbol π is typically not used for this purpose.) Lower case Roman letters ($a, b, c, \dots$) are also used, as are upper case Roman letters in the context of polygons. See the figures in this article for examples.

In geometric figures, angles may also be identified by the labels attached to the three points that define them. For example, the angle at vertex A enclosed by the rays AB and AC (i.e. the lines from point A to point B and point A to point C) is denoted $\angle$BAC or $\overset{\frown}{\text{BAC}}$. Sometimes, where there is no risk of confusion, the angle may be referred to simply by its vertex ("angle A").

Potentially, an angle denoted, say, $\angle$BAC might refer to any of four angles: the clockwise angle from B to C, the anticlockwise angle from B to C, the clockwise angle from C to B, or the anticlockwise angle from C to B, where the direction in which the angle is measured determines its sign (see Positive and negative angles). However, in many geometrical situations it is obvious from context that the positive angle less than or equal to 180 degrees is meant, and no ambiguity arises. Otherwise, a convention may be adopted so that $\angle$BAC always refers to the anticlockwise (positive) angle from B to C, and $\angle$CAB to the anticlockwise (positive) angle from C to B.

Types of Angles

Individual Angles

Right angle.

$$> 180°$$

Reflex angle.

$$b > 90°$$
$$a < 90°$$
$$c = 180°$$

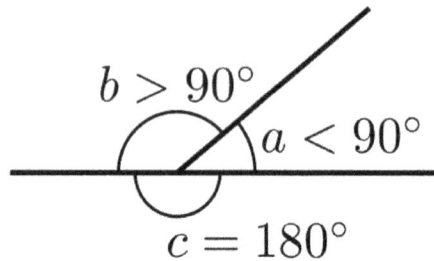

Acute (*a*), obtuse (*b*), and straight (*c*) angles. The acute and obtuse angles are also known as oblique angles.

- Angles smaller than a right angle (less than 90°) are called *acute angles* ("acute" meaning "sharp").

- An angle equal to 1/4 turn (90° or $\pi/2$ radians) is called a *right angle*. Two lines that form a right angle are said to be *normal, orthogonal,* or *perpendicular*.

- Angles larger than a right angle and smaller than a straight angle (between 90° and 180°) are called *obtuse angles* ("obtuse" meaning "blunt").

- An angle equal to 1/2 turn (180° or π radians) is called a *straight angle*.

- Angles larger than a straight angle but less than 1 turn (between 180° and 360°) are called *reflex angles*.

- An angle equal to 1 turn (360° or 2π radians) is called a *full angle, complete angle,* or a *perigon*.

- Angles that are not right angles or a multiple of a right angle are called *oblique angles*.

The names, intervals, and measured units are shown in a table below:

Name	acute	right angle	obtuse	straight	reflex	perigon
Units	**Interval**					
Turns	(0, 1/4)	1/4	(1/4, 1/2)	1/2	(1/2, 1)	1
Radians	$(0, 1/2\pi)$	$1/2\pi$	$(1/2\pi, \pi)$	π	$(\pi, 2\pi)$	2π
Degrees	(0, 90)°	90°	(90, 180)°	180°	(180, 360)°	360°
Gons	(0, 100)^g	100^g	(100, 200)^g	200^g	(200, 400)^g	400^g

Equivalence Angle Pairs

- Angles that have the same measure (i.e. the same magnitude) are said to be *equal* or *congruent*. An angle is defined by its measure and is not dependent upon the lengths of the sides of the angle (e.g. all *right angles* are equal in measure).

- Two angles which share terminal sides, but differ in size by an integer multiple of a turn, are called *coterminal angles.*

- A *reference angle* is the acute version of any angle determined by repeatedly subtracting or adding straight angle (1/2 turn, 180°, or π radians), to the results as necessary, until the magnitude of result is an acute angle, a value between 0 and 1/4 turn, 90°, or π/2 radians. For example, an angle of 30 degrees has a reference angle of 30 degrees, and an angle of 150 degrees also has a reference angle of 30 degrees (180 − 150). An angle of 750 degrees has a reference angle of 30 degrees (750 − 720).

Vertical and Adjacent Angle Pairs

When two straight lines intersect at a point, four angles are formed. Pairwise these angles are named according to their location relative to each other.

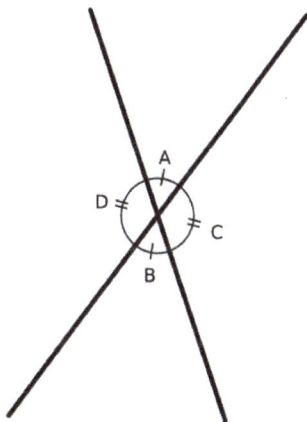

Angles A and B are a pair of vertical angles; angles C and D are a pair of vertical angles.

- A pair of angles opposite each other, formed by two intersecting straight lines that form an "X"-like shape, are called *vertical angles* or *opposite angles* or *vertically opposite angles.* They are abbreviated as *vert. opp. ∠s.*

The equality of vertically opposite angles is called the *vertical angle theorem.* Eudemus of Rhodes attributed the proof to Thales of Miletus. The proposition showed that since both of a pair of vertical angles are supplementary to both of the adjacent angles, the vertical angles are equal in measure. According to a historical Note, when Thales visited Egypt, he observed that whenever the Egyptians drew two intersecting lines, they would measure the vertical angles to make sure that they were equal. Thales concluded that one could prove that all vertical angles are equal if one accepted some general notions such as: all straight angles are equal, equals added to equals are equal, and equals subtracted from equals are equal.

In the figure, assume the measure of Angle $A = x$. When two adjacent angles form a straight line, they are supplementary. Therefore, the measure of Angle $C = 180 - x$. Similarly, the measure of Angle $D = 180 - x$. Both Angle C and Angle D have measures equal to $180 - x$ and are congruent. Since Angle B is supplementary to both Angles C and D, either of these angle measures may be used to determine the measure of Angle B. Using the measure of either Angle C or Angle D we find the measure of Angle $B = 180 - (180 - x) = 180 - 180 + x = x$. Therefore, both Angle A and Angle B have measures equal to x and are equal in measure.

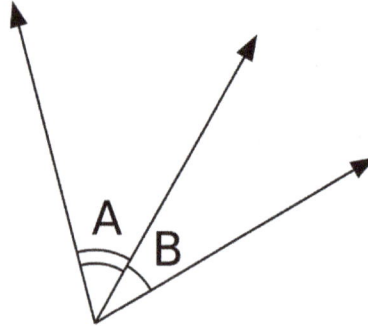

Angles A and B are adjacent.

- *Adjacent angles*, often abbreviated as *adj. $\angle s$*, are angles that share a common vertex and edge but do not share any interior points. In other words, they are angles that are side by side, or adjacent, sharing an "arm". Adjacent angles which sum to a right angle, straight angle or full angle are special and are respectively called *complementary*, *supplementary* and *explementary* angles.

A transversal is a line that intersects a pair of (often parallel) lines and is associated with *alternate interior angles*, *corresponding angles*, *interior angles*, and *exterior angles*.

Combining Angle Pairs

There are three special angle pairs which involve the summation of angles:

- Complementary angles are angle pairs whose measures sum to one right angle (1/4 turn, 90°, or $\pi/2$ radians). If the two complementary angles are adjacent their non-shared sides form a right angle. In Euclidean geometry, the two acute angles in a right triangle are complementary, because the sum of internal angles of a triangle is 180 degrees, and the right angle itself accounts for ninety degrees.

  The adjective complementary is from Latin *complementum*, associated with the verb *complere*, "to fill up". An acute angle is "filled up" by its complement to form a right angle.

  The difference between an angle and a right angle is termed the *complement* of the angle.

  If angles A and B are complementary, the following relationships hold:

$$\sin^2 A + \sin^2 B = 1. \quad \cos^2 A + \cos^2 B = 1.$$

$$\tan A = \cot B. \qquad \sec A = \csc B.$$

(The tangent of an angle equals the cotangent of its complement and its secant equals the cosecant of its complement.)

The prefix "co-" in the names of some trigonometric ratios refers to the word "complementary".

- Two angles that sum to a straight angle (1/2 turn, 180°, or π radians) are called supplementary angles.

If the two supplementary angles are adjacent (i.e. have a common vertex and share just one side), their non-shared sides form a straight line. Such angles are called a linear pair of angles. However, supplementary angles do not have to be on the same line, and can be separated in space. For example, adjacent angles of a parallelogram are supplementary, and opposite angles of a cyclic quadrilateral (one whose vertices all fall on a single circle) are supplementary.

If a point P is exterior to a circle with center O, and if the tangent lines from P touch the circle at points T and Q, then $\angle$TPQ and $\angle$TOQ are supplementary.

The sines of supplementary angles are equal. Their cosines and tangents (unless undefined) are equal in magnitude but have opposite signs.

In Euclidean geometry, any sum of two angles in a triangle is supplementary to the third, because the sum of internal angles of a triangle is a straight angle.

- Two angles that sum to a complete angle (1 turn, 360°, or 2π radians) are called explementary angles or conjugate angles.

The difference between an angle and a complete angle is termed the *explement* of the angle or *conjugate* of an angle.

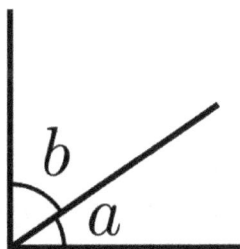

The *complementary* angles *a* and *b* (*b* is the *complement* of *a*, and *a* is the complement of *b*).

A reflex angle and its conjugate are *explementary* angles, and their sum is a *complete* angle.

$$b > 90°$$

$$a < 90°$$

$$c = 180°$$

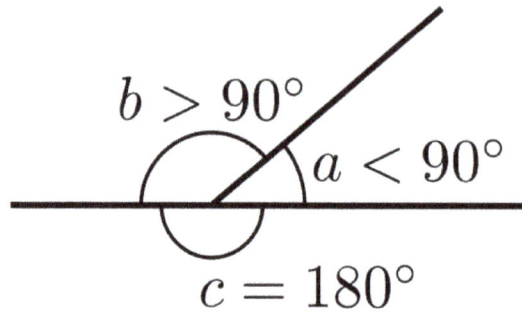

The angles *a* and *b* are *supplementary* angles.

Polygon Related Angles

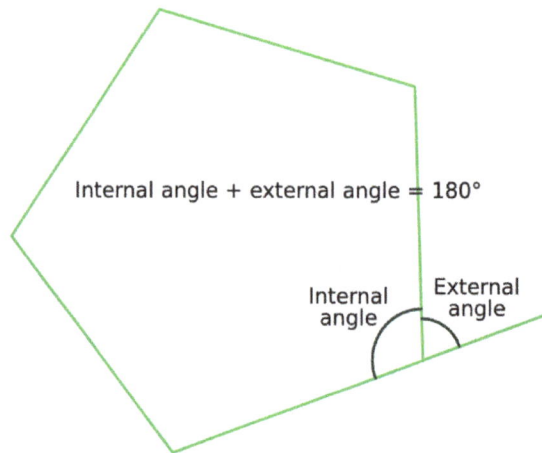

Internal angle + external angle = 180°

Internal
angle

External
angle

Internal and external angles.

- An angle that is part of a simple polygon is called an *interior angle* if it lies on the inside of that simple polygon. A simple concave polygon has at least one interior angle that is a reflex angle.

  In Euclidean geometry, the measures of the interior angles of a triangle add up to π radians, 180°, or 1/2 turn; the measures of the interior angles of a simple convex quadrilateral add up to 2π radians, 360°, or 1 turn. In general, the measures of the interior angles of a simple convex polygon with *n* sides add up to $(n - 2)\pi$ radians, or $180(n - 2)$ degrees, $(2n - 4)$ right angles, or $(n/2 - 1)$ turn.

- The supplement of an interior angle is called an *exterior angle*, that is, an interior angle and an exterior angle form a linear pair of angles. There are two exterior angles at each vertex of the polygon, each determined by extending one of the two sides of the polygon that meet at the vertex; these two angles are vertical angles and hence are equal. An exterior angle measures the amount of rotation one has to make at a vertex to trace out the polygon. If the corresponding interior angle is a reflex angle, the exterior angle should be considered negative. Even in a non-simple polygon it may be possible to define the exterior angle, but one will have to pick an orientation of the plane (or surface) to decide the sign of the exterior angle measure.

In Euclidean geometry, the sum of the exterior angles of a simple convex polygon will be one full turn (360°). The exterior angle here could be called a *supplementary exterior angle*. Exterior angles are commonly used in Logo Turtle Geometry when drawing regular polygons.

- In a triangle, the bisectors of two exterior angles and the bisector of the other interior angle are concurrent (meet at a single point).

- In a triangle, three intersection points, each of an external angle bisector with the opposite extended side, are collinear.

- In a triangle, three intersection points, two of them between an interior angle bisector and the opposite side, and the third between the other exterior angle bisector and the opposite side extended, are collinear.

- Some authors use the name *exterior angle* of a simple polygon to simply mean the *explement exterior angle* (*not* supplement!) of the interior angle. This conflicts with the above usage.

Plane Related Angles

- The angle between two planes (such as two adjacent faces of a polyhedron) is called a *dihedral angle*. It may be defined as the acute angle between two lines normal to the planes.

- The angle between a plane and an intersecting straight line is equal to ninety degrees minus the angle between the intersecting line and the line that goes through the point of intersection and is normal to the plane.

Measuring Angles

The size of a geometric angle is usually characterized by the magnitude of the smallest rotation that maps one of the rays into the other. Angles that have the same size are said to be *equal* or *congruent* or *equal in measure*.

In some contexts, such as identifying a point on a circle or describing the *orientation* of an object in two dimensions relative to a reference orientation, angles that differ by an exact multiple of a full turn are effectively equivalent. In other contexts, such as identifying a point on a spiral curve or describing the *cumulative rotation* of an object in two dimensions relative to a reference orientation, angles that differ by a non-zero multiple of a full turn are not equivalent.

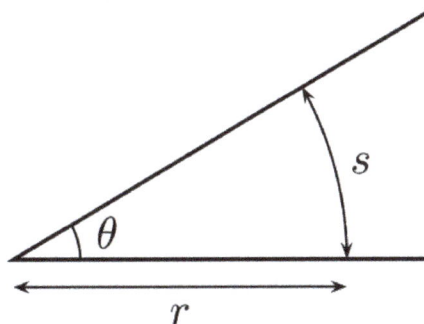

The measure of angle θ (in radians) is the quotient of s and r.

In order to measure an angle θ, a circular arc centered at the vertex of the angle is drawn, e.g. with a pair of compasses. The ratio of the length s of the arc by the radius r of the circle is the measure of the angle in radians.

The measure of the angle in another angular unit is then obtained by multiplying its measure in radians by the scaling factor $k/2\pi$, where k is the measure of a complete turn in the chosen unit (for example 360 for degrees or 400 for gradians):

$$\theta = k\frac{s}{2\pi r}.$$

The value of θ thus defined is independent of the size of the circle: if the length of the radius is changed then the arc length changes in the same proportion, so the ratio s/r is unaltered. (Proof. The formula above can be rewritten as $k = \theta r/s$. One turn, for which $\theta = n$ units, corresponds to an arc equal in length to the circle's circumference, which is $2\pi r$, so $s = 2\pi r$. Substituting n for θ and $2\pi r$ for s in the formula, results in $k = nr/2\pi r = n/2\pi$.)

Angle Addition Postulate

The angle addition postulate states that if B is in the interior of angle AOC, then

$$m\angle AOC = m\angle AOB + m\angle BOC$$

The measure of the angle AOC is the sum of the measure of angle AOB and the measure of angle BOC. In this postulate it does not matter in which unit the angle is measured as long as each angle is measured in the same unit.

Units

Units used to represent angles are listed below in descending magnitude order. Of these units, the *degree* and the *radian* are by far the most commonly used. Angles expressed in radians are dimensionless for the purposes of dimensional analysis.

Most units of angular measurement are defined such that one *turn* (i.e. one full circle) is equal to n units, for some whole number n. The two exceptions are the radian and the diameter part.

Turn ($n = 1$)

> The *turn*, also *cycle, full circle, revolution*, and *rotation*, is complete circular movement or measure (as to return to the same point) with circle or ellipse. A turn is abbreviated τ, *cyc*, *rev*, or *rot* depending on the application, but in the acronym *rpm* (revolutions per minute), just *r* is used. A *turn* of n units is obtained by setting $k = 1/2\pi$ in the formula above. The equivalence of 1 *turn* is 360°, 2π rad, 400 grad, and 4 right angles. The symbol τ can also be used as a mathematical constant to represent 2π radians. Used in this way ($k = \tau/2\pi$) allows for radians to be expressed as a fraction of a turn. For example, half a turn is $\tau/2 = \pi$.

Quadrant ($n = 4$)

> The *quadrant* is 1/4 of a turn, i.e. a *right angle*. It is the unit used in Euclid's Elements.

1 quad. = 90° = π/2 rad = 1/4 turn = 100 grad. In German the symbol ∟ has been used to denote a quadrant.

Sextant ($n = 6$)

The *sextant* (*angle of the equilateral triangle*) is 1/6 of a turn. It was the unit used by the Babylonians, and is especially easy to construct with ruler and compasses. The degree, minute of arc and second of arc are sexagesimal subunits of the Babylonian unit. 1 Babylonian unit = 60° = π/3 rad ≈ 1.047197551 rad.

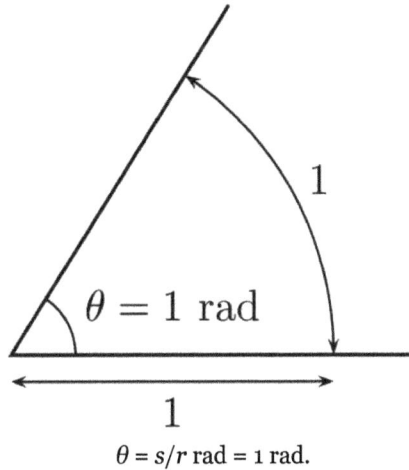

$\theta = s/r$ rad = 1 rad.

Radian ($n = 2\pi = 6.283\ldots$)

The *radian* is the angle subtended by an arc of a circle that has the same length as the circle's radius. The case of radian for the formula given earlier, a *radian* of $n = 2\pi$ units is obtained by setting $k = 2\pi/2\pi = 1$. One turn is 2π radians, and one radian is $180/\pi$ degrees, or about 57.2958 degrees. The radian is abbreviated *rad*, though this symbol is often omitted in mathematical texts, where radians are assumed unless specified otherwise. When radians are used angles are considered as dimensionless. The radian is used in virtually all mathematical work beyond simple practical geometry, due, for example, to the pleasing and "natural" properties that the trigonometric functions display when their arguments are in radians. The radian is the (derived) unit of angular measurement in the SI system.

Clock position ($n = 12$)

A clock position is the relative direction of an object described using the analogy of a 12-hour clock. One imagines a clock face lying either upright or flat in front of oneself, and identifies the twelve hour markings with the directions in which they point.

Hour angle ($n = 24$)

The astronomical *hour angle* is 1/24 of a turn. Since this system is amenable to measuring objects that cycle once per day (such as the relative position of stars), the sexagesimal subunits are called *minute of time* and *second of time*. Note that these are distinct from, and 15 times larger than, minutes and seconds of arc. 1 hour = 15° = π/12 rad = 1/6 quad. = 1/24 *turn* = 16 2/3 grad.

(Compass) point or wind ($n = 32$)

The *point*, used in navigation, is 1/32 of a turn. 1 point = 1/8 of a right angle = 11.25° = 12.5 grad. Each point is subdivided in four quarter-points so that 1 turn equals 128 quarter-points.

Hexacontade ($n = 60$)

The *hexacontade* is a unit of 6° that Eratosthenes used, so that a whole turn was divided into 60 units.

Pechus ($n = 144$–180)

–The *pechus* was a Babylonian unit equal to about 2° or 2 1/2°.

Binary degree ($n = 256$)

The *binary degree*, also known as the *binary radian* (or *brad*), is 1/256 of a turn. The binary degree is used in computing so that an angle can be efficiently represented in a single byte (albeit to limited precision). Other measures of angle used in computing may be based on dividing one whole turn into 2^n equal parts for other values of n.

Degree ($n = 360$)

The *degree*, denoted by a small superscript circle (°), is 1/360 of a turn, so one *turn* is 360°. The case of degrees for the formula given earlier, a *degree* of $n = 360°$ units is obtained by setting $k = 360°/2\pi$. One advantage of this old sexagesimal subunit is that many angles common in simple geometry are measured as a whole number of degrees. Fractions of a degree may be written in normal decimal notation (e.g. 3.5° for three and a half degrees), but the "minute" and "second" sexagesimal subunits of the "degree-minute-second" system are also in use, especially for geographical coordinates and in astronomy and ballistics:

Diameter part ($n = 376.99\ldots$)

The *diameter part* (occasionally used in Islamic mathematics) is 1/60 radian. One "diameter part" is approximately 0.95493°. There are about 376.991 diameter parts per turn.

Grad ($n = 400$)

The *grad*, also called *grade*, *gradian*, or *gon*, is 1/400 of a turn, so a right angle is 100 grads. It is a decimal subunit of the quadrant. A kilometre was historically defined as a centi-grad of arc along a great circle of the Earth, so the kilometer is the decimal analog to the sexagesimal nautical mile. The grad is used mostly in triangulation.

Mil ($n = 6000$–6400)

The *mil* is any of several units that are *approximately* equal to a milliradian. There are several definitions ranging from 0.05625 to 0.06 degrees (3.375 to 3.6 minutes), with the milliradian being approximately 0.05729578 degrees (3.43775 minutes). In NATO countries, it is defined as 1/6400 of a circle. Its value is approximately equal to the angle subtended by a width of 1 metre as seen from 1 km away ($2\pi/6400 = 0.0009817\ldots \approx 1/1000$).

Minute of arc (n = 21,600)

> The *minute of arc* (or *MOA*, *arcminute*, or just *minute*) is 1/60 of a degree = 1/21,600 turn. It is denoted by a single prime ('). For example, 3° 30' is equal to 3 × 60 + 30 = 210 minutes or 3 + 30/60 = 3.5 degrees. A mixed format with decimal fractions is also sometimes used, e.g. 3° 5.72' = 3 + 5.72/60 degrees. A nautical mile was historically defined as a minute of arc along a great circle of the Earth.

Second of arc (n = 1,296,000)

> The *second of arc* (or *arcsecond*, or just *second*) is 1/60 of a minute of arc and 1/3600 of a degree. It is denoted by a double prime (″). For example, 3° 7' 30″ is equal to 3 + 7/60 + 30/3600 degrees, or 3.125 degrees.

Positive and Negative Angles

Although the definition of the measurement of an angle does not support the concept of a negative angle, it is frequently useful to impose a convention that allows positive and negative angular values to represent orientations and/or rotations in opposite directions relative to some reference.

In a two-dimensional Cartesian coordinate system, an angle is typically defined by its two sides, with its vertex at the origin. The *initial side* is on the positive x-axis, while the other side or *terminal side* is defined by the measure from the initial side in radians, degrees, or turns. With *positive angles* representing rotations toward the positive y-axis and *negative angles* representing rotations toward the negative y-axis. When Cartesian coordinates are represented by *standard position*, defined by the x-axis rightward and the y-axis upward, positive rotations are anticlockwise and negative rotations are clockwise.

In many contexts, an angle of $-\theta$ is effectively equivalent to an angle of "one full turn minus θ". For example, an orientation represented as −45° is effectively equivalent to an orientation represented as 360° − 45° or 315°. However, a rotation of −45° would not be the same as a rotation of 315°.

In three-dimensional geometry, "clockwise" and "anticlockwise" have no absolute meaning, so the direction of positive and negative angles must be defined relative to some reference, which is typically a vector passing through the angle's vertex and perpendicular to the plane in which the rays of the angle lie.

In navigation, bearings are measured relative to north. By convention, viewed from above, bearing angle are positive clockwise, so a bearing of 45° corresponds to a north-east orientation. Negative bearings are not used in navigation, so a north-west orientation corresponds to a bearing of 315°.

Alternative Ways of Measuring the Size of an Angle

There are several alternatives to measuring the size of an angle by the angle of rotation. The *grade of a slope*, or *gradient* is equal to the tangent of the angle, or sometimes (rarely) the sine. A gradient is often expressed as a percentage. For very small values (less than 5%), the grade of a slope is approximately the measure of the angle in radians.

In rational geometry the *spread* between two lines is defined at the square of the sine of the angle between the lines. Since the sine of an angle and the sine of its supplementary angle are the same, any angle of rotation that maps one of the lines into the other leads to the same value for the spread between the lines.

Astronomical Approximations

Astronomers measure angular separation of objects in degrees from their point of observation.

- 0.5° is approximately the width of the sun or moon.

- 1° is approximately the width of a little finger at arm's length.

- 10° is approximately the width of a closed fist at arm's length.

- 20° is approximately the width of a handspan at arm's length.

These measurements clearly depend on the individual subject, and the above should be treated as rough rule of thumb approximations only.

Angles Between Curves

The angle between a line and a curve (mixed angle) or between two intersecting curves (curvilinear angle) is defined to be the angle between the tangents at the point of intersection.

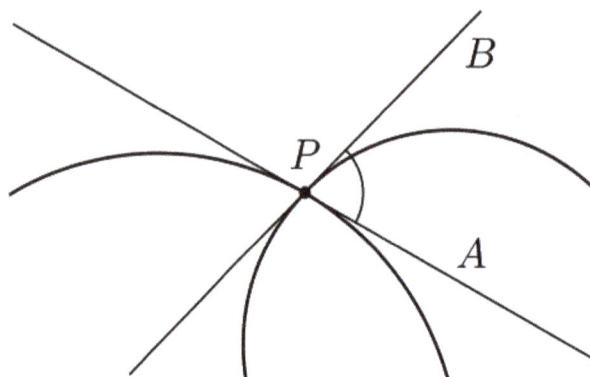

The angle between the two curves at P is defined as the angle between the tangents A and B at P.

Bisecting and Trisecting Angles

The ancient Greek mathematicians knew how to bisect an angle (divide it into two angles of equal measure) using only a compass and straightedge, but could only trisect certain angles. In 1837 Pierre Wantzel showed that for most angles this construction cannot be performed.

Dot Product and Generalisations

In the Euclidean space, the angle θ between two Euclidean vectors u and v is related to their dot product and their lengths by the formula

$$\mathbf{u} \cdot \mathbf{v} = \cos(\theta) \|\mathbf{u}\| \|\mathbf{v}\|.$$

This formula supplies an easy method to find the angle between two planes (or curved surfaces) from their normal vectors and between skew lines from their vector equations.

Inner Product

To define angles in an abstract real inner product space, we replace the Euclidean dot product (·) by the inner product $\langle \cdot, \cdot \rangle$, i.e.

$$\langle \mathbf{u}, \mathbf{v} \rangle = \cos(\theta) \, \|\mathbf{u}\| \, \|\mathbf{v}\|.$$

In a complex inner product space, the expression for the cosine above may give non-real values, so it is replaced with

$$\mathrm{Re}\left(\langle \mathbf{u}, \mathbf{v} \rangle\right) = \cos(\theta) \, \|\mathbf{u}\| \, \|\mathbf{v}\|.$$

or, more commonly, using the absolute value, with

$$\left|\langle \mathbf{u}, \mathbf{v} \rangle\right| = |\cos(\theta)| \, \|\mathbf{u}\| \, \|\mathbf{v}\|.$$

The latter definition ignores the direction of the vectors and thus describes the angle between one-dimensional subspaces $\mathrm{span}(\mathbf{u})$ and $\mathrm{span}(\mathbf{v})$ spanned by the vectors $\mathbf{u}$ and $\mathbf{v}$ correspondingly.

Angles between Subspaces

The definition of the angle between one-dimensional subspaces $\mathrm{span}(\mathbf{u})$ and $\mathrm{span}(\mathbf{v})$ given by

$$\left|\langle \mathbf{u}, \mathbf{v} \rangle\right| = |\cos(\theta)| \, \|\mathbf{u}\| \, \|\mathbf{v}\|$$

in a Hilbert space can be extended to subspaces of any finite dimensions. Given two subspaces $\mathcal{U}$, $\mathcal{W}$ with $\dim(\mathcal{U}) := k \leq \dim(\mathcal{W}) := l$, this leads to a definition of k angles called canonical or principal angles between subspaces.

Angles in Riemannian Geometry

In Riemannian geometry, the metric tensor is used to define the angle between two tangents. Where U and V are tangent vectors and g_{ij} are the components of the metric tensor G,

$$\cos \theta = \frac{g_{ij} U^i V^j}{\sqrt{\left|g_{ij} U^i U^j\right| \left|g_{ij} V^i V^j\right|}}.$$

Angles in Geography and Astronomy

In geography, the location of any point on the Earth can be identified using a *geographic coordinate system*. This system specifies the latitude and longitude of any location in terms of angles subtended at the centre of the Earth, using the equator and (usually) the Greenwich meridian as references.

In astronomy, a given point on the celestial sphere (that is, the apparent position of an astronomical object) can be identified using any of several *astronomical coordinate systems*, where the references vary according to the particular system. Astronomers measure the *angular separation* of two stars by imagining two lines through the centre of the Earth, each intersecting one of the stars. The angle between those lines can be measured, and is the angular separation between the two stars.

In both geography and astronomy, a sighting direction can be specified in terms of a vertical angle such as altitude /elevation with respect to the horizon as well as the azimuth with respect to north.

Astronomers also measure the *apparent size* of objects as an angular diameter. For example, the full moon has an angular diameter of approximately 0.5°, when viewed from Earth. One could say, "The Moon's diameter subtends an angle of half a degree." The small-angle formula can be used to convert such an angular measurement into a distance/size ratio.

Polygon

In elementary geometry, a polygon /ˈpɒlɪɡɒn/ is a plane figure that is bounded by a finite chain of straight line segments closing in a loop to form a closed chain or *circuit*. These segments are called its *edges* or *sides*, and the points where two edges meet are the polygon's *vertices* (singular: vertex) or *corners*. The interior of the polygon is sometimes called its *body*. An **n**-gon is a polygon with n sides; for example, a triangle is a 3-gon. A polygon is a 2-dimensional example of the more general polytope in any number of dimensions.

Some polygons of different kinds: open (excluding its boundary), boundary only (excluding interior), closed (including both boundary and interior), and self-intersecting.

The basic geometrical notion of a polygon has been adapted in various ways to suit particular purposes. Mathematicians are often concerned only with the bounding closed polygonal chain and with simple polygons which do not self-intersect, and they often define a polygon accordingly. A polygonal boundary may be allowed to intersect itself, creating star polygons and other self-intersecting polygons. These and other generalizations of polygons are described below.

Classification

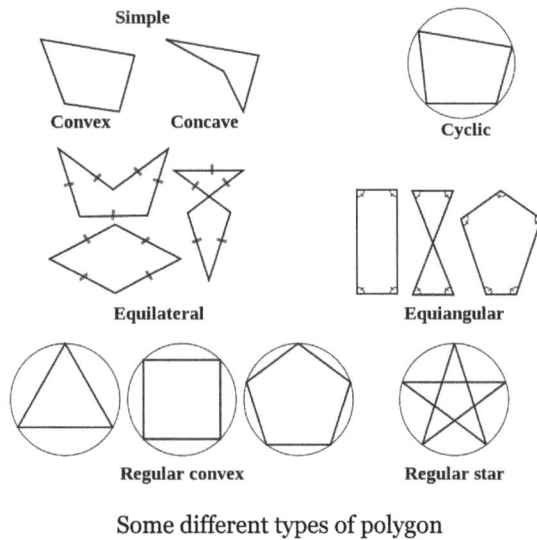

Some different types of polygon

Number of Sides

Polygons are primarily classified by the number of sides.

Convexity and Non-convexity

Polygons may be characterized by their convexity or type of non-convexity:

- Convex: any line drawn through the polygon (and not tangent to an edge or corner) meets its boundary exactly twice. As a consequence, all its interior angles are less than 180°. Equivalently, any line segment with endpoints on the boundary passes through only interior points between its endpoints.

- Non-convex: a line may be found which meets its boundary more than twice. Equivalently, there exists a line segment between two boundary points that passes outside the polygon.

- Simple: the boundary of the polygon does not cross itself. All convex polygons are simple.

- Concave. Non-convex and simple. There is at least one interior angle greater than 180°.

- Star-shaped: the whole interior is visible from at least one point, without crossing any edge. The polygon must be simple, and may be convex or concave.

- Self-intersecting: the boundary of the polygon crosses itself. Branko Grünbaum calls these coptic, though this term does not seem to be widely used. The term *complex* is sometimes used in contrast to *simple*, but this usage risks confusion with the idea of a *complex polygon* as one which exists in the complex Hilbert plane consisting of two complex dimensions.

- Star polygon: a polygon which self-intersects in a regular way. A polygon cannot be both a star and star-shaped.

Equality and Symmetry

- Equiangular: all corner angles are equal.

- Cyclic: all corners lie on a single circle, called the circumcircle.

- Isogonal or vertex-transitive: all corners lie within the same symmetry orbit. The polygon is also cyclic and equiangular.

- Equilateral: all edges are of the same length. The polygon need not be convex.

- Tangential: all sides are tangent to an inscribed circle.

- Isotoxal or edge-transitive: all sides lie within the same symmetry orbit. The polygon is also equilateral and tangential.

- Regular: the polygon is both *isogonal* and *isotoxal*. Equivalently, it is both *cyclic* and *equilateral*, or both *equilateral* and *equiangular*. A non-convex regular polygon is called a *regular star polygon*.

Miscellaneous

- Rectilinear: the polygon's sides meet at right angles, i.e., all its interior angles are 90 or 270 degrees.

- Monotone with respect to a given line *L*: every line orthogonal to L intersects the polygon not more than twice.

Properties and Formulas

Euclidean geometry is assumed throughout.

Angles

Any polygon has as many corners as it has sides. Each corner has several angles. The two most important ones are:

- Interior angle – The sum of the interior angles of a simple n-gon is $(n-2)\pi$ radians or $(n-2) \times 180$ degrees. This is because any simple n-gon (having n sides) can be considered to be made up of $(n-2)$ triangles, each of which has an angle sum of π radians or 180 degrees. The measure of any interior angle of a convex regular n-gon is $\left(1-\frac{2}{n}\right)\pi$ radians or $180 - \frac{360}{n}$ degrees. The interior angles of regular star polygons were first studied by Poinsot, in the same paper in which he describes the four regular star polyhedra: for a regular $\frac{p}{q}$-gon (a p-gon with central density q), each interior angle is $\frac{\pi(p-2q)}{p}$ radians or $\frac{180(p-2q)}{p}$ degrees.

- Exterior angle – The exterior angle is the supplementary angle to the interior angle. Tracing around a convex n-gon, the angle "turned" at a corner is the exterior or external angle. Tracing all the way around the polygon makes one full turn, so the sum of the exterior angles must be 360°. This argument can be generalized to concave simple polygons, if external angles that turn in the opposite direction are subtracted from the total turned. Tracing around an n-gon in general, the sum of the exterior angles (the total amount one rotates at the vertices) can be any integer multiple d of $360°$, e.g. $720°$ for a pentagram and $0°$ for an angular "eight" or antiparallelogram, where d is the density or starriness of the polygon.

Area and Centroid

Simple Polygons

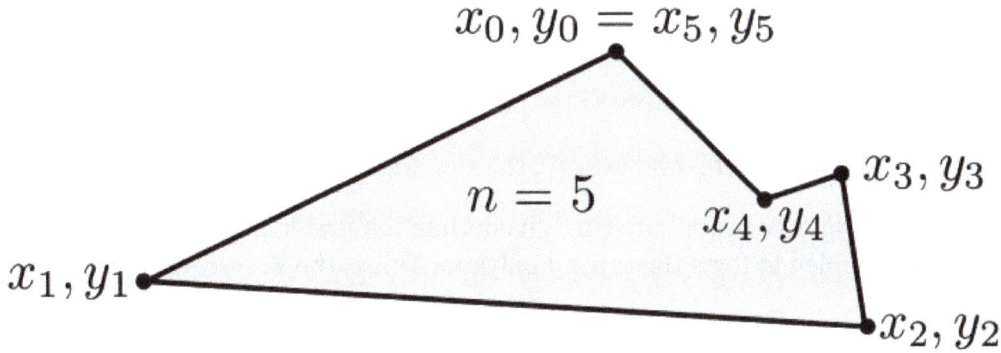

Coordinates of a non-convex pentagon.

For a non-self-intersecting (simple) polygon with n vertices x_i, y_i ($i = 1$ to n), the signed area and the Cartesian coordinates of the centroid are given by:

$$A = \frac{1}{2}\sum_{i=0}^{n-1}(x_i y_{i+1} - x_{i+1}y_i),$$

$$16A^2 = \sum_{i=1}^{n}\sum_{j=1}^{n}\begin{vmatrix} Q_{i,j} & Q_{i,j+1} \\ Q_{i+1,j} & Q_{i+1,j+1} \end{vmatrix},$$

where $Q_{i,j}$ is the squared distance between (x_i, y_i) and (x_j, y_j); and

$$C_x = \frac{1}{6A}\sum_{i=0}^{n-1}(x_i + x_{i+1})(x_i y_{i+1} - x_{i+1}y_i),$$

$$C_y = \frac{1}{6A}\sum_{i=0}^{n-1}(y_i + y_{i+1})(x_i y_{i+1} - x_{i+1}y_i).$$

To close the polygon, the first and last vertices are the same, i.e., $x_n, y_n = x_0, y_0$. The vertices must be ordered according to positive or negative orientation (counterclockwise or clockwise, respectively); if they are ordered negatively, the value given by the area formula will be negative but correct in absolute value, but when calculating C_x and C_y, the signed value of A (which in this case is negative) should be used. This is commonly called the shoelace formula or Surveyor's formula.

The area A of a simple polygon can also be computed if the lengths of the sides, $a_1, a_2, ..., a_n$ and the exterior angles, $\theta_1, \theta_2, ..., \theta_n$ are known, from:

$$A = \frac{1}{2}(a_1[a_2 \sin(\theta_1) + a_3 \sin(\theta_1 + \theta_2) + \cdots + a_{n-1} \sin(\theta_1 + \theta_2 + \cdots + \theta_{n-2})]$$

$$+ a_2[a_3 \sin(\theta_2) + a_4 \sin(\theta_2 + \theta_3) + \cdots + a_{n-1} \sin(\theta_2 + \cdots + \theta_{n-2})]$$

$$+ \cdots + a_{n-2}[a_{n-1} \sin(\theta_{n-2})]).$$

The formula was described by Lopshits in 1963.

If the polygon can be drawn on an equally spaced grid such that all its vertices are grid points, Pick's theorem gives a simple formula for the polygon's area based on the numbers of interior and boundary grid points: the former number plus one-half the latter number, minus 1.

In every polygon with perimeter p and area A, the isoperimetric inequality $p^2 > 4\pi A$ holds.

If any two simple polygons of equal area are given, then the first can be cut into polygonal pieces which can be reassembled to form the second polygon. This is the Bolyai–Gerwien theorem.

The area of a regular polygon is also given in terms of the radius r of its inscribed circle and its perimeter p by

$$A = \tfrac{1}{2} p \cdot r.$$

This radius is also termed its apothem and is often represented as a.

The area of a regular n-gon with side s inscribed in a unit circle is

$$A = \frac{ns}{4}\sqrt{4 - s^2}.$$

The area of a regular n-gon in terms of the radius R of its circumscribed circle and its perimeter p is given by

$$A = \frac{R}{2} \cdot p \cdot \sqrt{1 - \frac{p^2}{4n^2 R^2}}.$$

The area of a regular n-gon inscribed in a unit-radius circle, with side s and interior angle α, can also be expressed trigonometrically as

$$\frac{ns^2}{4}\cot\frac{\pi}{n} = \frac{ns^2}{4}\cot\frac{\alpha}{n-2} = n \cdot \sin\frac{\pi}{n} \cdot \cos\frac{\pi}{n} = n \cdot \sin\frac{\alpha}{n-2} \cdot \cos\frac{\alpha}{n-2}.$$

The lengths of the sides of a polygon do not in general determine the area. However, if the polygon is cyclic the sides *do* determine the area.

Of all n-gons with given sides, the one with the largest area is cyclic. Of all n-gons with a given perimeter, the one with the largest area is regular (and therefore cyclic).

Self-intersecting Polygons

The area of a self-intersecting polygon can be defined in two different ways, each of which gives a different answer:

- Using the above methods for simple polygons, we allow that particular regions within the polygon may have their area multiplied by a factor which we call the *density* of the region. For example, the central convex pentagon in the center of a pentagram has density 2. The two triangular regions of a cross-quadrilateral (like a figure 8) have opposite-signed densities, and adding their areas together can give a total area of zero for the whole figure.

- Considering the enclosed regions as point sets, we can find the area of the enclosed point set. This corresponds to the area of the plane covered by the polygon, or to the area of one or more simple polygons having the same outline as the self-intersecting one. In the case of the cross-quadrilateral, it is treated as two simple triangles.

Generalizations of Polygons

The idea of a polygon has been generalized in various ways. Some of the more important include:

- A spherical polygon is a circuit of arcs of great circles (sides) and vertices on the surface of a sphere. It allows the digon, a polygon having only two sides and two corners, which is impossible in a flat plane. Spherical polygons play an important role in cartography (map making) and in Wythoff's construction of the uniform polyhedra.

- A skew polygon does not lie in a flat plane, but zigzags in three (or more) dimensions. The Petrie polygons of the regular polytopes are well known examples.

- An apeirogon is an infinite sequence of sides and angles, which is not closed but has no ends because it extends indefinitely in both directions.

- A skew apeirogon is an infinite sequence of sides and angles that do not lie in a flat plane.

- A complex polygon is a configuration analogous to an ordinary polygon, which exists in the complex plane of two real and two imaginary dimensions.

- An abstract polygon is an algebraic partially ordered set representing the various elements (sides, vertices, etc.) and their connectivity. A real geometric polygon is said to be a *realization* of the associated abstract polygon. Depending on the mapping, all the generalizations described here can be realized.

- A polyhedron is a three-dimensional solid bounded by flat polygonal faces, analogous to a polygon in two dimensions. The corresponding shapes in four or higher dimensions are called polytopes.

Naming Polygons

Individual polygons are named (and sometimes classified) according to the number of sides, combining a Greek-derived numerical prefix with the suffix *-gon*, e.g. *pentagon*, *dodecagon*. The triangle, quadrilateral and nonagon are exceptions.

Beyond decagons (10-sided) and dodecagons (12-sided), mathematicians generally use numerical notation, for example 17-gon and 257-gon.

Exceptions exist for side counts that are more easily expressed in verbal form (e.g. 20 and 30), or are used by non-mathematicians. Some special polygons also have their own names; for example the regular star pentagon is also known as the pentagram.

Polygon names and miscellaneous properties		
Name	**Edges**	**Properties**
monogon	1	Not generally recognised as a polygon, although some disciplines such as graph theory sometimes use the term.
digon	2	Not generally recognised as a polygon in the Euclidean plane, although it can exist as a spherical polygon.
triangle (or trigon)	3	The simplest polygon which can exist in the Euclidean plane. Can tile the plane.
quadrilateral (or tetragon)	4	The simplest polygon which can cross itself; the simplest polygon which can be concave; the simplest polygon which can be non-cyclic. Can tile the plane.
pentagon	5	The simplest polygon which can exist as a regular star. A star pentagon is known as a pentagram or pentacle.
hexagon	6	Can tile the plane.
heptagon	7	The simplest polygon such that the regular form is not constructible with compass and straightedge. However, it can be constructed using a Neusis construction.
octagon	8	
nonagon (or enneagon)	9	"Nonagon" mixes Latin [*novem* = 9] with Greek, "enneagon" is pure Greek.
decagon	10	
hendecagon (or undecagon)	11	The simplest polygon such that the regular form cannot be constructed with compass, straightedge, and angle trisector.
dodecagon (or duodecagon)	12	
tridecagon (or triskaidecagon)	13	
tetradecagon (or tetrakaidecagon)	14	
pentadecagon (or pentakaidecagon)	15	
hexadecagon (or hexakaidecagon)	16	
heptadecagon (or heptakaidecagon)	17	Constructible polygon
octadecagon (or octakaidecagon)	18	
enneadecagon (or enneakaidecagon)	19	
icosagon	20	

icositetragon (or icosikaitetragon)	24	
triacontagon	30	
tetracontagon (or tessaracontagon)	40	
pentacontagon (or pentecontagon)	50	
hexacontagon (or hexecontagon)	60	
heptacontagon (or hebdomecontagon)	70	
octacontagon (or ogdoëcontagon)	80	
enneacontagon (or enenecontagon)	90	
hectogon (or hecatontagon)	100	
	257	Constructible polygon
chiliagon	1000	Philosophers including René Descartes, Immanuel Kant, David Hume, have used the chiliagon as an example in discussions.
myriagon	10,000	Used as an example in some philosophical discussions, for example in Descartes' *Meditations on First Philosophy*
	65,537	Constructible polygon
megagon	1,000,000	As with René Descartes' example of the chiliagon, the million-sided polygon has been used as an illustration of a well-defined concept that cannot be visualised. The megagon is also used as an illustration of the convergence of regular polygons to a circle.
apeirogon	∞	A degenerate polygon of infinitely many sides.

Constructing Higher Names

To construct the name of a polygon with more than 20 and less than 100 edges, combine the prefixes as follows. The "kai" term applies to 13-gons and higher was used by Kepler, and advocated by John H. Conway for clarity to concatenated prefix numbers in the naming of quasiregular polyhedra.

Tens		*and*		Ones		final suffix
	-kai-	1	-hena-	-gon		
20	icosi- (icosa- when alone)		2	-di-		
30	triaconta- (or triconta-)		3	-tri-		
40	tetraconta- (or tessaraconta-)		4	-tetra-		
50	pentaconta- (or penteconta-)		5	-penta-		
60	hexaconta- (or hexeconta-)		6	-hexa-		
70	heptaconta- (or hebdomeconta-)		7	-hepta-		
80	octaconta- (or ogdoëconta-)		8	-octa-		
90	enneaconta- (or eneneconta-)		9	-ennea-		

History

Polygons have been known since ancient times. The regular polygons were known to the ancient Greeks, with the pentagram, a non-convex regular polygon (star polygon), appearing as early as the 7th century B.C. on a krater by Aristonothos, found at Caere and now in the Capitoline Museum.

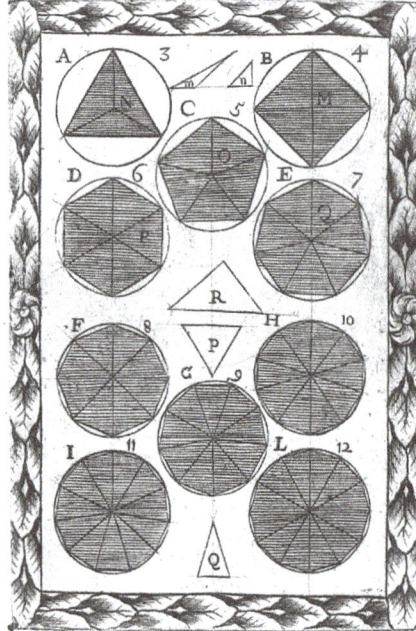

Historical image of polygons (1699)

The first known systematic study of non-convex polygons in general was made by Thomas Bradwardine in the 14th century.

In 1952, Geoffrey Colin Shephard generalized the idea of polygons to the complex plane, where each real dimension is accompanied by an imaginary one, to create complex polygons.

Polygons in Nature

Polygons appear in rock formations, most commonly as the flat facets of crystals, where the angles between the sides depend on the type of mineral from which the crystal is made.

The Giant's Causeway, in Northern Ireland

Regular hexagons can occur when the cooling of lava forms areas of tightly packed columns of basalt, which may be seen at the Giant's Causeway in Northern Ireland, or at the Devil's Postpile in California.

In biology, the surface of the wax honeycomb made by bees is an array of hexagons, and the sides and base of each cell are also polygons.

Polygons in Computer Graphics

A polygon in a computer graphics (image generation) system is a two-dimensional shape that is modelled and stored within its database. A polygon can be colored, shaded and textured, and its position in the database is defined by the coordinates of its vertices (corners).

Naming conventions differ from those of mathematicians:

- A simple polygon does not cross itself.

- a concave polygon is a simple polygon having at least one interior angle greater than 180°.

- A complex polygon does cross itself.

Any surface is modelled as a tessellation called polygon mesh. If a square mesh has $n + 1$ points (vertices) per side, there are n squared squares in the mesh, or $2n$ squared triangles since there are two triangles in a square. There are $(n + 1)^2 / 2(n^2)$ vertices per triangle. Where n is large, this approaches one half. Or, each vertex inside the square mesh connects four edges (lines).

The imaging system calls up the structure of polygons needed for the scene to be created from the database. This is transferred to active memory and finally, to the display system (screen, TV monitors etc.) so that the scene can be viewed. During this process, the imaging system renders polygons in correct perspective ready for transmission of the processed data to the display system. Although polygons are two-dimensional, through the system computer they are placed in a visual scene in the correct three-dimensional orientation.

In computer graphics and computational geometry, it is often necessary to determine whether a given point $P = (x_0, y_0)$ lies inside a simple polygon given by a sequence of line segments. This is called the Point in polygon test.

Curve

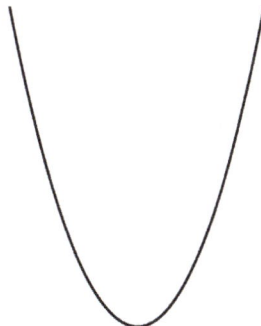

A parabola, a simple example of a curve

In mathematics, a curve (also called a curved line in older texts) is, generally speaking, an object similar to a line but that need not be straight. Thus, a curve is a generalization of a line, in that curvature is not necessarily zero.

Various disciplines within mathematics have given the term different meanings depending on the area of study, so the precise meaning depends on context. However, many of these meanings are special instances of the definition which follows. A curve is a topological space which is locally homeomorphic to a line. In everyday language, this means that a curve is a set of points which, near each of its points, looks like a line, up to a deformation. A simple example of a curve is the parabola, shown to the right. A large number of other curves have been studied in multiple mathematical fields.

A closed curve is a curve that forms a path whose starting point is also its ending point—that is, a path from any of its points to the same point.

Closely related meanings include the graph of a function (as in Phillips curve) and a two-dimensional graph.

History

Interest in curves began long before they were the subject of mathematical study. This can be seen in numerous examples of their decorative use in art and on everyday objects dating back to prehistoric times. Curves, or at least their graphical representations, are simple to create, for example by a stick in the sand on a beach.

Megalithic art from Newgrange showing an early interest in curves

Historically, the term "line" was used in place of the more modern term "curve". Hence the phrases "straight line" and "right line" were used to distinguish what are today called lines from "curved lines". For example, in Book I of Euclid's Elements, a line is defined as a "breadthless length" (Def. 2), while a *straight* line is defined as "a line that lies evenly with the points on itself" (Def. 4). Euclid's idea of a line is perhaps clarified by the statement "The extremities of a line are points," (Def. 3). Later commentators further classified lines according to various schemes. For example:

- Composite lines (lines forming an angle)

- Incomposite lines

    o Determinate (lines that do not extend indefinitely, such as the circle) o   Indeterminate (lines that extend indefinitely, such as the straight line and the parabola)

The curves created by slicing a cone (conic sections) were among the curves studied in ancient Greece.

The Greek geometers had studied many other kinds of curves. One reason was their interest in solving geometrical problems that could not be solved using standard compass and straightedge construction. These curves include:

- The conic sections, deeply studied by Apollonius of Perga

- The cissoid of Diocles, studied by Diocles and used as a method to double the cube.

- The conchoid of Nicomedes, studied by Nicomedes as a method to both double the cube and to trisect an angle.

- The Archimedean spiral, studied by Archimedes as a method to trisect an angle and square the circle.

- The spiric sections, sections of tori studied by Perseus as sections of cones had been studied by Apollonius.

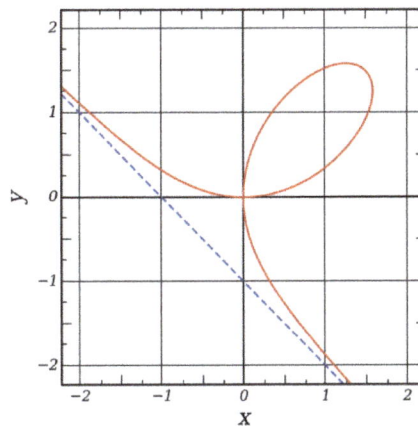

Analytic geometry allowed curves, such as the Folium of Descartes, to be defined using equations instead of geometrical construction.

A fundamental advance in the theory of curves was the advent of analytic geometry in the seventeenth century. This enabled a curve to be described using an equation rather than an elaborate geometrical construction. This not only allowed new curves to be defined and studied, but it enabled a formal distinction to be made between curves that can be defined using algebraic equations, algebraic curves, and those that cannot, transcendental curves. Previously, curves had been described as "geometrical" or "mechanical" according to how they were, or supposedly could be, generated.

Conic sections were applied in astronomy by Kepler. Newton also worked on an early example in the calculus of variations. Solutions to variational problems, such as the brachistochrone and tautochrone questions, introduced properties of curves in new ways (in this case, the cycloid). The catenary gets its name as the solution to the problem of a hanging chain, the sort of question that became routinely accessible by means of differential calculus.

In the eighteenth century came the beginnings of the theory of plane algebraic curves, in general. Newton had studied the cubic curves, in the general description of the real points into 'ovals'. The statement of Bézout's theorem showed a number of aspects which were not directly accessible to the geometry of the time, to do with singular points and complex solutions.

From the nineteenth century there is not a separate curve theory, but rather the appearance of curves as the one-dimensional aspect of projective geometry, and differential geometry; and later topology, when for example the Jordan curve theorem was understood to lie quite deep, as well as being required in complex analysis. The era of the space-filling curves finally provoked the modern definitions of curve.

Definition

In general, a curve is defined through a continuous function $\gamma : I \to X$ from an interval I of the real numbers into a topological space X. Depending on the context, it is either γ or its image $\gamma(I)$ which is called a curve.

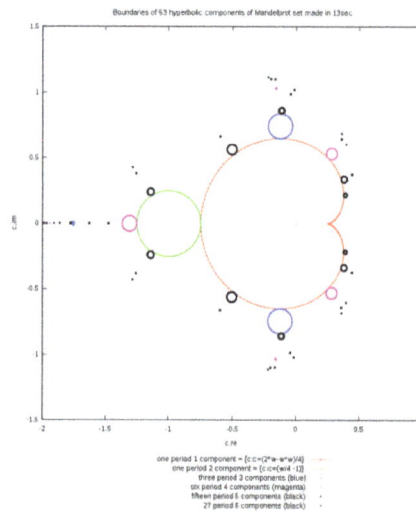

Boundaries of hyperbolic components of Mandelbrot set as closed curves

In general topology, when non-differentiable functions are considered, it is the map γ, which is called a curve, because its image may look very differently from what is commonly called a curve. For example, the image of the Peano curve completely fills the square. On the other hand, when one considers curves defined by a differentiable function (or, at least, a piecewise differentiable function), this is commonly the image of the function which is called a curve.

- The curve is said to be simple, or a Jordan arc, if γ is injective, i.e. if for all x, y in I, we have $\gamma(x) = \gamma(y)$ implies $x = y$. If I is a closed bounded interval $[a,b]$, we also allow the possibility $\gamma(a) = \gamma(b)$ (this convention makes it possible to talk about "closed" simple curves). In other words, this curve "does not cross itself and has no missing points".

- If $\gamma(x) = \gamma(y)$ for some $x \neq y$ (other than the extremities of I), then $\gamma(x)$ is called a double (or multiple) point of the curve. This is a special case of a singular point of a curve.

- A curve γ is said to be closed or a loop if $I = [a,b]$ and if $\gamma(a) = \gamma(b)$. A closed curve is thus the image of a continuous mapping of the circle S^1; a simple closed curve is also called a Jordan curve. The Jordan curve theorem states that such curves divide the plane into an "interior" and an "exterior".

A plane curve is a curve for which X is the Euclidean plane—these are the examples first encountered—or in some cases the projective plane. A space curve is a curve for which X is of three dimensions, usually Euclidean space; a skew curve is a space curve which lies in no plane. These definitions of plane, space and skew curves apply also to real algebraic curves, although the above definition of a curve does not applies (a real algebraic curve may be disconnected).

This definition of curve captures our intuitive notion of a curve as a connected, continuous geometric figure that is "like" a line, without thickness and drawn without interruption, although it also includes figures that can hardly be called curves in common usage. For example, the image of a curve can cover a square in the plane (space-filling curve). The image of simple plane curve can have Hausdorff dimension bigger than one and even positive Lebesgue measure (the last example can be obtained by small variation of the Peano curve construction). The dragon curve is another unusual example.

Differentiable Curve

Roughly speaking a differentiable curve is a curve that is defined as being locally the image of an injective differentiable function $\gamma : I \to X$ from an interval I of the real numbers into a differentiable manifold X, often $\mathbb{R}^n$.

More precisely, a differentiable curve is a subset C of X such every point of C has a neighborhood U such $C \cap U$ is diffeomorphic to an interval of the real numbers. In other words, a differentiable curve differentiable manifold of dimension one.

Length of a Curve

If $X = \mathbb{R}^n$ is an Euclidean space and $\gamma : [a,b] \to \mathbb{R}^n$ is an injective differentiable function, then the image of γ is a curve of length

$$\text{length}(\gamma) = \int_a^b |\gamma'(t)| \, dt.$$

This length is independent of the choice of the function γ that has been chosen for parameterizing the curve.

In particular, the length s of the graph of a differentiable function $y = f(x)$ defined on the interval $[a, b]$ is

$$s = \int_a^b \sqrt{1 + f'(x)^2}\, dx.$$

More generally, if X is a metric space with metric d, then we can define the length of a curve defined by $\gamma : [a, b] \to X$ by

$$\text{length}(\gamma) = \sup\left\{ \sum_{i=1}^n d(\gamma(t_i), \gamma(t_{i-1})) : n \in \mathbb{N} \text{ and } a = t_0 < t_1 < \cdots < t_n = b \right\}.$$

where the sup is over all n and all partitions $t_0 < t_1 < \cdots < t_n$ of $[a, b]$.

A rectifiable curve is a curve with finite length. A parametrization of γ is called natural (or unit speed or parametrised by arc length) if for any $t_1, t_2 \in [a, b]$, we have

$$\text{length}(\gamma\,|_{[t_1, t_2]}) = |t_2 - t_1|.$$

If γ is a Lipschitz-continuous function, then it is automatically rectifiable. Moreover, in this case, one can define the speed (or metric derivative) of γ at t_0 as

$$\text{speed}(t_0) = \limsup_{t \to t_0} \frac{d(\gamma(t), \gamma(t_0))}{|t - t_0|}$$

and then

$$\text{length}(\gamma) = \int_a^b \text{speed}(t)dt.$$

Differential Geometry

While the first examples of curves that are met are mostly plane curves (that is, in everyday words, *curved lines* in *two-dimensional space*), there are obvious examples such as the helix which exist naturally in three dimensions. The needs of geometry, and also for example classical mechanics are to have a notion of curve in space of any number of dimensions. In general relativity, a world line is a curve in spacetime.

If X is a differentiable manifold, then we can define the notion of *differentiable curve* in X. This general idea is enough to cover many of the applications of curves in mathematics. From a local point of view one can take X to be Euclidean space. On the other hand, it is useful to be more general, in that (for example) it is possible to define the tangent vectors to X by means of this notion of curve.

If X is a smooth manifold, a *smooth curve* in X is a smooth map

$$\gamma : I \to X.$$

This is a basic notion. There are less and more restricted ideas, too. If X is a C^k manifold (i.e., a manifold whose charts are k times continuously differentiable), then a C^k curve in X is such a curve which is only assumed to be C^k (i.e. k times continuously differentiable). If X is an analytic manifold (i.e. infinitely differentiable and charts are expressible as power series), and γ is an analytic map, then γ is said to be an *analytic curve*.

A differentiable curve is said to be *regular* if its derivative never vanishes. (In words, a regular curve never slows to a stop or backtracks on itself.) Two C^k differentiable curves

$$\gamma_1 : I \to X \text{ and}$$

$$\gamma_2 : J \to X$$

are said to be *equivalent* if there is a bijective C^k map

$$p : J \to I$$

such that the inverse map

$$p^{-1} : I \to J$$

is also C^k, and

$$\gamma_2(t) = \gamma_1(p(t))$$

for all . The map γ_2 is called a *reparametrisation* of γ_1; and this makes an equivalence relation on the set of all C^k differentiable curves in X. A C^k *arc* is an equivalence class of C^k curves under the relation of reparametrisation.

Algebraic Curve

Algebraic curves are the curves considered in algebraic geometry. A plane algebraic curve is the locus of the points of coordinates x, y such that $f(x, y) = 0$, where f is a polynomial in two variables defined over some field F. Algebraic geometry normally looks not only on points with coordinates in F but on all the points with coordinates in an algebraically closed field K. If C is a curve defined by a polynomial f with coefficients in F, the curve is said defined over F. The points of the curve C with coordinates in a field G are said rational over G and can be denoted $C(G)$). When G is the field of the rational numbers, one simply talks of *rational points*. For example, Fermat's Last Theorem may be restated as: *For n > 2, every rational point of the Fermat curve of degree n has a zero coordinate.*

Algebraic curves can also be space curves, or curves in a space of higher dimension, say n. They are defined as algebraic varieties of dimension one. They may be obtained as the common solutions of at least $n-1$ polynomial equations in n variables. If $n-1$ polynomials are sufficient to define a curve in a space of dimension n, the curve is said to be a complete intersection. By eliminating variables (by any tool of elimination theory), an algebraic curve may be projected onto a plane algebraic curve, which however may introduce new singularities such as cusps or double points.

A plane curve may also be completed in a curve in the projective plane: if a curve is defined by a polynomial f of total degree d, then $w^d f(u/w, v/w)$ simplifies to a homogeneous polynomial $g(u, v, w)$ of degree d. The values of u, v, w such that $g(u, v, w) = 0$ are the homogeneous coordinates of the points of the completion of the curve in the projective plane and the points of the initial curve are those such w is not zero. An example is the Fermat curve $u^n + v^n = w^n$, which has an affine form $x^n + y^n = 1$. A similar process of homogenization may be defined for curves in higher dimensional spaces

Important examples of algebraic curves are the conics, which are nonsingular curves of degree two and genus zero, and elliptic curves, which are nonsingular curves of genus one studied in number theory and which have important applications to cryptography. Because algebraic curves in fields of characteristic zero are most often studied over the complex numbers, algebraic curves in algebraic geometry may be considered as real surfaces. In particular, the non-singular complex projective algebraic curves are called Riemann surfaces.

Geometric Topology

In mathematics, geometric topology is the study of manifolds and maps between them, particularly embeddings of one manifold into another.

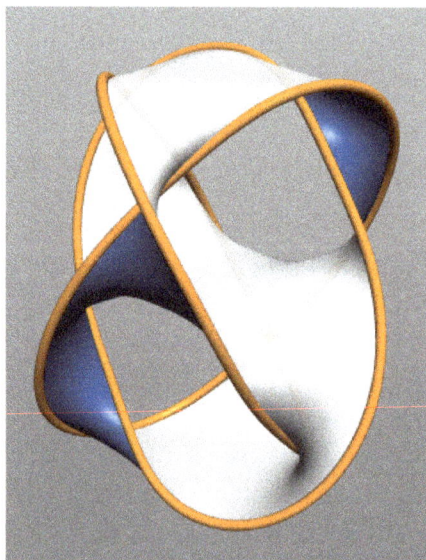

A Seifert surface bounded by a set of Borromean rings. Seifert surfaces for links are a useful tool in geometric topology.

History

Geometric topology as an area distinct from algebraic topology may be said to have originated in the 1935 classification of lens spaces by Reidemeister torsion, which required distinguishing spaces that are homotopy equivalent but not homeomorphic. This was the origin of *simple* homotopy theory.

Differences between Low-dimensional and High-dimensional topology

Manifolds differ radically in behavior in high and low dimension.

High-dimensional topology refers to manifolds of dimension 5 and above, or in relative terms, embeddings in codimension 3 and above. Low-dimensional topology is concerned with questions in dimensions up to 4, or embeddings in codimension up to 2.

Dimension 4 is special, in that in some respects (topologically), dimension 4 is high-dimensional, while in other respects (differentiably), dimension 4 is low-dimensional; this overlap yields phenomena exceptional to dimension 4, such as exotic differentiable structures on R^4. Thus the topological classification of 4-manifolds is in principle easy, and the key questions are: does a topological manifold admit a differentiable structure, and if so, how many? Notably, the smooth case of dimension 4 is the last open case of the generalized Poincaré conjecture.

The distinction is because surgery theory works in dimension 5 and above (in fact, it works topologically in dimension 4, though this is very involved to prove), and thus the behavior of manifolds in dimension 5 and above is controlled algebraically by surgery theory. In dimension 4 and below (topologically, in dimension 3 and below), surgery theory does not work, and other phenomena occur. Indeed, one approach to discussing low-dimensional manifolds is to ask "what would surgery theory predict to be true, were it to work?" – and then understand low-dimensional phenomena as deviations from this.

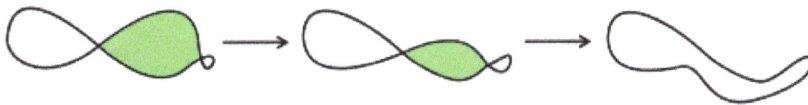

The Whitney trick requires 2+1 dimensions, hence surgery theory requires 5 dimensions.

The precise reason for the difference at dimension 5 is because the Whitney embedding theorem, the key technical trick which underlies surgery theory, requires 2+1 dimensions. Roughly, the Whitney trick allows one to "unknot" knotted spheres – more precisely, remove self-intersections of immersions; it does this via a homotopy of a disk – the disk has 2 dimensions, and the homotopy adds 1 more – and thus in codimension greater than 2, this can be done without intersecting itself; hence embeddings in codimension greater than 2 can be understood by surgery. In surgery theory, the key step is in the middle dimension, and thus when the middle dimension has codimension more than 2 (loosely, 2½ is enough, hence total dimension 5 is enough), the Whitney trick works. The key consequence of this is Smale's h-cobordism theorem, which works in dimension 5 and above, and forms the basis for surgery theory.

A modification of the Whitney trick can work in 4 dimensions, and is called Casson handles – because there are not enough dimensions, a Whitney disk introduces new kinks, which can be resolved by another Whitney disk, leading to a sequence ("tower") of disks. The limit of this tower yields a topological but not differentiable map, hence surgery works topologically but not differentiably in dimension 4.

Important Tools in Geometric Topology

Fundamental Group

In all dimensions, the fundamental group of a manifold is a very important invariant, and determines much of the structure; in dimensions 1, 2 and 3, the possible fundamental groups are restricted, while in every dimension 4 and above every finitely presented group is the fundamental group of a manifold (note that it is sufficient to show this for 4- and 5-dimensional manifolds, and then to take products with spheres to get higher ones).

Orientability

A manifold is orientable if it has a consistent choice of orientation, and a connected orientable manifold has exactly two different possible orientations. In this setting, various equivalent formulations of orientability can be given, depending on the desired application and level of generality. Formulations applicable to general topological manifolds often employ methods of homology theory, whereas for differentiable manifolds more structure is present, allowing a formulation in terms of differential forms. An important generalization of the notion of orientability of a space is that of orientability of a family of spaces parameterized by some other space (a fiber bundle) for which an orientation must be selected in each of the spaces which varies continuously with respect to changes in the parameter values.

Handle Decompositions

A handle decomposition of an m-manifold M is a union

$$M_{-1} \subset M_0 \subset M_1 \subset M_2 \subset \cdots \subset M_{m-1} \subset M_m = M$$

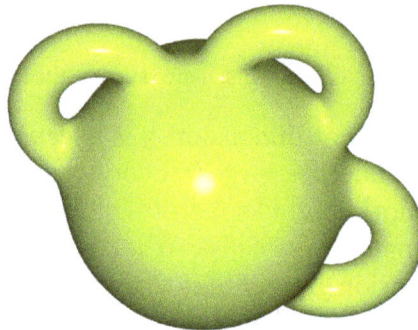

A 3-ball with three 1-handles attached.

where each M_i is obtained from M_{i-1} by the attaching of i-handles. A handle decomposition is to a manifold what a CW-decomposition is to a topological space—in many regards the purpose of a handle decomposition is to have a language analogous to CW-complexes, but adapted to the world of smooth manifolds. Thus an i-handle is the smooth analogue of an i-cell. Handle decompositions of manifolds arise naturally via Morse theory. The modification of handle structures is closely linked to Cerf theory.

Local Flatness

Local flatness is a property of a submanifold in a topological manifold of larger dimension. In the category of topological manifolds, locally flat submanifolds play a role similar to that of embedded submanifolds in the category of smooth manifolds.

Suppose a d dimensional manifold N is embedded into an n dimensional manifold M (where $d < n$). If $x \in N$, we say N is locally flat at x if there is a neighborhood $U \subset M$ of x such that the topological pair $(U, U \cap N)$ is homeomorphic to the pair $(\mathbb{R}^n, \mathbb{R}^d)$, with a standard inclusion of $\mathbb{R}^d$ as a subspace of $\mathbb{R}^n$. That is, there exists a homeomorphism $U \to R^n$ such that the image of $U \cap N$ coincides with $\mathbb{R}^d$.

Schönflies Theorems

The generalized Schoenflies theorem states that, if an $(n-1)$-dimensional sphere S is embedded into the n-dimensional sphere S^n in a locally flat way (that is, the embedding extends to that of a thickened sphere), then the pair (S^n, S) is homeomorphic to the pair (S^n, S^{n-1}), where S^{n-1} is the equator of the n-sphere. Brown and Mazur received the Veblen Prize for their independent proofs of this theorem.

Branches of Geometric Topology

Low-dimensional Topology

Low-dimensional topology includes:

- Surface (topology)s (2-manifolds)

- 3-manifolds

- 4-manifolds

each have their own theory, where there are some connections.

Low-dimensional topology is strongly geometric, as reflected in the uniformization theorem in 2 dimensions – every surface admits a constant curvature metric; geometrically, it has one of 3 possible geometries: positive curvature/spherical, zero curvature/flat, negative curvature/hyperbolic – and the geometrization conjecture (now theorem) in 3 dimensions – every 3-manifold can be cut into pieces, each of which has one of 8 possible geometries.

2-dimensional topology can be studied as complex geometry in one variable (Riemann surfaces are complex curves) – by the uniformization theorem every conformal class of metrics is equivalent to a unique complex one, and 4-dimensional topology can be studied from the point of view of complex geometry in two variables (complex surfaces), though not every 4-manifold admits a complex structure.

Knot Theory

Knot theory is the study of mathematical knots. While inspired by knots which appear in daily life in shoelaces and rope, a mathematician's knot differs in that the ends are joined together so that it cannot be undone. In mathematical language, a knot is an embedding of a circle in 3-dimensional Euclidean space, R^3 (since we're using topology, a circle isn't bound to the classical geometric concept, but to all of its homeomorphisms). Two mathematical knots are equivalent if one can be transformed into the other via a deformation of R^3 upon itself (known as an ambient isotopy); these transformations correspond to manipulations of a knotted string that do not involve cutting the string or passing the string through itself.

To gain further insight, mathematicians have generalized the knot concept in several ways. Knots can be considered in other three-dimensional spaces and objects other than circles can be used;

see *knot (mathematics)*. Higher-dimensional knots are n-dimensional spheres in m-dimensional Euclidean space.

High-dimensional Geometric Topology

In high-dimensional topology, characteristic classes are a basic invariant, and surgery theory is a key theory.

A characteristic class is a way of associating to each principal bundle on a topological space X a cohomology class of X. The cohomology class measures the extent to which the bundle is "twisted" — particularly, whether it possesses sections or not. In other words, characteristic classes are global invariants which measure the deviation of a local product structure from a global product structure. They are one of the unifying geometric concepts in algebraic topology, differential geometry and algebraic geometry.

Surgery theory is a collection of techniques used to produce one manifold from another in a 'controlled' way, introduced by Milnor (1961). Surgery refers to cutting out parts of the manifold and replacing it with a part of another manifold, matching up along the cut or boundary. This is closely related to, but not identical with, handlebody decompositions. It is a major tool in the study and classification of manifolds of dimension greater than 3.

More technically, the idea is to start with a well-understood manifold M and perform surgery on it to produce a manifold M' having some desired property, in such a way that the effects on the homology, homotopy groups, or other interesting invariants of the manifold are known.

The classification of exotic spheres by Kervaire and Milnor (1963) led to the emergence of surgery theory as a major tool in high-dimensional topology.

References

- Faber, Richard L. (1983). Foundations of Euclidean and Non-Euclidean Geometry. New York: Marcel Dekker. ISBN 0-8247-1748-1.

- Richards, Joan L. (1988), Mathematical Visions: The Pursuit of Geometry in Victorian England, Boston: Academic Press, ISBN 0-12-587445-6

- Harry F. Davis & Arthur David Snider (1988) Introduction to Vector Analysis, 5th edition, page 1, Wm. C. Brown Publishers ISBN 0-697-06814-5

- Brannan, David A.; Esplen, Matthew F.; Gray, Jeremy J. (1998), Geometry, Cambridge: Cambridge University Press, ISBN 0-521-59787-0

- Dembowski, Peter (1968), Finite geometries, Ergebnisse der Mathematik und ihrer Grenzgebiete, Band 44, Berlin, New York: Springer-Verlag, ISBN 3-540-61786-8, MR 0233275

- Henderson, David W.; Taimina, Daina (2005), Experiencing Geometry/Euclidean and Non-Euclidean with History (3rd ed.), Pearson Prentice-Hall, ISBN 978-0-13-143748-7

- Parr, H. E. (1970). Revision Course in School mathematics. Mathematics Textbooks Second Edition. G Bell and Sons Ltd. ISBN 0-7135-1717-4.

- Cornel, Antonio (2002). Geometry for Secondary Schools. Mathematics Textbooks Second Edition. Bookmark Inc. ISBN 971-569-441-1.

- Wong, TW; Wong, MS. "Angles in Intersecting and Parallel Lines". New Century Mathematics. 1B (1 ed.). Hong

Kong: Oxford University Press. pp. 161–163. ISBN 978-0-19-800176-8.

- Henderson, David W.; Taimina, Daina (2005), Experiencing Geometry / Euclidean and Non-Euclidean with History (3rd ed.), Pearson Prentice Hall, p. 104, ISBN 9780131437487

- Kappraff, Jay (2002). Beyond measure: a guided tour through nature, myth, and number. World Scientific. p. 258. ISBN 978-981-02-4702-7.

- Salomon, David (2011). The Computer Graphics Manual. Springer Science & Business Media. pp. 88–90. ISBN 978-0-85729-886-7.

- Darling, David J., The universal book of mathematics: from Abracadabra to Zeno's paradoxes, John Wiley & Sons, 2004. Page 249. ISBN 0-471-27047-4.

- Potter, Vincent G., On Understanding Understanding: A Philosophy of Knowledge, 2nd ed, Fordham University Press, 1993, p. 86, ISBN 0-8232-1486-9.

- A.S. Parkhomenko (2001), "Line (curve)", in Hazewinkel, Michiel, Encyclopedia of Mathematics, Springer, ISBN 978-1-55608-010-4

- B.I. Golubov (2001), "Rectifiable curve", in Hazewinkel, Michiel, Encyclopedia of Mathematics, Springer, ISBN 978-1-55608-010-4

Triangles: An Overview

Triangles are plain figures which have three angles and the three angles always add to 180°. Triangles on the bases of their sides can be categorized into three types, equilateral, isosceles and scalene. Equilateral triangles have all sides of the same length, an isosceles triangle has two sides of the same length and a scalene triangle has 3 different lengths.

Triangle

In Euclidean geometry any three points, when non-collinear, determine a unique triangle and a unique plane (i.e. a two-dimensional Euclidean space). This article is about triangles in Euclidean geometry except where otherwise noted.

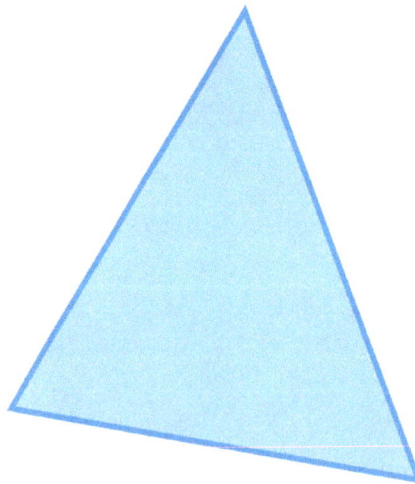

A triangle

Types of Triangle

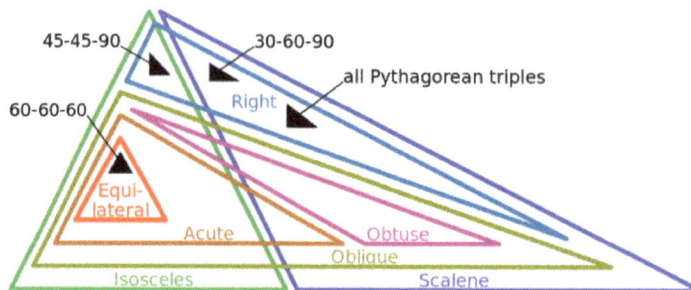

Euler diagram of types of triangles, using the definition that isosceles triangles have *at least* 2 equal sides, i.e. equilateral triangles are isosceles.

By Lengths of Sides

Triangles can be classified according to the lengths of their sides:

- An *equilateral triangle* has all sides the same length. An equilateral triangle is also a regular polygon with all angles measuring 60°.

- An *isosceles triangle* has two sides of equal length. An isosceles triangle also has two angles of the same measure, namely the angles opposite to the two sides of the same length; this fact is the content of the isosceles triangle theorem, which was known by Euclid. Some mathematicians define an isosceles triangle to have exactly two equal sides, whereas others define an isosceles triangle as one with *at least* two equal sides. The latter definition would make all equilateral triangles isosceles triangles. The 45–45–90 right triangle, which appears in the tetrakis square tiling, is isosceles.

- A *scalene triangle* has all its sides of different lengths. Equivalently, it has all angles of different measure.

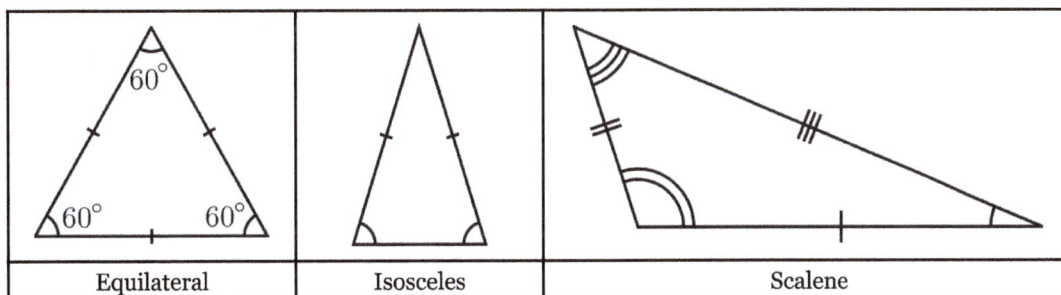

| Equilateral | Isosceles | Scalene |

Hatch marks, also called tick marks, are used in diagrams of triangles and other geometric figures to identify sides of equal lengths. A side can be marked with a pattern of "ticks", short line segments in the form of tally marks; two sides have equal lengths if they are both marked with the same pattern. In a triangle, the pattern is usually no more than 3 ticks. An equilateral triangle has the same pattern on all 3 sides, an isosceles triangle has the same pattern on just 2 sides, and a scalene triangle has different patterns on all sides since no sides are equal. Similarly, patterns of 1, 2, or 3 concentric arcs inside the angles are used to indicate equal angles. An equilateral triangle has the same pattern on all 3 angles, an isosceles triangle has the same pattern on just 2 angles, and a scalene triangle has different patterns on all angles since no angles are equal.

By Internal Angles

Triangles can also be classified according to their internal angles, measured here in degrees.

- A *right triangle* (or *right-angled triangle*, formerly called a *rectangled triangle*) has one of its interior angles measuring 90° (a right angle). The side opposite to the right angle is the hypotenuse, the longest side of the triangle. The other two sides are called the *legs* or *catheti* (singular: *cathetus*) of the triangle. Right triangles obey the Pythagorean theorem: the sum of the squares of the lengths of the two legs is equal to the square of the length of the hypotenuse: $a^2 + b^2 = c^2$, where a and b are the lengths of the legs and c is the length of the hypotenuse. Special right triangles are right triangles with additional properties that make calculations involving them easier. One of the two most famous is the 3–4–5 right triangle, where $3^2 + 4^2 = 5^2$. In this situation, 3, 4, and 5 are a Pythagorean triple. The other one is an isosceles triangle that has 2 angles that each measure 45 degrees.

- Triangles that do not have an angle measuring 90° are called oblique triangles.

- A triangle with all interior angles measuring less than 90° is an acute triangle or *acute-angled triangle*. If c is the length of the longest side, then $a^2 + b^2 > c^2$, where a and b are the lengths of the other sides.

- A triangle with one interior angle measuring more than 90° is an obtuse triangle or *obtuse-angled triangle*. If c is the length of the longest side, then $a^2 + b^2 < c^2$, where a and b are the lengths of the other sides.

- A triangle with an interior angle of 180° (and collinear vertices) is degenerate.

- A right degenerate triangle has collinear vertices, two of which are coincident.

A triangle that has two angles with the same measure also has two sides with the same length, and therefore it is an isosceles triangle. It follows that in a triangle where all angles have the same measure, all three sides have the same length, and such a triangle is therefore equilateral.

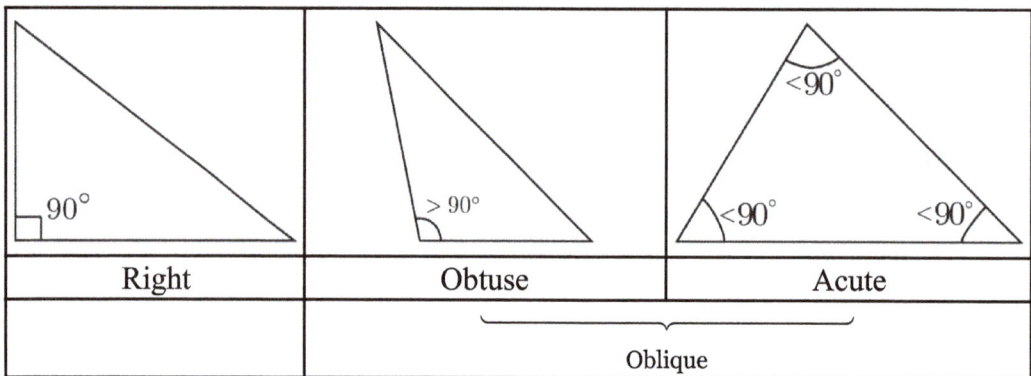

Basic Facts

Triangles are assumed to be two-dimensional plane figures, unless the context provides otherwise. In rigorous treatments, a triangle is therefore called a *2-simplex*. Elementary facts about triangles were presented by Euclid in books 1–4 of his *Elements*, around 300 BC.

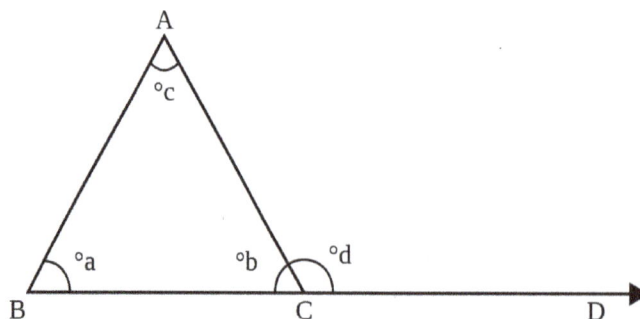

A triangle, showing exterior angle d.

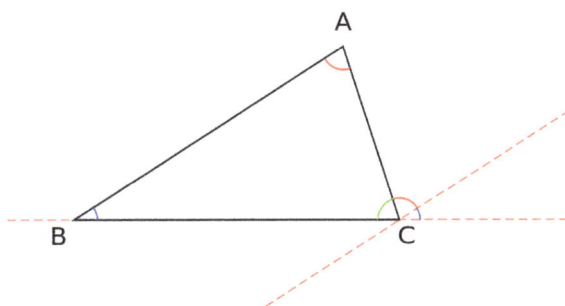

The measures of the interior angles of the triangle always add up to 180 degrees (same color to point out they are equal).

The sum of the measures of the interior angles of a triangle in Euclidean space is always 180 degrees. This fact is equivalent to Euclid's parallel postulate. This allows determination of the measure of the third angle of any triangle given the measure of two angles. An *exterior angle* of a triangle is an angle that is a linear pair (and hence supplementary) to an interior angle. The measure of an exterior angle of a triangle is equal to the sum of the measures of the two interior angles that are not adjacent to it; this is the exterior angle theorem. The sum of the measures of the three exterior angles (one for each vertex) of any triangle is 360 degrees.

Similarity and Congruence

Two triangles are said to be *similar* if every angle of one triangle has the same measure as the corresponding angle in the other triangle. The corresponding sides of similar triangles have lengths that are in the same proportion, and this property is also sufficient to establish similarity.

Some basic theorems about similar triangles are:

-   If and only if one pair of internal angles of two triangles have the same measure as each other, and another pair also have the same measure as each other, the triangles are similar.

-   If and only if one pair of corresponding sides of two triangles are in the same proportion as are another pair of corresponding sides, and their included angles have the same measure, then the triangles are similar. (The *included angle* for any two sides of a polygon is the internal angle between those two sides.)

-   If and only if three pairs of corresponding sides of two triangles are all in the same proportion, then the triangles are similar.

Two triangles that are congruent have exactly the same size and shape: all pairs of corresponding interior angles are equal in measure, and all pairs of corresponding sides have the same length. (This is a total of six equalities, but three are often sufficient to prove congruence.)

Some individually necessary and sufficient conditions for a pair of triangles to be congruent are:

-   SAS Postulate: Two sides in a triangle have the same length as two sides in the other triangle, and the included angles have the same measure.

-   ASA: Two interior angles and the included side in a triangle have the same measure and length, respectively, as those in the other triangle. (The *included side* for a pair of angles is the side that is common to them.)

- SSS: Each side of a triangle has the same length as a corresponding side of the other triangle.

- AAS: Two angles and a corresponding (non-included) side in a triangle have the same measure and length, respectively, as those in the other triangle. (This is sometimes referred to as *AAcorrS* and then includes ASA above.)

Some individually sufficient conditions are:

- Hypotenuse-Leg (HL) Theorem: The hypotenuse and a leg in a right triangle have the same length as those in another right triangle. This is also called RHS (right-angle, hypotenuse, side).

- Hypotenuse-Angle Theorem: The hypotenuse and an acute angle in one right triangle have the same length and measure, respectively, as those in the other right triangle. This is just a particular case of the AAS theorem.

An important condition is:

- Side-Side-Angle (or Angle-Side-Side) condition: If two sides and a corresponding non-included angle of a triangle have the same length and measure, respectively, as those in another triangle, then this is *not* sufficient to prove congruence; but if the angle given is opposite to the longer side of the two sides, then the triangles are congruent. The Hypotenuse-Leg Theorem is a particular case of this criterion. The Side-Side-Angle condition does not by itself guarantee that the triangles are congruent because one triangle could be obtuse-angled and the other acute-angled.

Using right triangles and the concept of similarity, the trigonometric functions sine and cosine can be defined. These are functions of an angle which are investigated in trigonometry.

Right Triangles

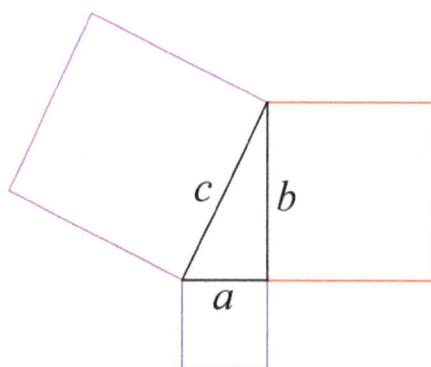

The Pythagorean theorem

A central theorem is the Pythagorean theorem, which states in any right triangle, the square of the length of the hypotenuse equals the sum of the squares of the lengths of the two other sides. If the hypotenuse has length c, and the legs have lengths a and b, then the theorem states that

$$a^2 + b^2 = c^2.$$

The converse is true: if the lengths of the sides of a triangle satisfy the above equation, then the triangle has a right angle opposite side c.

Some other facts about right triangles:

- The acute angles of a right triangle are complementary.

$$a + b + 90° = 180° \Rightarrow a + b = 90° \Rightarrow a = 90° - b$$

- If the legs of a right triangle have the same length, then the angles opposite those legs have the same measure. Since these angles are complementary, it follows that each measures 45 degrees. By the Pythagorean theorem, the length of the hypotenuse is the length of a leg times $\sqrt{2}$.

- In a right triangle with acute angles measuring 30 and 60 degrees, the hypotenuse is twice the length of the shorter side, and the longer side is equal to the length of the shorter side times $\sqrt{3}$:

$$c = 2a$$

$$b = a \times \sqrt{3}.$$

For all triangles, angles and sides are related by the law of cosines and law of sines (also called the *cosine rule* and *sine rule*).

Existence of a Triangle

Condition on the Sides

The triangle inequality states that the sum of the lengths of any two sides of a triangle must be greater than or equal to the length of the third side. That sum can equal the length of the third side only in the case of a degenerate triangle, one with collinear vertices. It is not possible for that sum to be less than the length of the third side. A triangle with three given positive side lengths exists if and only if those side lengths satisfy the triangle inequality.

Conditions on the Angles

Three given angles form a non-degenerate triangle (and indeed an infinitude of them) if and only if both of these conditions hold: (a) each of the angles is positive, and (b) the angles sum to 180°. If degenerate triangles are permitted, angles of 0° are permitted.

Trigonometric Conditions

Three positive angles α, β, and γ, each of them less than 180°, are the angles of a triangle if and only if any one of the following conditions holds:

$$\tan\frac{\alpha}{2}\tan\frac{\beta}{2} + \tan\frac{\beta}{2}\tan\frac{\gamma}{2} + \tan\frac{\gamma}{2}\tan\frac{\alpha}{2} = 1,$$

$$\sin^2\frac{\alpha}{2}+\sin^2\frac{\beta}{2}+\sin^2\frac{\gamma}{2}+2\sin\frac{\alpha}{2}\sin\frac{\beta}{2}\sin\frac{\gamma}{2}=1,$$

$$(2\alpha)+\sin(2\beta)+\sin(2\gamma)=4\sin(\alpha)\sin(\beta)\sin(\gamma),$$

$$\cos^2\alpha+\cos^2\beta+\cos^2\gamma+2\cos(\alpha)\cos(\beta)\cos(\gamma)=1,$$

$$\tan(\alpha)+\tan(\beta)+\tan(\gamma)=\tan(\alpha)\tan(\beta)\tan(\gamma),$$

the latter equality applying only if none of the angles is 90° (so the tangent function's value is always finite).

Points, Lines, and Circles Associated with a Triangle

There are thousands of different constructions that find a special point associated with (and often inside) a triangle, satisfying some unique property: Encyclopedia of Triangle Centers for a catalogue of them. Often they are constructed by finding three lines associated in a symmetrical way with the three sides (or vertices) and then proving that the three lines meet in a single point: an important tool for proving the existence of these is Ceva's theorem, which gives a criterion for determining when three such lines are concurrent. Similarly, lines associated with a triangle are often constructed by proving that three symmetrically constructed points are collinear: here Menelaus' theorem gives a useful general criterion.

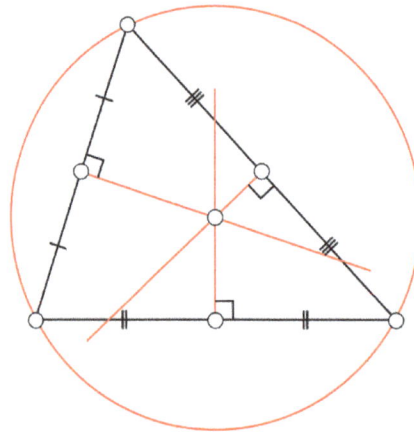

The circumcenter is the center of a circle passing through the three vertices of the triangle.

A perpendicular bisector of a side of a triangle is a straight line passing through the midpoint of the side and being perpendicular to it, i.e. forming a right angle with it. The three perpendicular bisectors meet in a single point, the triangle's circumcenter, usually denoted by O; this point is the center of the circumcircle, the circle passing through all three vertices. The diameter of this circle, called the *circumdiameter*, can be found from the law of sines stated above. The circumcircle's radius is called the *circumradius*.

Thales' theorem implies that if the circumcenter is located on one side of the triangle, then the opposite angle is a right one. If the circumcenter is located inside the triangle, then the triangle is acute; if the circumcenter is located outside the triangle, then the triangle is obtuse.

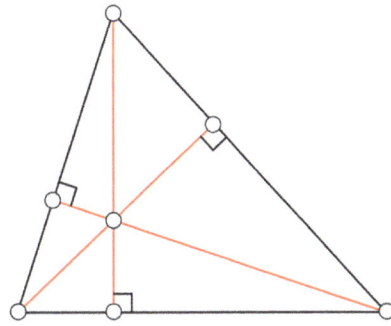

The intersection of the altitudes is the orthocenter.

An altitude of a triangle is a straight line through a vertex and perpendicular to (i.e. forming a right angle with) the opposite side. This opposite side is called the *base* of the altitude, and the point where the altitude intersects the base (or its extension) is called the *foot* of the altitude. The length of the altitude is the distance between the base and the vertex. The three altitudes intersect in a single point, called the orthocenter of the triangle, usually denoted by H. The orthocenter lies inside the triangle if and only if the triangle is acute.

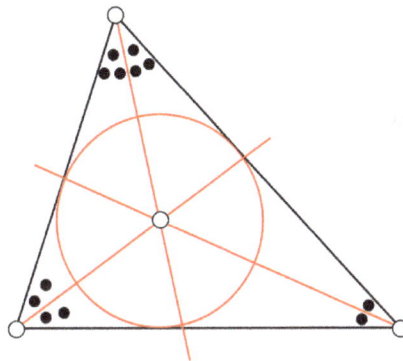

The intersection of the angle bisectors is the center of the incircle.

An angle bisector of a triangle is a straight line through a vertex which cuts the corresponding angle in half. The three angle bisectors intersect in a single point, the incenter, usually denoted by I, the center of the triangle's incircle. The incircle is the circle which lies inside the triangle and touches all three sides. Its radius is called the *inradius*. There are three other important circles, the excircles; they lie outside the triangle and touch one side as well as the extensions of the other two. The centers of the in- and excircles form an orthocentric system.

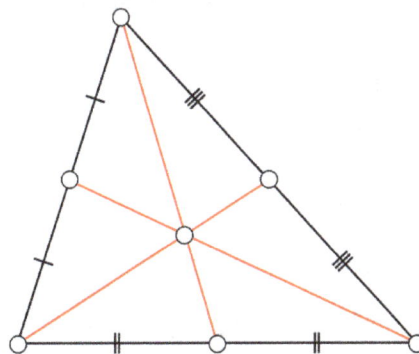

The intersection of the medians is the centroid.

A median of a triangle is a straight line through a vertex and the midpoint of the opposite side, and divides the triangle into two equal areas. The three medians intersect in a single point, the triangle's centroid or geometric barycenter, usually denoted by G. The centroid of a rigid triangular object (cut out of a thin sheet of uniform density) is also its center of mass: the object can be balanced on its centroid in a uniform gravitational field. The centroid cuts every median in the ratio 2:1, i.e. the distance between a vertex and the centroid is twice the distance between the centroid and the midpoint of the opposite side.

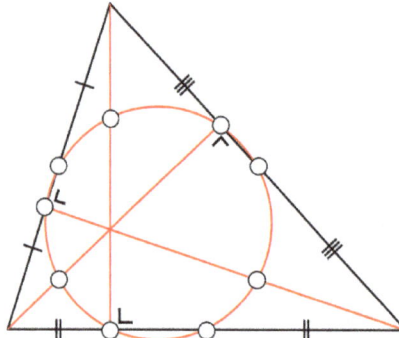

Nine-point circle demonstrates a symmetry where six points lie on the edge of the triangle.

The midpoints of the three sides and the feet of the three altitudes all lie on a single circle, the triangle's nine-point circle. The remaining three points for which it is named are the midpoints of the portion of altitude between the vertices and the orthocenter. The radius of the nine-point circle is half that of the circumcircle. It touches the incircle (at the Feuerbach point) and the three excircles.

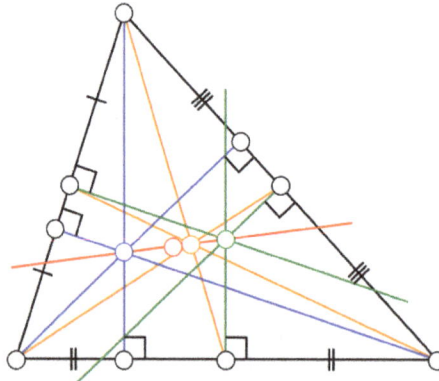

Euler's line is a straight line through the centroid (orange), orthocenter (blue), circumcenter (green) and center of the nine-point circle (red).

The centroid (yellow), orthocenter (blue), circumcenter (green) and center of the nine-point circle (red point) all lie on a single line, known as Euler's line (red line). The center of the nine-point circle lies at the midpoint between the orthocenter and the circumcenter, and the distance between the centroid and the circumcenter is half that between the centroid and the orthocenter.

The center of the incircle is not in general located on Euler's line.

If one reflects a median in the angle bisector that passes through the same vertex, one obtains a symmedian. The three symmedians intersect in a single point, the symmedian point of the triangle.

Computing the Sides and Angles

There are various standard methods for calculating the length of a side or the measure of an angle. Certain methods are suited to calculating values in a right-angled triangle; more complex methods may be required in other situations.

Trigonometric Ratios in Right Triangles

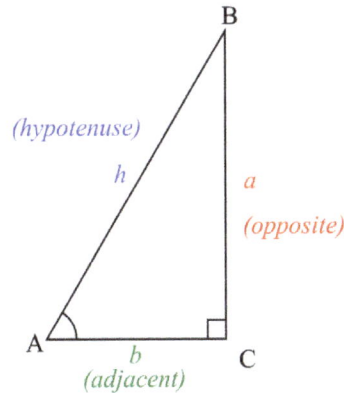

A right triangle always includes a 90° ($\pi/2$ radians) angle, here with label C. Angles A and B may vary. Trigonometric functions specify the relationships among side lengths and interior angles of a right triangle.

In right triangles, the trigonometric ratios of sine, cosine and tangent can be used to find unknown angles and the lengths of unknown sides. The sides of the triangle are known as follows:

- The *hypotenuse* is the side opposite the right angle, or defined as the longest side of a right-angled triangle, in this case h.

- The *opposite side* is the side opposite to the angle we are interested in, in this case a.

- The *adjacent side* is the side that is in contact with the angle we are interested in and the right angle, hence its name. In this case the adjacent side is b.

Sine, Cosine and Tangent

The *sine* of an angle is the ratio of the length of the opposite side to the length of the hypotenuse. In our case

$$\sin A = \frac{\text{opposite side}}{\text{hypotenuse}} = \frac{a}{h}.$$

Note that this ratio does not depend on the particular right triangle chosen, as long as it contains the angle A, since all those triangles are similar.

The *cosine* of an angle is the ratio of the length of the adjacent side to the length of the hypotenuse. In our case

$$\cos A = \frac{\text{adjacent side}}{\text{hypotenuse}} = \frac{b}{h}.$$

The *tangent* of an angle is the ratio of the length of the opposite side to the length of the adjacent side. In our case

$$\tan A = \frac{\text{opposite side}}{\text{adjacent side}} = \frac{a}{b} = \frac{\sin A}{\cos A}.$$

The acronym "SOH-CAH-TOA" is a useful mnemonic for these ratios.

Inverse Functions

The inverse trigonometric functions can be used to calculate the internal angles for a right angled triangle with the length of any two sides.

Arcsin can be used to calculate an angle from the length of the opposite side and the length of the hypotenuse.

$$\theta = \arcsin\left(\frac{\text{opposite side}}{\text{hypotenuse}}\right)$$

Arccos can be used to calculate an angle from the length of the adjacent side and the length of the hypotenuse.

$$\theta = \arccos\left(\frac{\text{adjacent side}}{\text{hypotenuse}}\right)$$

Arctan can be used to calculate an angle from the length of the opposite side and the length of the adjacent side.

$$\theta = \arctan\left(\frac{\text{opposite side}}{\text{adjacent side}}\right)$$

In introductory geometry and trigonometry courses, the notation $\sin^{-1}$, $\cos^{-1}$, etc., are often used in place of arcsin, arccos, etc. However, the arcsin, arccos, etc., notation is standard in higher mathematics where trigonometric functions are commonly raised to powers, as this avoids confusion between multiplicative inverse and compositional inverse.

Sine, cosine and Tangent Rules

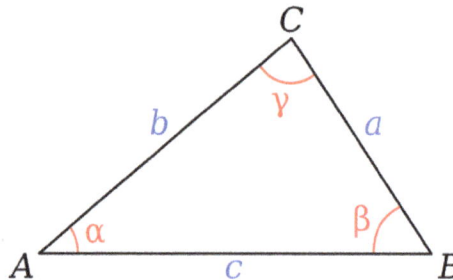

A triangle with sides of length a, b and c and angles of α, β and γ respectively.

The law of sines, or sine rule, states that the ratio of the length of a side to the sine of its corresponding opposite angle is constant, that is

$$\frac{a}{\sin \alpha} = \frac{b}{\sin \beta} = \frac{c}{\sin \gamma}.$$

This ratio is equal to the diameter of the circumscribed circle of the given triangle. Another interpretation of this theorem is that every triangle with angles α, β and γ is similar to a triangle with side lengths equal to $\sin \alpha$, $\sin \beta$ and $\sin \gamma$. This triangle can be constructed by first constructing a circle of diameter 1, and inscribing in it two of the angles of the triangle. The length of the sides of that triangle will be $\sin \alpha$, $\sin \beta$ and $\sin \gamma$. The side whose length is $\sin \alpha$ is opposite to the angle whose measure is α, etc.

The law of cosines, or cosine rule, connects the length of an unknown side of a triangle to the length of the other sides and the angle opposite to the unknown side. As per the law:

For a triangle with length of sides a, b, c and angles of α, β, γ respectively, given two known lengths of a triangle a and b, and the angle between the two known sides γ (or the angle opposite to the unknown side c), to calculate the third side c, the following formula can be used:

$$c^2 = a^2 + b^2 - 2ab\cos(\gamma)$$

$$b^2 = a^2 + c^2 - 2ac\cos(\beta)$$

$$a^2 = b^2 + c^2 - 2bc\cos(\alpha)$$

If the lengths of all three sides of any triangle are known the three angles can be calculated:

$$\alpha = \arccos\left(\frac{b^2 + c^2 - a^2}{2bc}\right)$$

$$\beta = \arccos\left(\frac{a^2 + c^2 - b^2}{2ac}\right)$$

$$\gamma = \arccos\left(\frac{a^2 + b^2 - c^2}{2ab}\right)$$

The law of tangents or tangent rule, can be used to find a side or an angle when you know two sides and an angle or two angles and a side. It states that:

$$\frac{a-b}{a+b} = \frac{\tan[\frac{1}{2}(\alpha - \beta)]}{\tan[\frac{1}{2}(\alpha + \beta)]}.$$

Solution of Triangles

"Solution of triangles" is the main trigonometric problem: to find missing characteristics of a triangle (three angles, the lengths of the three sides etc.) when at least three of these characteristics are given. The triangle can be located on a plane or on a sphere. This problem often occurs in various trigonometric applications, such as geodesy, astronomy, construction, navigation etc.

Computing the Area of a Triangle

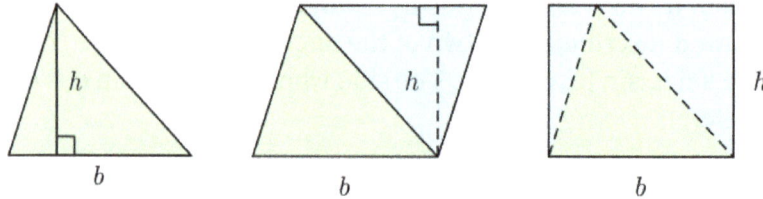

The area of a triangle can be demonstrated as half of the area of a parallelogram which has the same base length and height.

Calculating the area T of a triangle is an elementary problem encountered often in many different situations. The best known and simplest formula is:

$$T = \frac{1}{2}bh$$

where b is the length of the base of the triangle, and h is the height or altitude of the triangle. The term "base" denotes any side, and "height" denotes the length of a perpendicular from the vertex opposite the side onto the line containing the side itself. In 499 CE Aryabhata, a great mathematician-astronomer from the classical age of Indian mathematics and Indian astronomy, used this method in the *Aryabhatiya* (section 2.6).

Although simple, this formula is only useful if the height can be readily found, which is not always the case. For example, the surveyor of a triangular field might find it relatively easy to measure the length of each side, but relatively difficult to construct a 'height'. Various methods may be used in practice, depending on what is known about the triangle. The following is a selection of frequently used formulae for the area of a triangle.

Using Trigonometry

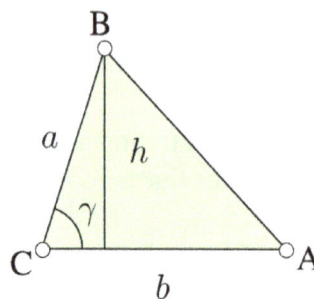

Applying trigonometry to find the altitude h.

The height of a triangle can be found through the application of trigonometry.

Knowing SAS: Using the labels in the image on the right, the altitude is $h = a \sin \gamma$. Substituting this in the formula $T = \frac{1}{2} bh$ derived above, the area of the triangle can be expressed as:

$$T = \frac{1}{2} ab \sin \gamma = \frac{1}{2} bc \sin \alpha = \frac{1}{2} ca \sin \beta$$

(where α is the interior angle at A, β is the interior angle at B, γ is the interior angle at C and c is the line AB).

Furthermore, since $\sin \alpha = \sin (\pi - \alpha) = \sin (\beta + \gamma)$, and similarly for the other two angles:

$$T = \frac{1}{2} ab \sin(\alpha + \beta) = \frac{1}{2} bc \sin(\beta + \gamma) = \frac{1}{2} ca \sin(\gamma + \alpha).$$

Knowing AAS:

$$T = \frac{b^2 (\sin \alpha)(\sin(\alpha + \beta))}{2 \sin \beta},$$

and analogously if the known side is a or c.

Knowing ASA:

$$T = \frac{a^2}{2(\cot \beta + \cot \gamma)} = \frac{a^2 (\sin \beta)(\sin \gamma)}{2 \sin(\beta + \gamma)},$$

and analogously if the known side is b or c.

Using Heron's Formula

The shape of the triangle is determined by the lengths of the sides. Therefore, the area can also be derived from the lengths of the sides. By Heron's formula:

$$T = \sqrt{s(s-a)(s-b)(s-c)}$$

where $s = \frac{a+b+c}{2}$ is the semiperimeter, or half of the triangle's perimeter.

Three other equivalent ways of writing Heron's formula are

$$T = \frac{1}{4} \sqrt{(a^2 + b^2 + c^2)^2 - 2(a^4 + b^4 + c^4)}$$

$$T = \frac{1}{4} \sqrt{2(a^2 b^2 + a^2 c^2 + b^2 c^2) - (a^4 + b^4 + c^4)}$$

$$T = \frac{1}{4} \sqrt{(a+b-c)(a-b+c)(-a+b+c)(a+b+c)}.$$

Using Vectors

The area of a parallelogram embedded in a three-dimensional Euclidean space can be calculated using vectors. Let vectors AB and AC point respectively from A to B and from A to C. The area of parallelogram $ABDC$ is then

$$|\mathbf{AB} \times \mathbf{AC}|,$$

which is the magnitude of the cross product of vectors AB and AC. The area of triangle ABC is half of this,

$$\frac{1}{2}|\mathbf{AB} \times \mathbf{AC}|.$$

The area of triangle ABC can also be expressed in terms of dot products as follows:

$$\frac{1}{2}\sqrt{(\mathbf{AB} \cdot \mathbf{AB})(\mathbf{AC} \cdot \mathbf{AC}) - (\mathbf{AB} \cdot \mathbf{AC})^2} = \frac{1}{2}\sqrt{|\mathbf{AB}|^2|\mathbf{AC}|^2 - (\mathbf{AB} \cdot \mathbf{AC})^2}.$$

In two-dimensional Euclidean space, expressing vector AB as a free vector in Cartesian space equal to (x_1, y_1) and AC as (x_2, y_2), this can be rewritten as:

$$\frac{1}{2}|x_1 y_2 - x_2 y_1|.$$

Using Coordinates

If vertex A is located at the origin (0, 0) of a Cartesian coordinate system and the coordinates of the other two vertices are given by $B = (x_B, y_B)$ and $C = (x_C, y_C)$, then the area can be computed as $\frac{1}{2}$ times the absolute value of the determinant

$$T = \frac{1}{2}\left|\det\begin{pmatrix} x_B & x_C \\ y_B & y_C \end{pmatrix}\right| = \frac{1}{2}|x_B y_C - x_C y_B|.$$

For three general vertices, the equation is:

$$T = \frac{1}{2}\left|\det\begin{pmatrix} x_A & x_B & x_C \\ y_A & y_B & y_C \\ 1 & 1 & 1 \end{pmatrix}\right| = \frac{1}{2}|x_A y_B - x_A y_C + x_B y_C - x_B y_A + x_C y_A - x_C y_B|,$$

which can be written as

$$T = \frac{1}{2}|(x_A - x_C)(y_B - y_A) - (x_A - x_B)(y_C - y_A)|.$$

If the points are labeled sequentially in the counterclockwise direction, the above determinant expressions are positive and the absolute value signs can be omitted. The above formula is known as the shoelace formula or the surveyor's formula.

If we locate the vertices in the complex plane and denote them in counterclockwise sequence as $a = x_A + y_A i$, $b = x_B + y_B i$, and $c = x_C + y_C i$, and denote their complex conjugates as $\bar{a}$, $\bar{b}$, and $\bar{c}$, then the formula

$$T = \frac{i}{4} \begin{vmatrix} a & \bar{a} & 1 \\ b & \bar{b} & 1 \\ c & \bar{c} & 1 \end{vmatrix}$$

is equivalent to the shoelace formula.

In three dimensions, the area of a general triangle $A = (x_A, y_A, z_A)$, $B = (x_B, y_B, z_B)$ and $C = (x_C, y_C, z_C)$ is the Pythagorean sum of the areas of the respective projections on the three principal planes (i.e. $x = 0$, $y = 0$ and $z = 0$):

$$T = \frac{1}{2} \sqrt{ \begin{vmatrix} x_A & x_B & x_C \\ y_A & y_B & y_C \\ 1 & 1 & 1 \end{vmatrix}^2 + \begin{vmatrix} y_A & y_B & y_C \\ z_A & z_B & z_C \\ 1 & 1 & 1 \end{vmatrix}^2 + \begin{vmatrix} z_A & z_B & z_C \\ x_A & x_B & x_C \\ 1 & 1 & 1 \end{vmatrix}^2 }.$$

Using Line Integrals

The area within any closed curve, such as a triangle, is given by the line integral around the curve of the algebraic or signed distance of a point on the curve from an arbitrary oriented straight line L. Points to the right of L as oriented are taken to be at negative distance from L, while the weight for the integral is taken to be the component of arc length parallel to L rather than arc length itself.

This method is well suited to computation of the area of an arbitrary polygon. Taking L to be the x-axis, the line integral between consecutive vertices (x_i, y_i) and (x_{i+1}, y_{i+1}) is given by the base times the mean height, namely $(x_{i+1} - x_i)(y_i + y_{i+1})/2$. The sign of the area is an overall indicator of the direction of traversal, with negative area indicating counterclockwise traversal. The area of a triangle then falls out as the case of a polygon with three sides.

While the line integral method has in common with other coordinate-based methods the arbitrary choice of a coordinate system, unlike the others it makes no arbitrary choice of vertex of the triangle as origin or of side as base. Furthermore, the choice of coordinate system defined by L commits to only two degrees of freedom rather than the usual three, since the weight is a local distance (e.g. $x_{i+1} - x_i$ in the above) whence the method does not require choosing an axis normal to L.

When working in polar coordinates it is not necessary to convert to Cartesian coordinates to use line integration, since the line integral between consecutive vertices (r_i, θ_i) and (r_{i+1}, θ_{i+1}) of a polygon is given directly by $r_i r_{i+1} \sin(\theta_{i+1} - \theta_i)/2$. This is valid for all values of θ, with some decrease in numerical accuracy when $|\theta|$ is many orders of magnitude greater than π. With this formulation negative area indicates clockwise traversal, which should be kept in mind when mixing polar and cartesian coordinates. Just as the choice of y-axis ($x = 0$) is immaterial for line integration in cartesian coordinates, so is the choice of zero heading ($\theta = 0$) immaterial here.

Formulas Resembling Heron's Formula

Three formulas have the same structure as Heron's formula but are expressed in terms of different variables. First, denoting the medians from sides a, b, and c respectively as m_a, m_b, and m_c and their semi-sum $(m_a + m_b + m_c)/2$ as σ, we have

$$T = \frac{4}{3}\sqrt{\sigma(\sigma - m_a)(\sigma - m_b)(\sigma - m_c)}.$$

Next, denoting the altitudes from sides a, b, and c respectively as h_a, h_b, and h_c, and denoting the semi-sum of the reciprocals of the altitudes as $H = (h_a^{-1} + h_b^{-1} + h_c^{-1}) >$ we have

$$T^{-1} = 4\sqrt{H(H - h_a^{-1})(H - h_b^{-1})(H - h_c^{-1})}.$$

And denoting the semi-sum of the angles' sines as $S = [(\sin \alpha) + (\sin \beta) + (\sin \gamma)]/2$, we have

$$T = D^2\sqrt{S(S - \sin \alpha)(S - \sin \beta)(S - \sin \gamma)}$$

where D is the diameter of the circumcircle:

$$D = \frac{a}{\sin \alpha} = \frac{b}{\sin \beta} = \frac{c}{\sin \gamma}.$$

Using Pick's Theorem

Pick's theorem for a technique for finding the area of any arbitrary lattice polygon (one drawn on a grid with vertically and horizontally adjacent lattice points at equal distances, and with vertices on lattice points).

The theorem states:

$$T = I + \frac{1}{2}B - 1$$

where I is the number of internal lattice points and B is the number of lattice points lying on the border of the polygon.

Other Area Formulas

Numerous other area formulas exist, such as

$$T = r \cdot s,$$

where r is the inradius, and s is the semiperimeter (in fact this formula holds for *all* tangential polygons), and

$$T = r_a(s - a) = r_b(s - b) = r_c(s - c)$$

where r_a, r_b, r_c are the radii of the excircles tangent to sides a, b, c respectively.

We also have

$$T = \frac{1}{2}D^2(\sin\alpha)(\sin\beta)(\sin\gamma)$$

and

$$T = \frac{abc}{2D} = \frac{abc}{4R}$$

for circumdiameter D; and

$$T = \frac{\tan\alpha}{4}(b^2 + c^2 - a^2)$$

for angle $\alpha \neq 90°$.

The area can also be expressed as

$$T = \sqrt{rr_a r_b r_c}.$$

In 1885, Baker gave a collection of over a hundred distinct area formulas for the triangle. These include:

$$T = \frac{1}{2}[abch_a h_b h_c]^{1/3},$$

$$T = \frac{1}{2}\sqrt{abh_a h_b},$$

$$T = \frac{a+b}{2(h_a^{-1} + h_b^{-1})},$$

$$T = \frac{Rh_b h_c}{a}$$

for circumradius (radius of the circumcircle) R, and

$$T = \frac{h_a h_b}{2\sin\gamma}.$$

Upper Bound on the Area

The area T of any triangle with perimeter p satisfies

$$T \leq \frac{p^2}{12\sqrt{3}},$$

with equality holding if and only if the triangle is equilateral.

Other upper bounds on the area T are given by

$$4\sqrt{3}T \le a^2 + b^2 + c^2$$

and

$$4\sqrt{3}T \le \frac{9abc}{a+b+c},$$

both again holding if and only if the triangle is equilateral.

Bisecting the Area

There are infinitely many lines that bisect the area of a triangle. Three of them are the medians, which are the only area bisectors that go through the centroid. Three other area bisectors are parallel to the triangle's sides.

Any line through a triangle that splits both the triangle's area and its perimeter in half goes through the triangle's incenter. There can be one, two, or three of these for any given triangle.

Medians, Angle Bisectors, Perpendicular Side Bisectors, and Altitudes

The medians and the sides are related by

$$\frac{3}{4}(a^2 + b^2 + c^2) = m_a^2 + m_b^2 + m_c^2$$

and

$$m_a = \frac{1}{2}\sqrt{2b^2 + 2c^2 - a^2} = \sqrt{\frac{1}{2}(a^2 + b^2 + c^2) - \frac{3}{4}a^2},$$

and equivalently for m_b and m_c.

For angle A opposite side a, the length of the internal angle bisector is given by

$$w_A = \frac{2\sqrt{bcs(s-a)}}{b+c} = \sqrt{bc[1 - \frac{a^2}{(b+c)^2}]} = \frac{2bc}{b+c}\cos\frac{A}{2},$$

for semiperimeter s, where the bisector length is measured from the vertex to where it meets the opposite side.

The interior perpendicular bisectors are given by

$$p_a = \frac{2aT}{a^2 + b^2 - c^2},$$

$$p_b = \frac{2bT}{a^2 + b^2 - c^2},$$

$$p_c = \frac{2cT}{a^2 - b^2 + c^2},$$

where the sides are $a \geq b \geq c$ and the area is T.

The altitude from, for example, the side of length a is

$$h_a = \frac{2T}{a}.$$

Circumradius and Inradius

The following formulas involve the circumradius R and the inradius r:

$$R = \sqrt{\frac{a^2 b^2 c^2}{(a+b+c)(-a+b+c)(a-b+c)(a+b-c)}};$$

$$r = \sqrt{\frac{(-a+b+c)(a-b+c)(a+b-c)}{4(a+b+c)}};$$

$$\frac{1}{r} = \frac{1}{h_a} + \frac{1}{h_b} + \frac{1}{h_c}$$

where h_a etc. are the altitudes to the subscripted sides;

$$\frac{r}{R} = \frac{4T^2}{sabc} = \cos\alpha + \cos\beta + \cos\gamma - 1;$$

and

$$2Rr = \frac{abc}{a+b+c}.$$

The product of two sides of a triangle equals the altitude to the third side times the diameter of the circumcircle:

$$ab = h_c D, bc = h_a D, ca = h_b D.$$

Adjacent Triangles

Suppose two adjacent but non-overlapping triangles share the same side of length f and share the same circumcircle, so that the side of length f is a chord of the circumcircle and the triangles have side lengths (a, b, f) and (c, d, f), with the two triangles together forming a cyclic quadrilateral with side lengths in sequence (a, b, c, d). Then

$$f^2 = \frac{(ac+bd)(ad+bc)}{(ab+cd)}.$$

Centroid

Let G be the centroid of a triangle with vertices A, B, and C, and let P be any interior point. Then the distances between the points are related by

$$(PA)^2 + (PB)^2 + (PC)^2 = (GA)^2 + (GB)^2 + (GC)^2 + 3(PG)^2.$$

The sum of the squares of the triangle's sides equals three times the sum of the squared distances of the centroid from the vertices:

$$AB^2 + BC^2 + CA^2 = 3(GA^2 + GB^2 + GC^2).$$

Let q_a, q_b, and q_c be the distances from the centroid to the sides of lengths a, b, and c. Then

$$\frac{q_a}{q_b} = \frac{b}{a}, \quad \frac{q_b}{q_c} = \frac{c}{b}, \quad \frac{q_a}{q_c} = \frac{c}{a}$$

and

$$q_a \cdot a = q_b \cdot b = q_c \cdot c = \frac{2}{3}T.$$

Circumcenter, Incenter, and Orthocenter

Carnot's Theorem states that the sum of the distances from the circumcenter to the three sides equals the sum of the circumradius and the inradius. Here a segment's length is considered to be negative if and only if the segment lies entirely outside the triangle. This method is especially useful for deducing the properties of more abstract forms of triangles, such as the ones induced by Lie algebras, that otherwise have the same properties as usual triangles.

Euler's theorem states that the distance d between the circumcenter and the incenter is given by

$$d^2 = R(R - 2r)$$

or equivalently

$$\frac{1}{R-d} + \frac{1}{R+d} = \frac{1}{r},$$

where R is the circumradius and r is the inradius. Thus for all triangles $R \geq 2r$, with equality holding for equilateral triangles.

If we denote that the orthocenter divides one altitude into segments of lengths u and v, another altitude into segment lengths w and x, and the third altitude into segment lengths y and z, then $uv = wx = yz$.

The distance from a side to the circumcenter equals half the distance from the opposite vertex to the orthocenter.

The sum of the squares of the distances from the vertices to the orthocenter plus the sum of the squares of the sides equals twelve times the square of the circumradius.

Angles

For any triangle,

$$a = b\cos C + c\cos B, \quad b = c\cos A + a\cos C, \quad c = a\cos B + b\cos A.$$

Morley's Trisector Theorem

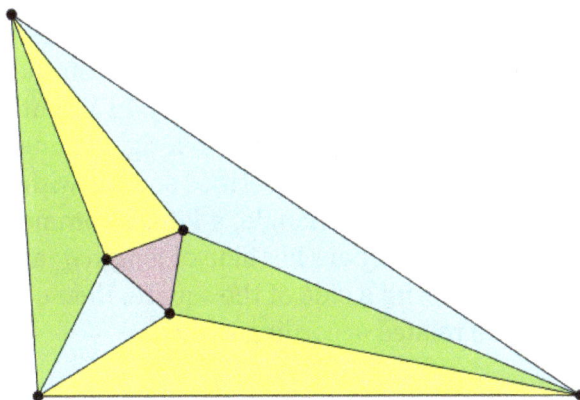

The Morley triangle, resulting from the trisection of each interior angle. This is an example of a finite subdivision rule.

Morley's trisector theorem states that in any triangle, the three points of intersection of the adjacent angle trisectors form an equilateral triangle, called the Morley triangle.

Figures Inscribed in a Triangle

Conics

As discussed above, every triangle has a unique inscribed circle (incircle) that is interior to the triangle and tangent to all three sides.

Every triangle has a unique Steiner inellipse which is interior to the triangle and tangent at the midpoints of the sides. Marden's theorem shows how to find the foci of this ellipse. This ellipse has the greatest area of any ellipse tangent to all three sides of the triangle.

The Mandart inellipse of a triangle is the ellipse inscribed within the triangle tangent to its sides at the contact points of its excircles.

For any ellipse inscribed in a triangle ABC, let the foci be P and Q. Then

$$\frac{\overline{PA}{\cdot}\overline{QA}}{\overline{CA}{\cdot}\overline{AB}} + \frac{\overline{PB}{\cdot}\overline{QB}}{\overline{AB}{\cdot}\overline{BC}} + \frac{\overline{PC}{\cdot}\overline{QC}}{\overline{BC}{\cdot}\overline{CA}} = 1.$$

Convex Polygon

Every convex polygon with area T can be inscribed in a triangle of area at most equal to $2T$. Equality holds (exclusively) for a parallelogram.

Hexagon

The Lemoine hexagon is a cyclic hexagon with vertices given by the six intersections of the sides of a triangle with the three lines that are parallel to the sides and that pass through its symmedian point. In either its simple form or its self-intersecting form, the Lemoine hexagon is interior to the triangle with two vertices on each side of the triangle.

Squares

Every acute triangle has three inscribed squares (squares in its interior such that all four of a square's vertices lie on a side of the triangle, so two of them lie on the same side and hence one side of the square coincides with part of a side of the triangle). In a right triangle two of the squares coincide and have a vertex at the triangle's right angle, so a right triangle has only two *distinct* inscribed squares. An obtuse triangle has only one inscribed square, with a side coinciding with part of the triangle's longest side. Within a given triangle, a longer common side is associated with a smaller inscribed square. If an inscribed square has side of length q_a and the triangle has a side of length a, part of which side coincides with a side of the square, then q_a, a, the altitude h_a from the side a, and the triangle's area T are related according to

$$q_a = \frac{2Ta}{a^2 + 2T} = \frac{ah_a}{a + h_a}.$$

The largest possible ratio of the area of the inscribed square to the area of the triangle is 1/2, which occurs when $a^2 = 2T$, $q = a/2$, and the altitude of the triangle from the base of length a is equal to a. The smallest possible ratio of the side of one inscribed square to the side of another in the same non-obtuse triangle is $2\sqrt{2}/3 = 0.94\dots$ Both of these extreme cases occur for the isosceles right triangle.

Triangles

From an interior point in a reference triangle, the nearest points on the three sides serve as the vertices of the pedal triangle of that point. If the interior point is the circumcenter of the reference triangle, the vertices of the pedal triangle are the midpoints of the reference triangle's sides, and so the pedal triangle is called the midpoint triangle or medial triangle. The midpoint triangle subdivides the reference triangle into four congruent triangles which are similar to the reference triangle.

The Gergonne triangle or intouch triangle of a reference triangle has its vertices at the three points of tangency of the reference triangle's sides with its incircle. The extouch triangle of a reference triangle has its vertices at the points of tangency of the reference triangle's excircles with its sides (not extended).

Figures Circumscribed About a Triangle

The tangential triangle of a reference triangle (other than a right triangle) is the triangle whose sides are on the tangent lines to the reference triangle's circumcircle at its vertices.

As mentioned above, every triangle has a unique circumcircle, a circle passing through all three vertices, whose center is the intersection of the perpendicular bisectors of the triangle's sides.

Further, every triangle has a unique Steiner circumellipse, which passes through the triangle's vertices and has its center at the triangle's centroid. Of all ellipses going through the triangle's vertices, it has the smallest area.

The Kiepert hyperbola is the unique conic which passes through the triangle's three vertices, its centroid, and its circumcenter.

Of all triangles contained in a given convex polygon, there exists a triangle with maximal area whose vertices are all vertices of the given polygon.

Specifying the Location of a Point in a Triangle

One way to identify locations of points in (or outside) a triangle is to place the triangle in an arbitrary location and orientation in the Cartesian plane, and to use Cartesian coordinates. While convenient for many purposes, this approach has the disadvantage of all points' coordinate values being dependent on the arbitrary placement in the plane.

Two systems avoid that feature, so that the coordinates of a point are not affected by moving the triangle, rotating it, or reflecting it as in a mirror, any of which give a congruent triangle, or even by rescaling it to give a similar triangle:

- Trilinear coordinates specify the relative distances of a point from the sides, so that coordinates $x:y:z$ indicate that the ratio of the distance of the point from the first side to its distance from the second side is $x:y$, etc.

- Barycentric coordinates of the form $\alpha:\beta:\gamma$ specify the point's location by the relative weights that would have to be put on the three vertices in order to balance the otherwise weightless triangle on the given point.

Non-planar Triangles

A non-planar triangle is a triangle which is not contained in a (flat) plane. Some examples of non-planar triangles in non-Euclidean geometries are spherical triangles in spherical geometry and hyperbolic triangles in hyperbolic geometry.

While the measures of the internal angles in planar triangles always sum to 180°, a hyperbolic triangle has measures of angles that sum to less than 180°, and a spherical triangle has measures of angles that sum to more than 180°. A hyperbolic triangle can be obtained by drawing on a negatively curved surface, such as a saddle surface, and a spherical triangle can be obtained by drawing on a positively curved surface such as a sphere. Thus, if one draws a giant triangle on the surface of the Earth, one will find that the sum of the measures of its angles is greater than 180°; in fact it will be between 180° and 540°. In particular it is possible to draw a triangle on a sphere such that the measure of each of its internal angles is equal to 90°, adding up to a total of 270°.

Specifically, on a sphere the sum of the angles of a triangle is

$$180° \times (1 + 4f),$$

where f is the fraction of the sphere's area which is enclosed by the triangle. For example, suppose that we draw a triangle on the Earth's surface with vertices at the North Pole, at a point on the equator at 0° longitude, and a point on the equator at 90° West longitude. The great circle line between the latter two points is the equator, and the great circle line between either of those points and the North Pole is a line of longitude; so there are right angles at the two points on the equator. Moreover, the angle at the North Pole is also 90° because the other two vertices differ by 90° of longitude. So the sum of the angles in this triangle is 90° + 90° + 90° = 270°. The triangle encloses 1/4 of the northern hemisphere (90°/360° as viewed from the North Pole) and therefore 1/8 of the Earth's surface, so in the formula f = 1/8; thus the formula correctly gives the sum of the triangle's angles as 270°.

From the above angle sum formula we can also see that the Earth's surface is locally flat: If we draw an arbitrarily small triangle in the neighborhood of one point on the Earth's surface, the fraction f of the Earth's surface which is enclosed by the triangle will be arbitrarily close to zero. In this case the angle sum formula simplifies to 180°, which we know is what Euclidean geometry tells us for triangles on a flat surface.

Triangles in Construction

The Flatiron Building in New York is shaped like a triangular prism

Rectangles have been the most popular and common geometric form for buildings since the shape is easy to stack and organize; as a standard, it is easy to design furniture and fixtures to fit inside rectangularly shaped buildings. But triangles, while more difficult to use conceptually, provide a great deal of strength. As computer technology helps architects design creative new buildings, triangular shapes are becoming increasingly prevalent as parts of buildings and as the primary shape for some types of skyscrapers as well as building materials. In Tokyo in 1989, architects had wondered whether it was possible to build a 500-story tower to provide affordable office space for this densely packed city, but with the danger to buildings from earthquakes, architects considered that a triangular shape would have been necessary if such a building was ever to have been built (it hasn't by 2011).

In New York City, as Broadway crisscrosses major avenues, the resulting blocks are cut like triangles, and buildings have been built on these shapes; one such building is the triangularly shaped Flatiron Building which real estate people admit has a "warren of awkward spaces that do not easily accommodate modern office furniture" but that has not prevented the structure from becoming a landmark icon. Designers have made houses in Norway using triangular themes. Triangle shapes have appeared in churches as well as public buildings including colleges as well as supports for innovative home designs.

Triangles are sturdy; while a rectangle can collapse into a parallelogram from pressure to one of its points, triangles have a natural strength which supports structures against lateral pressures. A triangle will not change shape unless its sides are bent or extended or broken or if its joints break; in essence, each of the three sides supports the other two. A rectangle, in contrast, is more dependent on the strength of its joints in a structural sense. Some innovative designers have proposed making bricks not out of rectangles, but with triangular shapes which can be combined in three dimensions. It is likely that triangles will be used increasingly in new ways as architecture increases in complexity. It is important to remember that triangles are strong in terms of rigidity, but while packed in a tessellating arrangement triangles are not as strong as hexagons under compression (hence the prevalence of hexagonal forms in nature). Tessellated triangles still maintain superior strength for cantilevering however, and this is the basis for one of the strongest man made structures, the tetrahedral truss.

Types of Triangles

Equilateral Triangle

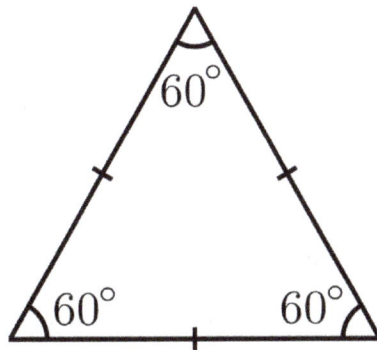

Type   Regular polygon

In geometry, an equilateral triangle is a triangle in which all three sides are equal. In the familiar Euclidean geometry, equilateral triangles are also equiangular; that is, all three internal angles are also congruent to each other and are each 60°. They are regular polygons, and can therefore also be referred to as regular triangles.

Principal Properties

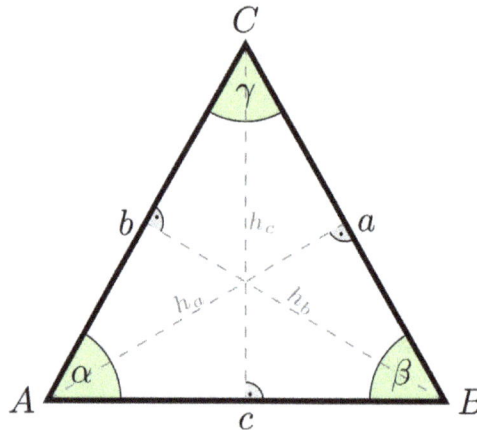

An equilateral triangle. It has equal sides ($a=b=c$), equal angles ($\alpha = \beta = \gamma$), and equal altitudes ($h_a=h_b=h_c$).

Denoting the common length of the sides of the equilateral triangle as a, we can determine using the Pythagorean theorem that:

- The area is $A = \dfrac{\sqrt{3}}{4}a^2$

- The perimeter is $p = 3a$

- The radius of the circumscribed circle is $R = \dfrac{a}{\sqrt{3}}$

- The radius of the inscribed circle is $r = \dfrac{\sqrt{3}}{6}a$ or $r = \dfrac{R}{2}$

- The geometric center of the triangle is the center of the circumscribed and inscribed circles

- And the altitude (height) from any side is $h = \dfrac{\sqrt{3}}{2}a.$

Denoting the radius of the circumscribed circle as R, we can determine using trigonometry that:

- The area of the triangle is $A = \dfrac{3\sqrt{3}}{4}R^2$

Many of these quantities have simple relationships to the altitude ("h") of each vertex from the opposite side:

- The area is $A = \dfrac{h^2}{\sqrt{3}}$

- The height of the center from each side is $\dfrac{h}{3}$

- The radius of the circle circumscribing the three vertices is $R = \dfrac{2h}{3}$

- The radius of the inscribed circle is $r = \dfrac{h}{3}$

In an equilateral triangle, the altitudes, the angle bisectors, the perpendicular bisectors and the medians to each side coincide.

Characterizations

A triangle ABC that has the sides a, b, c, semiperimeter s, area T, exradii r_a, r_b, r_c (tangent to a, b, c respectively), and where R and r are the radii of the circumcircle and incircle respectively, is equilateral if and only if any one of the statements in the following nine categories is true. Thus these are properties that are unique to equilateral triangles.

Sides

- $a = b = c$

- $a^2 + b^2 + c^2 = ab + bc + ca$

- $abc = (a+b-c)(a-b+c)(-a+b+c)$ (Lehmus)

- $(a+b+c)\left(\dfrac{1}{a}+\dfrac{1}{b}+\dfrac{1}{c}\right) = 9$

- $\dfrac{1}{a}+\dfrac{1}{b}+\dfrac{1}{c} = \dfrac{\sqrt{25Rr-2r^2}}{4Rr}$

Semiperimeter

- $s = 2R + (3\sqrt{3}-4)r$ (Blundon)

- $s^2 = 3r^2 + 12Rr$

- $s^2 = 3\sqrt{3}T$

- $s = 3\sqrt{3}r$

- $s = \dfrac{3\sqrt{3}}{2}R$

Angles

- $A = B = C = 60°$

- $\cos A + \cos B + \cos C = \dfrac{3}{2}$

- $\sin\dfrac{A}{2}\sin\dfrac{B}{2}\sin\dfrac{C}{2} = \dfrac{1}{8}$

Area

- $A = \dfrac{a^2+b^2+c^2}{4\sqrt{3}}$ (Weizenbock)

$$A = \frac{\sqrt{3}}{4}(abc)^{\frac{2}{3}}$$

Circumradius, Inradius and Exradii

- $R = 2r$ (Chapple-Euler)

- $9R^2 = a^2 + b^2 + c^2$

- $r = \dfrac{r_a + r_b + r_c}{9}$

- $r_a = r_b = r_c$

Equal Cevians

Three kinds of cevians are equal for (and only for) equilateral triangles:

- The three altitudes have equal lengths.

- The three medians have equal lengths.

- The three angle bisectors have equal lengths.

Coincident Triangle Centers

Every triangle center of an equilateral triangle coincides with its centroid, which implies that the equilateral triangle is the only triangle with no Euler line connecting some of the centers. For some pairs of triangle centers, the fact that they coincide is enough to ensure that the triangle is equilateral. In particular:

- A triangle is equilateral if any two of the circumcenter, incenter, centroid, or orthocenter coincide.

- It is also equilateral if its circumcenter coincides with the Nagel point, or if its incenter coincides with its nine-point center.

Six Triangles Formed by Partitioning by the Medians

For any triangle, the three medians partition the triangle into six smaller triangles.

- A triangle is equilateral if and only if any three of the smaller triangles have either the same perimeter or the same inradius.

- A triangle is equilateral if and only if the circumcenters of any three of the smaller triangles have the same distance from the centroid.

Points in the Plane

- A triangle is equilateral if and only if, for *every* point P in the plane, with distances p, q, and r to the triangle's sides and distances x, y, and z to its vertices,

$$4(p^2 + q^2 + r^2) \geq x^2 + y^2 + z^2.$$

Notable Theorems

Morley's trisector theorem states that, in any triangle, the three points of intersection of the adjacent angle trisectors form an equilateral triangle.

Napoleon's theorem states that, if equilateral triangles are constructed on the sides of any triangle, either all outward, or all inward, the centers of those equilateral triangles themselves form an equilateral triangle.

A version of the isoperimetric inequality for triangles states that the triangle of greatest area among all those with a given perimeter is equilateral.

Viviani's theorem states that, for any interior point P in an equilateral triangle with distances d, e, and f from the sides and altitude h,

$$d + e + f = h,$$

independent of the location of P.

Pompeiu's theorem states that, if P is an arbitrary point in an equilateral triangle ABC, then there exists a triangle with sides of lengths PA, PB, and PC. That is, PA, PB, and PC satisfy the triangle inequality that any two of them sum to at least as great as the third.

Other Properties

By Euler's inequality, the equilateral triangle has the smallest ratio R/r of the circumradius to the inradius of any triangle: specifically, $R/r = 2$.

The triangle of largest area of all those inscribed in a given circle is equilateral; and the triangle of smallest area of all those circumscribed around a given circle is equilateral.

The ratio of the area of the incircle to the area of an equilateral triangle, $\dfrac{\pi}{3\sqrt{3}}$, is larger than that of any non-equilateral triangle.

The ratio of the area to the square of the perimeter of an equilateral triangle, $\dfrac{1}{12\sqrt{3}}$, is larger than that for any other triangle.

If a segment splits an equilateral triangle into two regions with equal perimeters and with areas A_1 and A_2, then

$$\frac{7}{9} \leq \frac{A_1}{A_2} \leq \frac{9}{7}.$$

If a triangle is placed in the complex plane with complex vertices z_1, z_2, and z_3, then for either non-real cube root ω of 1 the triangle is equilateral if and only if

$$z_1 + \omega z_2 + \omega^2 z_3 = 0.$$

Given a point P in the interior of an equilateral triangle, the ratio of the sum of its distances from the vertices to the sum of its distances from the sides is greater than or equal to 2, equality holding when P is the centroid. In no other triangle is there a point for which this ratio is as small as 2. This is the Erdős–Mordell inequality; a stronger variant of it is Barrow's inequality, which replaces the perpendicular distances to the sides with the distances from P to the points where the angle bisectors of $\angle APB$, $\angle BPC$, and $\angle CPA$ cross the sides (A, B, and C being the vertices).

For any point P in the plane, with distances p, q, and t from the vertices A, B, and C respectively,

$$3(p^4 + q^4 + t^4 + a^4) = (p^2 + q^2 + t^2 + a^2)^2.$$

For any point P on the inscribed circle of an equilateral triangle, with distances p, q, and t from the vertices,

$$4(p^2 + q^2 + t^2) = 5a^2$$

and

$$16(p^4 + q^4 + t^4) = 11a^4.$$

For any point P on the minor arc BC of the circumcircle, with distances p, q, and t from A, B, and C respectively,

$$p = q + t$$

and

$$q^2 + qt + t^2 = a^2;$$

moreover, if point D on side BC divides PA into segments PD and DA with DA having length z and PD having length y, then

$$z = \frac{t^2 + tq + q^2}{t + q},$$

which also equals $\frac{t^3 - q^3}{t^2 - q^2}$ if $t \neq q$; and

$$\frac{1}{q} + \frac{1}{t} = \frac{1}{y},$$

which is the optic equation.

There are numerous triangle inequalities that hold with equality if and only if the triangle is equilateral.

An equilateral triangle is the most symmetrical triangle, having 3 lines of reflection and rotational symmetry of order 3 about its center. Its symmetry group is the dihedral group of order 6 D_3.

Equilateral triangles are the only triangles whose Steiner inellipse is a circle (specifically, it is the incircle).

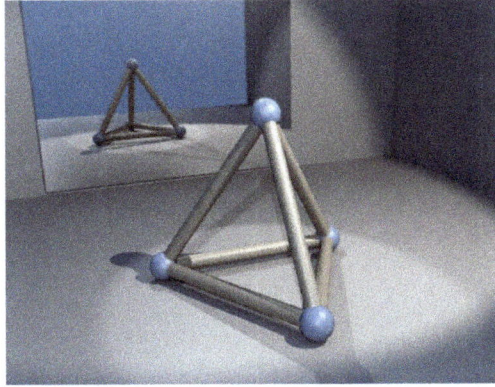

A regular tetrahedron is made of four equilateral triangles.

Equilateral triangles are found in many other geometric constructs. The intersection of circles whose centers are a radius width apart is a pair of equilateral arches, each of which can be inscribed with an equilateral triangle. They form faces of regular and uniform polyhedra. Three of the five Platonic solids are composed of equilateral triangles. In particular, the regular tetrahedron has four equilateral triangles for faces and can be considered the three-dimensional analogue of the shape. The plane can be tiled using equilateral triangles giving the triangular tiling.

Geometric Construction

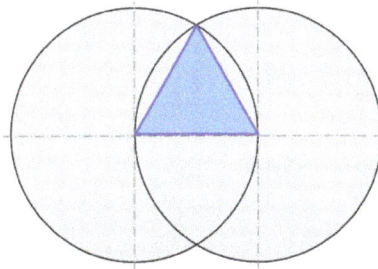

Construction of equilateral triangle with compass and straightedge

An equilateral triangle is easily constructed using a compass and straightedge, as 3 is a Fermat prime. Draw a straight line, and place the point of the compass on one end of the line, and swing an arc from that point to the other point of the line segment. Repeat with the other side of the line. Finally, connect the point where the two arcs intersect with each end of the line segment

An alternative method is to draw a circle with radius r, place the point of the compass on the circle and draw another circle with the same radius. The two circles will intersect in two points. An equilateral triangle can be constructed by taking the two centers of the circles and either of the points of intersection.

In both methods a by-product is the formation of vesica piscis.

The proof that the resulting figure is an equilateral triangle is the first proposition in Book I of Euclid's Elements.

Derivation of Area Formula

The area formula $A = \dfrac{\sqrt{3}}{4}a^2$ in terms of side length a can be derived directly using the Pythagorean theorem or using trigonometry.

Using the Pythagorean Theorem

The area of a triangle is half of one side a times the height h from that side:

$$A = \frac{1}{2}ah.$$

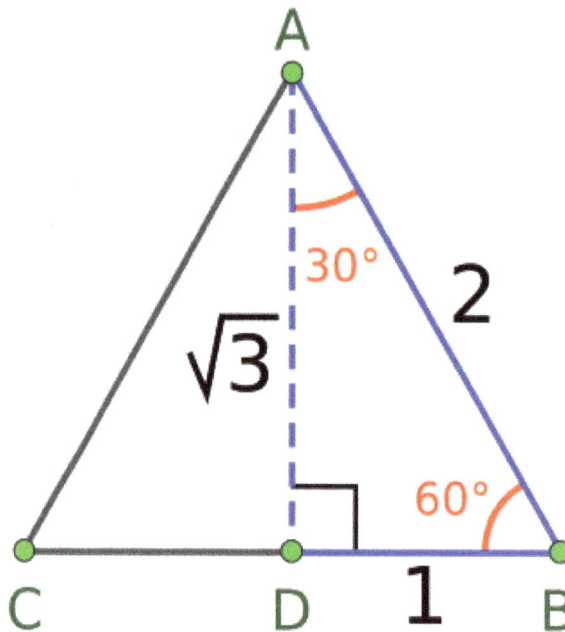

An equilateral triangle with a side of 2 has a height of $\sqrt{3}$ as the sine of $60°$ is $\sqrt{3}/2$.

The legs of either right triangle formed by an altitude of the equilateral triangle are half of the base a, and the hypotenuse is the side a of the equilateral triangle. The height of an equilateral triangle can be found using the Pythagorean theorem

$$\left(\frac{a}{2}\right)^2 + h^2 = a^2$$

so that

$$h = \frac{\sqrt{3}}{2}a.$$

Substituting h into the area formula $(1/2)ah$ gives the area formula for the equilateral triangle:

$$A = \frac{\sqrt{3}}{4}a^2.$$

Using Trigonometry

Using trigonometry, the area of a triangle with any two sides a and b, and an angle C between them is

$$A = \frac{1}{2}ab\sin C.$$

Each angle of an equilateral triangle is 60°, so

$$A = \frac{1}{2}ab\sin 60°.$$

The sine of 60° is $\frac{\sqrt{3}}{2}$. Thus

$$A = \frac{1}{2}ab \times \frac{\sqrt{3}}{2} = \frac{\sqrt{3}}{4}ab = \frac{\sqrt{3}}{4}a$$

since all sides of an equilateral triangle are equal.

In Culture and Society

Equilateral triangles have frequently appeared in man made constructions:

- Some archaeological sites have equilateral triangles as part of their construction, for example Lepenski Vir in Serbia.

- The shape also occurs in modern architecture such as Randhurst Mall and the Jefferson National Expansion Memorial.

- The Flag of Nicaragua, the Flag of the Philippines, the Seal of the President of the Philippines, and the Flag of Junqueirópolis contain equilateral triangles.

- It is a shape of a variety of road signs, including the Yield sign.

- The fraternity Tau Kappa Epsilon uses the equilateral triangle as its primary symbol.

Isosceles Triangle

Isosceles triangle with vertical axis of symmetry

In geometry, an isosceles triangle is a triangle that has two sides of equal length. Sometimes it is specified as having two *and only two* sides of equal length, and sometimes as having *at least* two sides of equal length, the latter version thus including the equilateral triangle as a special case.

By the isosceles triangle theorem, the two angles opposite the equal sides are themselves equal, while if the third side is different then the third angle is different.

By the Steiner–Lehmus theorem, every triangle with two angle bisectors of equal length is isosceles.

Terminology

In an isosceles triangle that has exactly two equal sides, the equal sides are called legs and the third side is called the base. The angle included by the legs is called the *vertex angle* and the angles that have the base as one of their sides are called the *base angles*. The vertex opposite the base is called the apex.

Euclid defined an isosceles triangle as one having exactly two equal sides, but modern treatments prefer to define them as having at least two equal sides, making *equilateral triangles* (with three equal sides) a special case of isosceles triangles. In the equilateral triangle case, since all sides are equal, any side can be called the base, if needed, and the term leg is not generally used.

Symmetry

A triangle with exactly two equal sides has exactly one axis of symmetry, which goes through the vertex angle and also goes through the midpoint of the base. Thus the axis of symmetry coincides with (1) the angle bisector of the vertex angle, (2) the median drawn to the base, (3) the altitude drawn from the vertex angle, and (4) the perpendicular bisector of the base.

Acute, Right and Obtuse

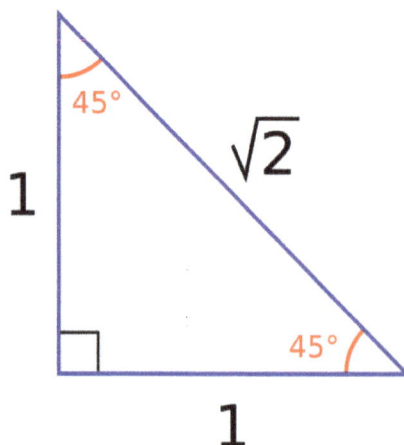

Isosceles right triangle

Whether the isosceles triangle is acute, right or obtuse depends on the vertex angle. In Euclidean geometry, the base angles cannot be obtuse (greater than 90°) or right (equal to 90°) because their measures would sum to at least 180°, the total of all angles in any Euclidean triangle. Since a triangle is obtuse (resp. right) if and only if one of its angles is obtuse (resp. right), an isosceles triangle is obtuse, right or acute if and only if its vertex angle is respectively obtuse, right or acute.

Euler Line

The Euler line of any triangle goes through the triangle's orthocenter (the intersection of its three altitudes), its centroid (the intersection of its three medians), and its circumcenter (the intersection of its three sides' perpendicular bisectors, which is the center of the circumcircle that passes through the three vertices). In an isosceles triangle with exactly two equal sides, the Euler line coincides with the axis of symmetry. This can be seen as follows. Since as pointed out in the previous section the axis of symmetry coincides with an altitude, the intersection of the altitudes, which must lie on that altitude, must therefore lie on the axis of symmetry; since the axis coincides with a median, the intersection of the medians, which must lie on that median, must therefore lie on the axis of symmetry; and since the axis coincides with a perpendicular bisector, the intersection of the perpendicular bisectors, which must lie on that perpendicular bisector, must therefore lie on the axis of symmetry.

If the vertex angle is acute (so the isosceles triangle is an acute triangle), then the orthocenter, the centroid, and the circumcenter all fall inside the triangle. If the vertex angle, and therefore the triangle, is obtuse, then the centroid still falls in the triangle's interior, but the circumcenter falls outside it (beyond the base), and the orthocenter also falls outside the triangle (beyond the apex).

In an isosceles triangle the incenter (the intersection of its angle bisectors, which is the center of the incircle, that is, the circle which is internally tangent to the triangle's three sides) lies on the Euler line.

Steiner Inellipse

The Steiner inellipse of any triangle is the unique ellipse that is internally tangent to the triangle's three sides at their midpoints. In an isosceles triangle, if the legs are longer than the base then the Steiner inellipse's major axis coincides with the triangle's axis of symmetry; if the legs are shorter than the base, then the ellipse's minor axis coincides with the triangle's axis of symmetry.

Formulas

For an isosceles triangle with equal sides of length a and base of length b, the general triangle formulas for (1) the length of the triangle-interior portion of the angle bisector of the vertex angle, (2) the length of the median drawn to the base, (3) length of the altitude drawn to the base, and (4) the length of the triangle-interior portion of the perpendicular bisector of the base all simplify to $\frac{1}{2}\sqrt{4a^2 - b^2}$.

For any isosceles triangle with area T and perimeter p, we have

$$2pb^3 - p^2b^2 + 16T^2 = 0.$$

Area

The area of an isosceles triangle can be derived using the Pythagorean Theorem: The sum of the squares of half the base b and the height h is the square of either of the other two sides of length a:

$$\left(\frac{b}{2}\right)^2 + h^2 = a^2,$$

$$h = \frac{\sqrt{4a^2 - b^2}}{2}.$$

By substituting the height, the formula for the area of an isosceles triangle can be derived from the general formula one-half the base times the height:

$$T = \frac{b}{4}\sqrt{4a^2 - b^2}.$$

This is what Heron's formula reduces to in the isosceles case.

If the apex angle (θ) and leg lengths (a) of an isosceles triangle are known, then the area of that triangle is:

$$T = 2\left(\frac{1}{2}a\sin\left(\frac{\theta}{2}\right)a\cos\left(\frac{\theta}{2}\right)\right)$$

$$= a^2 \sin\left(\frac{\theta}{2}\right)\cos\left(\frac{\theta}{2}\right).$$

This is derived by drawing a perpendicular line from the base of the triangle, which bisects the vertex angle and creates two right triangles. The bases of these two right triangles are both equal to the hypotenuse times the sine of the bisected angle by definition of the term "sine". For the same reason, the heights of these triangles are equal to the hypotenuse times the cosine of the bisected angle. Using the trigonometric identity $\sin\theta = 2\sin\left(\frac{\theta}{2}\right)\cos\left(\frac{\theta}{2}\right)$, we get

$$T = \frac{1}{2}a^2 \sin\theta,$$

which is a special case of the general triangle area formula one-half the product of two sides times the sine of the included angle.

The Isosceles Triangle Theorem

The theorem which states that the base angles of an isosceles triangle are equal appears as Proposition I.5 in Euclid. This result has been called the *pons asinorum* (the bridge of asses). Some say that this is probably because of the diagram used by Euclid in his demonstration of the result.

Others claim that the name stems from the fact that this is the first difficult result in Euclid, and acts to separate those who can understand Euclid's geometry from those who can't.

Partitioning Into Isosceles Triangles

For any integer $n \geq 4$, any triangle can be partitioned into n isosceles triangles.

In a right triangle, the median from the hypotenuse (that is, the line segment from the midpoint of the hypotenuse to the right-angled vertex) divides the right triangle into two isosceles triangles. This is because the midpoint of the hypotenuse is the center of the circumcircle of the right triangle, and each of the two triangles created by the partition has two equal radii as two of its sides.

The golden triangle is isosceles and has a ratio of either leg to the base equal to the golden ratio, and has angles 72°, 72°, and 36° in the ratios 2:2:1. It can be partitioned into another golden triangle and a golden gnomon, also isosceles, with ratio of base to leg equaling the golden ratio and with angles 36°, 36°, and 108° in the ratios 1:1:3.

Miscellaneous

If a cubic equation has two complex roots and one real root, then when these roots are plotted in the complex plane they are the vertices of an isosceles triangle whose axis of symmetry coincides with the horizontal (real) axis. This is because the complex roots are complex conjugates and hence are symmetric about the real axis.

Either diagonal of a rhombus divides it into two congruent isosceles triangles.

The Calabi triangle, which is isosceles, is the unique non-equilateral triangle in which the largest square that fits in its interior can be positioned in any of three different ways.

If an isosceles triangle ABC with equal legs AB and BC has a segment drawn from A to a point D on ray BC, and if the reflection of AD around AC intersects ray BC at E, then $BC^2 = BD \times BE$.

There are exactly two distinct isosceles triangles with given area T and perimeter p if the isoperimetric inequality holds strictly as $p^2 > 12\sqrt{3}T$. If the inequality is replaced by the corresponding equality, there is only one such triangle, which is equilateral.

If the two equal sides have length a and the other side has length c, then the internal angle bisector t from one of the two equal-angled vertices satisfies

$$\frac{2ac}{a+c} > t > \frac{ac\sqrt{2}}{a+c}$$

as well as

$$t < \frac{4a}{3} ;$$

and conversely, if the latter condition holds, an isosceles triangle parametrized by a and t exists.

Fallacy of the Isosceles Triangle

A well known fallacy is the false proof of the statement that *all triangles are isosceles*. This argument has been attributed to Lewis Carroll, but W.W. Rouse Ball claims priority in this matter. The fallacy is rooted in Euclid's lack of recognition of the concept of *betweenness* and the resulting ambiguity of *inside* versus *outside* of figures.

Right Triangle

A right triangle (American English) or right-angled triangle (British English) is a triangle in which one angle is a right angle (that is, a 90-degree angle). The relation between the sides and angles of a right triangle is the basis for trigonometry.

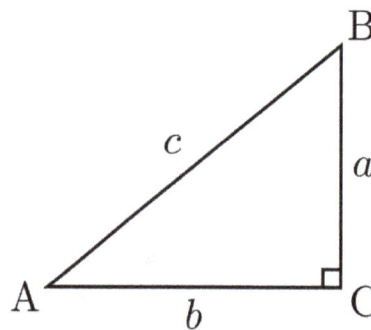

Right triangle

The side opposite the right angle is called the *hypotenuse* (side c in the figure). The sides adjacent to the right angle are called *legs* (or *catheti*, singular: *cathetus*). Side a may be identified as the side *adjacent to angle B* and *opposed to* (or *opposite*) *angle A*, while side b is the side *adjacent to angle A* and *opposed to angle B*.

If the lengths of all three sides of a right triangle are integers, the triangle is said to be a Pythagorean triangle and its side lengths are collectively known as a Pythagorean triple.

Principal Properties

Area

As with any triangle, the area is equal to one half the base multiplied by the corresponding height. In a right triangle, if one leg is taken as the base then the other is height, so the area of a right triangle is one half the product of the two legs. As a formula the area T is

$$T = \tfrac{1}{2}ab$$

where a and b are the legs of the triangle.

If the incircle is tangent to the hypotenuse AB at point P, then denoting the semi-perimeter $(a + b + c) / 2$ as s, we have PA = $s - a$ and PB = $s - b$, and the area is given by

$$T = \text{PA·PB} = (s - a)(s - b).$$

This formula only applies to right triangles.

Altitudes

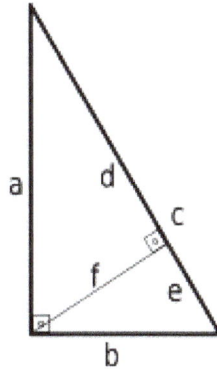

Altitude of a right triangle

If an altitude is drawn from the vertex with the right angle to the hypotenuse then the triangle is divided into two smaller triangles which are both similar to the original and therefore similar to each other. From this:

- The altitude to the hypotenuse is the geometric mean (mean proportional) of the two segments of the hypotenuse.

- Each leg of the triangle is the mean proportional of the hypotenuse and the segment of the hypotenuse that is adjacent to the leg.

In equations,

$$f \quad de \text{ (this is sometimes known as the right triangle altitude theorem)}$$

$$b^2 = ce,$$

$$a^2 = cd$$

where a, b, c, d, e, f are as shown in the diagram. Thus

$$f = \frac{ab}{c}.$$

Moreover, the altitude to the hypotenuse is related to the legs of the right triangle by

$$\frac{1}{a^2} + \frac{1}{b^2} = \frac{1}{f^2}.$$

For solutions of this equation in integer values of a, b, f, and c.

The altitude from either leg coincides with the other leg. Since these intersect at the right-angled vertex, the right triangle's orthocenter—the intersection of its three altitudes—coincides with the right-angled vertex.

Pythagorean Theorem

The Pythagorean theorem states that:

In any right triangle, the area of the square whose side is the hypotenuse (the side opposite the right angle) is equal to the sum of the areas of the squares whose sides are the two legs (the two sides that meet at a right angle).

This can be stated in equation form as

$$a^2 + b^2 = c^2$$

where c is the length of the hypotenuse, and a and b are the lengths of the remaining two sides.

Pythagorean triples are integer values of a, b, c satisfying this equation.

Inradius and Circumradius

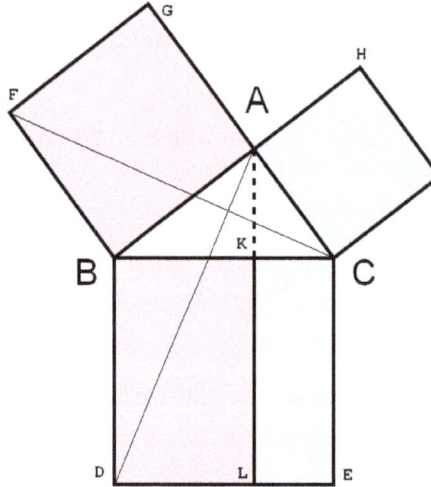

Illustration of the Pythagorean Theorem

The radius of the incircle of a right triangle with legs a and b and hypotenuse c is

$$r = \frac{a+b-c}{2} = \frac{ab}{a+b+c}.$$

The radius of the circumcircle is half the length of the hypotenuse,

$$R = \frac{c}{2}.$$

Thus the sum of the circumradius and the inradius is half the sum of the legs:

$$R + r = \frac{a+b}{2}.$$

One of the legs can be expressed in terms of the inradius and the other leg as

$$a = \frac{2r(b-r)}{b-2r}.$$

Characterizations

A triangle ABC with sides $a \le b < c$ semiperimeter s, area T, altitude h opposite the longest side, circumradius R, inradius r, exradii r_a, r_b, r_c (tangent to a, b, c respectively), and medians m_a, m_b, m_c is a right triangle if and only if any one of the statements in the following six categories is true. All of them are of course also properties of a right triangle, since characterizations are equivalences.

Sides and Semiperimeter

- $a^2 + b^2 = c^2$ (Pythagorean theorem)

- $(s-a)(s-b) = s(s-c)$

- $s = 2R + r.$

- $a^2 + b^2 + c^2 = 8R^2.$

Angles

- A and B are complementary.

- $\cos A \cos B \cos C = 0.$

- $\sin^2 A + \sin^2 B + \sin^2 C = 2.$

- $\cos^2 A + \cos^2 B + \cos^2 C = 1.$

- $\sin 2A = \sin 2B = 2 \sin A \sin B.$

Area

- $T = \dfrac{ab}{2}$

- $T = r_a r_b = r r_c$

- $T = r(2R + r)$

- $T = PA \cdot PB$, where P is the tangency point of the incircle at the longest side AB.

Inradius and Exradii

- $r = s - c = (a+b-c)/2$

- $r_a = s - b = (a-b+c)/2$

- $r_b = s - a = (-a+b+c)/2$

- $r_c = s = (a+b+c)/2$

- $r_a + r_b + r_c + r = a + b + c$

- $r_a^2 + r_b^2 + r_c^2 + r^2 = a^2 + b^2 + c^2$

- $r = \dfrac{r_a r_b}{r_c}$

Altitude and Medians

- $h = \dfrac{ab}{c}$
- $m_a^2 + m_b^2 + m_c^2 = 6R^2.$

- The length of one median is equal to the circumradius.

- The shortest altitude (the one from the vertex with the biggest angle) is the geometric mean of the line segments it divides the opposite (longest) side into. This is the right triangle altitude theorem.

Circumcircle and Incircle

- The triangle can be inscribed in a semicircle, with one side coinciding with the entirety of the diameter (Thales' theorem).

- The circumcenter is the midpoint of the longest side.

- The longest side is a diameter of the circumcircle ($c = 2R$).

- The circumcircle is tangent to the nine-point circle.

- The orthocenter lies on the circumcircle.

- The distance between the incenter and the orthocenter is equal to $\sqrt{2}r$.

Trigonometric Ratios

The trigonometric functions for acute angles can be defined as ratios of the sides of a right triangle. For a given angle, a right triangle may be constructed with this angle, and the sides labeled opposite, adjacent and hypotenuse with reference to this angle according to the definitions above. These ratios of the sides do not depend on the particular right triangle chosen, but only on the given angle, since all triangles constructed this way are similar. If, for a given angle α, the opposite side, adjacent side and hypotenuse are labeled O, A and H respectively, then the trigonometric functions are

$$\sin\alpha = \frac{O}{H}, \cos\alpha = \frac{A}{H}, \tan\alpha = \frac{O}{A}, \sec\alpha = \frac{H}{A}, \cot\alpha = \frac{A}{O}, \csc\alpha = \frac{H}{O}.$$

For the expression of hyperbolic functions as ratio of the sides of a right triangle, the hyperbolic triangle of a hyperbolic sector.

Special Right Triangles

The values of the trigonometric functions can be evaluated exactly for certain angles using right triangles with special angles. These include the *30-60-90 triangle* which can be used to evaluate the trigonometric functions for any multiple of $\pi/6$, and the *45-45-90 triangle* which can be used to evaluate the trigonometric functions for any multiple of $\pi/4$.

Kepler Triangle

Let H, G, and A be the harmonic mean, the geometric mean, and the arithmetic mean of two positive numbers a and b with $a > b$. If a right triangle has legs H and G and hypotenuse A, then

$$\frac{A}{H} = \frac{A^2}{G^2} = \frac{G^2}{H^2} = \phi$$

and

$$\frac{a}{b} = \phi^3,$$

where ϕ is the golden ratio $\frac{1+\sqrt{5}}{2}$. Since the sides of this right triangle are in geometric progression, this is the Kepler triangle.

Thales' Theorem

Thales' theorem states that if A is any point of the circle with diameter BC (except B or C themselves) ABC is a right triangle where A is the right angle. The converse states that if a right triangle is inscribed in a circle then the hypotenuse will be a diameter of the circle. A corollary is that the length of the hypotenuse is twice the distance from the right angle vertex to the midpoint of the hypotenuse. Also, the center of the circle that circumscribes a right triangle is the midpoint of the hypotenuse and its radius is one half the length of the hypotenuse.

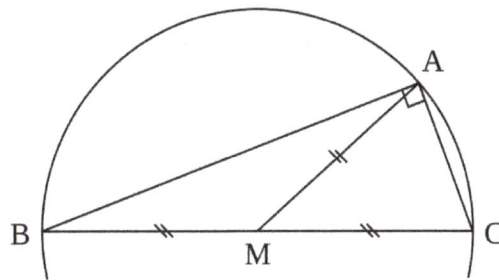

Median of a right angle of a triangle

Medians

The following formulas hold for the medians of a right triangle:

$$m_a^2 + m_b^2 = 5m_c^2 = \frac{5}{4}c^2.$$

The median on the hypotenuse of a right triangle divides the triangle into two isosceles triangles, because the median equals one-half the hypotenuse.

The medians m_a and m_b from the legs satisfy

$$4c^4 + 9a^2b^2 = 16m_a^2 m_b^2.$$

Euler Line

In a right triangle, the Euler line contains the median on the hypotenuse—that is, it goes through both the right-angled vertex and the midpoint of the side opposite that vertex. This is because the right triangle's orthocenter, the intersection of its altitudes, falls on the right-angled vertex while its circumcenter, the intersection of its perpendicular bisectors of sides, falls on the midpoint of the hypotenuse.

Inequalities

In any right triangle the diameter of the incircle is less than half the hypotenuse, and more strongly it is less than or equal to the hypotenuse times $(\sqrt{2} - 1)$.

In a right triangle with legs a, b and hypotenuse c,

$$c \geq \frac{\sqrt{2}}{2}(a + b)$$

with equality only in the isosceles case.

If the altitude from the hypotenuse is denoted h_c, then

$$h_c \leq \frac{\sqrt{2}}{4}(a + b)$$

with equality only in the isosceles case.

Other Properties

If segments of lengths p and q emanating from vertex C trisect the hypotenuse into segments of length $c/3$, then

$$p^2 + q^2 = 5\left(\frac{c}{3}\right)^2.$$

The right triangle is the only triangle having two, rather than one or three, distinct inscribed squares.

Let h and k ($h > k$) be the sides of the two inscribed squares in a right triangle with hypotenuse c. Then

$$\frac{1}{c^2} + \frac{1}{h^2} = \frac{1}{k^2}.$$

These sides and the incircle radius r are related by a similar formula:

$$\frac{1}{r} = -\frac{1}{c} + \frac{1}{h} + \frac{1}{k}.$$

The perimeter of a right triangle equals the sum of the radii of the incircle and the three excircles:

$$a + b + c = r + r_a + r_b + r_c.$$

Acute and Obtuse Triangles

An acute triangle is a triangle with all three angles acute (less than 90°). An obtuse triangle is one with one obtuse angle (greater than 90°) and two acute angles. Since a triangle's angles must sum to 180°, no triangle can have more than one obtuse angle.

Acute and obtuse triangles are the two different types of oblique triangles—triangles that are not right triangles because they have no 90° angle.

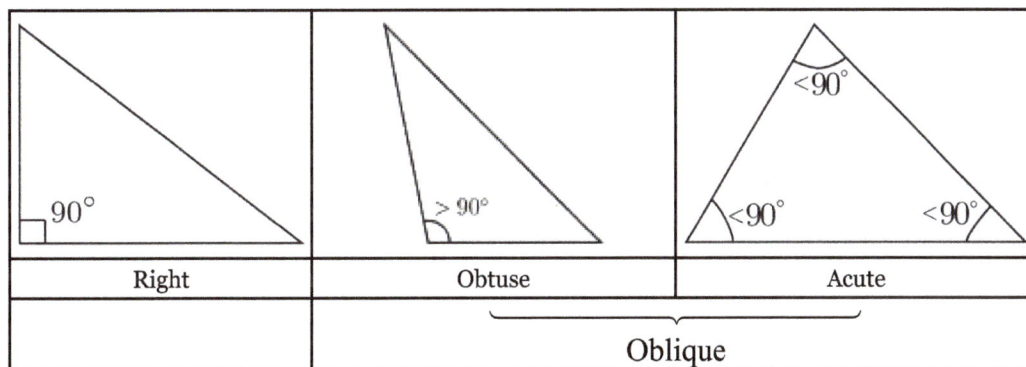

Right	Obtuse	Acute
	Oblique	

Properties

In all triangles, the centroid—the intersection of the medians, each of which connects a vertex with the midpoint of the opposite side—and the incenter—the center of the circle that is internally tangent to all three sides—are in the interior of the triangle. However, while the orthocenter and the circumcenter are in an acute triangle's interior, they are exterior to an obtuse triangle.

The orthocenter is the intersection point of the triangle's three altitudes, each of which perpendicularly connects a side to the opposite vertex. In the case of an acute triangle, all three of these segments lie entirely in the triangle's interior, and so they intersect in the interior. But for an obtuse triangle, the altitudes from the two acute angles intersect only the extensions of the opposite sides. These altitudes fall entirely outside the triangle, resulting in their intersection with each other (and hence with the extended altitude from the obtuse-angled vertex) occurring in the triangle's exterior.

Likewise, a triangle's circumcenter—the intersection of the three sides' perpendicular bisectors, which is the center of the circle that passes through all three vertices—falls inside an acute triangle but outside an obtuse triangle.

The right triangle is the in-between case: both its circumcenter and its orthocenter lie on its boundary.

In any triangle, any two angle measures A and B opposite sides a and b respectively are related according to

$A > B$ if and only if $a > b$.

This implies that the longest side in an obtuse triangle is the one opposite the obtuse-angled vertex.

An acute triangle has three inscribed squares, each with one side coinciding with part of a side of the triangle and with the square's other two vertices on the remaining two sides of the triangle. (In a right triangle two of these are merged into the same square, so there are only two distinct inscribed squares.) However, an obtuse triangle has only one inscribed square, one of whose sides coincides with part of the longest side of the triangle.

All triangles in which the Euler line is parallel to one side are acute. This property holds for side BC if and only if $(\tan B)(\tan C) = 3$.

Inequalities

Sides

If angle C is obtuse then for sides a, b, and c we have

$$\frac{c^2}{2} < a^2 + b^2 < c^2,$$

with the left inequality approaching equality in the limit only as the apex angle of an isosceles triangle approaches 180°, and with the right inequality approaching equality only as the obtuse angle approaches 90°.

If the triangle is acute then

$$a^2 + b^2 > c^2, \quad b^2 + c^2 > a^2, \quad c^2 + a^2 > b^2.$$

Altitude

If C is the greatest angle and h_c is the altitude from vertex C, then for an acute triangle

$$\frac{1}{h_c^2} < \frac{1}{a^2} + \frac{1}{b^2},$$

with the opposite inequality if C is obtuse.

Medians

With longest side c and medians m_a and m_b from the other sides,

$$4c^2 + 9a^2 b^2 > 16 m_a^2 m_b^2$$

for an acute triangle but with the inequality reversed for an obtuse triangle.

The median m_c from the longest side is greater or less than the circumradius for an acute or obtuse triangle respectively:

$$m_c > R$$

for acute triangles, with the opposite for obtuse triangles.

Area

Ono's inequality for the area A,

$$27(b^2 + c^2 - a^2)^2 (c^2 + a^2 - b^2)^2 (a^2 + b^2 - c^2)^2 \leq (4A)^6,$$

holds for all acute triangles but not for all obtuse triangles.

Trigonometric Functions

For an acute triangle we have, for angles A, B, and C,

$$\cos^2 A + \cos^2 B + \cos^2 C < 1,$$

with the reverse inequality holding for an obtuse triangle.

For an acute triangle with circumradius R,

$$a \cos^3 A + b \cos^3 B + c \cos^3 C \leq \frac{abc}{4R^2}$$

and

$$\cos^3 A + \cos^3 B + \cos^3 C + \cos A \cos B \cos C \geq \frac{1}{2}.$$

For an acute triangle,

$$\sin^2 A + \sin^2 B + \sin^2 C > 2,$$

with the reverse inequality for an obtuse triangle.

For an acute triangle,

$$A \cdot \sin B + \sin B \cdot \sin C + \sin C \cdot \sin A \leq (\cos A + \cos B + \cos C)^2.$$

For any triangle the triple tangent identity states that the sum of the angles' tangents equals their product. Since an acute angle has a positive tangent value while an obtuse angle has a negative one, the expression for the product of the tangents shows that

$$\tan A + \tan B + \tan C = \tan A \cdot \tan B \cdot \tan C > 0$$

for acute triangles, while the opposite direction of inequality holds for obtuse triangles.

We have

$$A + \tan B + \tan C \geq 2(\sin 2A + \sin 2B + \sin 2C)$$

for acute triangles, and the reverse for obtuse triangles.

For all acute triangles,

$$(\tan A + \tan B + \tan C)^2 \geq (\sec A + 1)^2 + (\sec B + 1)^2 + (\sec C + 1)^2.$$

For all acute triangles with inradius r and circumradius R,

$$a \tan A + b \tan B + c \tan C \geq 10R - 2r.$$

For an acute triangle with area K,

$$(\sqrt{\cot A} + \sqrt{\cot B} + \sqrt{\cot C})^2 \leq \frac{K}{r^2}.$$

Circumradius, Inradius, and Exradii

The sum of the circumradius R and the inradius r is less than or greater than half the sum of the shortest sides a and b as the triangle is acute or obtuse:

$$R + r < \frac{a+b}{2}.$$

For an acute triangle with medians m_a, m_b, and m_c and circumradius R, we have

$$m_a^2 + m_b^2 + m_c^2 > 6R^2$$

while the opposite inequality holds for an obtuse triangle.

Also, an acute triangle satisfies

$$r^2 + r_a^2 + r_b^2 + r_c^2 < 8R^2,$$

in terms of the excircle radii r_a, r_b, and r_c, again with the reverse inequality holding for an obtuse triangle.

For an acute triangle with semiperimeter s,

$$s - r > 2R,$$

and the reverse inequality holds for an obtuse triangle.

For an acute triangle with area K,

$$ab + bc + ca \geq 2R(R + r) + \frac{8K}{\sqrt{3}}.$$

Distances Involving Triangle Centers

For an acute triangle the distance between the circumcenter O and the orthocenter H satisfies

$$OH < R,$$

with the opposite inequality holding for an obtuse triangle.

For an acute triangle the distance between the incircle center I and orthocenter H satisfies

$$IH < r\sqrt{2},$$

where r is the inradius, with the reverse inequality for an obtuse triangle.

Inscribed Square

If one of the inscribed squares of an acute triangle has side length x_a and another has side length x_b with $x_a < x_b$, then

$$1 \geq \frac{x_a}{x_b} \geq \frac{2\sqrt{2}}{3} \approx 0.94.$$

Two Triangles

If two obtuse triangles have sides (a, b, c) and (p, q, r) with c and r being the respective longest sides, then

$$ap + bq < cr.$$

Examples

Triangles with Special Names

The Calabi triangle, which is the only non-equilateral triangle for which the largest square that fits in the interior can be positioned in any of three different ways, is obtuse and isosceles with base angles 39.1320261...° and third angle 101.7359477...°.

The equilateral triangle, with three 60° angles, is acute.

The Morley triangle, formed from any triangle by the intersections of its adjacent angle trisectors, is equilateral and hence acute.

The golden triangle is the isosceles triangle in which the ratio of the duplicated side to the base side equals the golden ratio. It is acute, with angles 36°, 72°, and 72°, making it the only triangle with angles in the proportions 1:2:2.

The heptagonal triangle, with sides coinciding with a side, the shorter diagonal, and the longer diagonal of a regular heptagon, is obtuse, with angles $\pi/7, 2\pi/7,$ and $4\ /7$.

Triangles with Integer Sides

The only triangle with consecutive integers for an altitude and the sides is acute, having sides (13,14,15) and altitude from side 14 equal to 12.

The smallest-perimeter triangle with integer sides in arithmetic progression, and the smallest-perimeter integer-sided triangle with distinct sides, is obtuse: namely the one with sides (2, 3, 4).

The only triangles with one angle being twice another and having integer sides in arithmetic progression are acute: namely, the (4,5,6) triangle and its multiples.

There are no acute integer-sided triangles with area = perimeter, but there are three obtuse ones, having sides (6,25,29), (7,15,20), and (9,10,17).

The smallest integer-sided triangle with three rational medians is acute, with sides (68, 85, 87).

Heron triangles have integer sides and integer area. The oblique Heron triangle with the smallest perimeter is acute, with sides (6, 5, 5). The two oblique Heron triangles that share the smallest area are the acute one with sides (6, 5, 5) and the obtuse one with sides (8, 5, 5), the area of each being 12.

List of Triangle Inequalities

In geometry, triangle inequalities are inequalities involving the parameters of triangles, that hold for every triangle, or for every triangle meeting certain conditions. The inequalities give an ordering of two different values: they are of the form "less than", "less than or equal to", "greater than", or "greater than or equal to". The parameters in a triangle inequality can be the side lengths, the semiperimeter, the angle measures, the values of trigonometric functions of those angles, the area of the triangle, the medians of the sides, the altitudes, the lengths of the internal angle bisectors from each angle to the opposite side, the perpendicular bisectors of the sides, the distance from an arbitrary point to another point, the inradius, the exradii, the circumradius, and/or other quantities.

Unless otherwise specified, this article deals with triangles in the Euclidean plane.

Main Parameters and Notation

The parameters most commonly appearing in triangle inequalities are:

- the side lengths a, b, and c;

- the semiperimeter $s = (a + b + c) / 2$ (half the perimeter p);

- the angle measures A, B, and C of the angles of the vertices opposite the respective sides a, b, and c (with the vertices denoted with the same symbols as their angle measures);

- the values of trigonometric functions of the angles;

- the area T of the triangle;

- the medians m_a, m_b, and m_c of the sides (each being the length of the line segment from the midpoint of the side to the opposite vertex);

- the altitudes h_a, h_b, and h_c (each being the length of a segment perpendicular to one side and reaching from that side (or possibly the extension of that side) to the opposite vertex);

- the lengths of the internal angle bisectors t_a, t_b, and t_c (each being a segment from a vertex to the opposite side and bisecting the vertex's angle);

- the perpendicular bisectors p_a, p_b, and p_c of the sides (each being the length of a segment perpendicular to one side at its midpoint and reaching to one of the other sides);

- the lengths of line segments with an endpoint at an arbitrary point P in the plane (for example, the length of the segment from P to vertex A is denoted PA or AP);

- the inradius r (radius of the circle inscribed in the triangle, tangent to all three sides), the exradii r_a, r_b, and r_c (each being the radius of an excircle tangent to side a, b, or c respectively and tangent to the extensions of the other two sides), and the circumradius R (radius of the circle circumscribed around the triangle and passing through all three vertices).

Side Lengths

The basic triangle inequality is

$$a < b + c, \quad b < c + a, \quad c < a + b$$

or equivalently

$$\max(a, b, c) < s.$$

In addition,

$$\frac{a}{b+c} + \frac{b}{a+c} + \frac{c}{a+b} < 2,$$

where the value of the right side is the lowest possible bound, approached asymptotically as certain classes of triangles approach the degenerate case of zero area.

We have

$$3\left(\frac{a}{b} + \frac{b}{c} + \frac{c}{a}\right) \geq 2\left(\frac{b}{a} + \frac{c}{b} + \frac{a}{c}\right) + 3.$$

$$abc \geq (a+b-c)(a-b+c)(-a+b+c).$$

$$\frac{1}{3} \leq \frac{a^2+b^2+c^2}{(a+b+c)^2} \leq \frac{1}{2}.$$

$$\sqrt{a+b-c} + \sqrt{a-b+c} + \sqrt{-a+b+c} \leq \sqrt{a} + \sqrt{b} + \sqrt{c}.$$

$$a^2 b(a-b) + b^2 c(b-c) + c^2 a(c-a) \geq 0.$$

If angle C is obtuse (greater than 90°) then

$$a^2 + b^2 < c^2;$$

if C is acute (less than 90°) then

$$a^2 + b^2 > c^2.$$

The in-between case of equality when C is a right angle is the Pythagorean theorem.

In general,

$$a^2 + b^2 > \frac{c^2}{2},$$

with equality approached in the limit only as the apex angle of an isosceles triangle approaches 180°.

If the centroid of the triangle is inside the triangle's incircle, then

$$a^2 < 4bc, \quad b^2 < 4ac, \quad c^2 < 4ab.$$

While all of the above inequalities are true because a, b, and c must follow the basic triangle inequality that the longest side is less than half the perimeter, the following relations hold for all positive a, b, and c:

$$\frac{3abc}{ab+bc+ca} \leq \sqrt[3]{abc} \leq \frac{a+b+c}{3},$$

each holding with equality only when $a = b = c$. This says that in the non-equilateral case the harmonic mean of the sides is less than their geometric mean which in turn is less than their arithmetic mean.

Angles

$$\cos A + \cos B + \cos C \leq \frac{3}{2}.$$

$$(1-\cos A)(1-\cos B)(1-\cos C) \ge \cos A \cdot \cos B \cdot \cos C.$$

$$\cos^4 \frac{A}{2} + \cos^4 \frac{B}{2} + \cos^4 \frac{C}{2} \le \frac{s^3}{2abc}$$

for semi-perimeter s, with equality only in the equilateral case.

$$a+b+c \ge 2\sqrt{bc}\,\cos A + 2\sqrt{ca}\,\cos B + 2\sqrt{ab}\,\cos C.$$

$$\sin A + \sin B + \sin C \le \frac{3\sqrt{3}}{2}.$$

$$\sin^2 A + \sin^2 B + \sin^2 C \le \frac{9}{4}.$$

$$\sin A \cdot \sin B \cdot \sin C \le \left(\frac{\sin A + \sin B + \sin C}{3}\right)^3 \le \left(\sin \frac{A+B+C}{3}\right)^3 = \sin^3\left(\frac{\pi}{3}\right) = \frac{3\sqrt{3}}{8}.$$

$$\sin A + \sin B \cdot \sin C \le \varphi$$

where $\varphi = \dfrac{1+\sqrt{5}}{2}$, the golden ratio.

$$\sin \frac{A}{2} \cdot \sin \frac{B}{2} \cdot \sin \frac{C}{2} \le \frac{1}{8}.$$

$$\tan^2 \frac{A}{2} + \tan^2 \frac{B}{2} + \tan^2 \frac{C}{2} \ge 1.$$

$$\cot A + \cot B + \cot C \ge \sqrt{3}.$$

$$\sin A \cdot \cos B + \sin B \cdot \cos C + \sin C \cdot \cos A \le \frac{3\sqrt{3}}{4}.$$

For circumradius R and inradius r we have

$$\max\left(\sin \frac{A}{2}, \sin \frac{B}{2}, \sin \frac{C}{2}\right) \le \frac{1}{2}\left(1 + \sqrt{1 - \frac{2r}{R}}\right),$$

with equality if and only if the triangle is isosceles with apex angle greater than or equal to 60°; and

$$\min\left(\sin \frac{A}{2}, \sin \frac{B}{2}, \sin \frac{C}{2}\right) \ge \frac{1}{2}\left(1 - \sqrt{1 - \frac{2r}{R}}\right),$$

with equality if and only if the triangle is isosceles with apex angle less than or equal to 60°.

We also have

$$\frac{r}{R} - \sqrt{1 - \frac{2r}{R}} \le \cos A \le \frac{r}{R} + \sqrt{1 - \frac{2r}{R}}$$

and likewise for angles B, C, with equality in the first part if the triangle is isosceles and the apex angle is at least 60° and equality in the second part if and only if the triangle is isosceles with apex angle no greater than 60°.

Further, any two angle measures A and B opposite sides a and b respectively are related according to

$$A > B \quad \text{if and only if} \quad a > b,$$

which is related to the isosceles triangle theorem and its converse, which state that $A = B$ if and only if $a = b$.

By Euclid's exterior angle theorem, any exterior angle of a triangle is greater than either of the interior angles at the opposite vertices:

$$180° - A > \max(B, C).$$

If a point D is in the interior of triangle ABC, then

$$\angle BDC > \angle A.$$

For an acute triangle we have

$$\cos^2 A + \cos^2 B + \cos^2 C < 1,$$

with the reverse inequality holding for an obtuse triangle.

Area

Weitzenböck's inequality is, in terms of area T,

$$a^2 + b^2 + c^2 \ge 4\sqrt{3} \cdot T,$$

with equality only in the equilateral case. This is a corollary of the Hadwiger–Finsler inequality, which is

$$a^2 + b^2 + c^2 \ge (a - b)^2 + (b - c)^2 + (c - a)^2 + 4\sqrt{3} \cdot T.$$

Also,

$$ab + bc + ca \ge 4\sqrt{3} \cdot T$$

and

$$\frac{bc}{2}\sqrt{\frac{a+b+c}{a^3+b^3+c^3+abc}} \le \frac{1}{4}\sqrt[6]{\frac{3(a+b+c)^3(abc)^4}{a^3+b^3+c^3}} \le \frac{\sqrt{3}}{4}(abc)^{2/3}.$$

From the last upper bound on T, using the arithmetic-geometric mean inequality, is obtained the isoperimetric inequality for triangles:

$$T \le \frac{\sqrt{3}}{36}(a+b+c)^2 = \frac{\sqrt{3}}{9}s^2$$

for semiperimeter s. This is sometimes stated in terms of perimeter p as

$$p^2 \ge 12\sqrt{3}\cdot T,$$

with equality for the equilateral triangle.

We also have

with equality only in the equilateral case;

$$38T^2 \le 2s^4 - a^4 - b^4 - c^4$$

for semiperimeter s; and

$$\frac{1}{a}+\frac{1}{b}+\frac{1}{c} < \frac{s}{T}.$$

Ono's inequality for acute triangles (those with all angles less than 90°) is

$$27(b^2+c^2-a^2)^2(c^2+a^2-b^2)^2(a^2+b^2-c^2)^2 \le (4T)^6.$$

The area of the triangle can be compared to the area of the incircle:

$$\frac{\text{Area of incircle}}{\text{Area of triangle}} \le \frac{\pi}{3\sqrt{3}}$$

with equality only for the equilateral triangle.

If an inner triangle is inscribed in a reference triangle so that the inner triangle's vertices partition the perimeter of the reference triangle into equal length segments, the ratio of their areas is bounded by

$$\frac{\text{Area of inscribed triangle}}{\text{Area of reference triangle}} \le \frac{1}{4}.$$

Let the interior angle bisectors of A, B, and C meet the opposite sides at D, E, and F. Then

$$\frac{3abc}{4(a^3+b^3+c^3)} \le \frac{\text{Area of triangle}\,DEF}{\text{Area of triangle}\,ABC} \le \frac{1}{4}.$$

Medians and Centroid

The three medians m_a, m_b, m_c of a triangle each connect a vertex with the midpoint of the opposite side, and the sum of their lengths satisfies

$$\frac{3}{4}(a+b+c) < m_a + m_b + m_c < a+b+c.$$

Moreover,

$$\left(\frac{m_a}{a}\right)^2 + \left(\frac{m_b}{b}\right)^2 + \left(\frac{m_c}{c}\right)^2 \geq \frac{9}{4},$$

with equality only in the equilateral case, and for inradius r,

If we further denote the lengths of the medians extended to their intersections with the circumcircle as M_a, M_b, and M_c, then

$$\frac{M_a}{m_a} + \frac{M_b}{m_b} + \frac{M_c}{m_c} \geq 4.$$

The centroid G is the intersection of the medians. Let AG, BG, and CG meet the circumcircle at U, V, and W respectively. Then both

$$GU + GV + GW \geq AG + BG + CG$$

and

$$GU \cdot GV \cdot GW \geq AG \cdot BG \cdot CG;$$

in addition,

$$\sin GBC + \sin GCA + \sin GAB \leq \frac{3}{2}.$$

For an acute triangle we have

$$m_a^2 + m_b^2 + m_c^2 > 6R^2$$

in terms of the circumradius R, while the opposite inequality holds for an obtuse triangle.

Denoting as IA, IB, IC the distances of the incenter from the vertices, the following holds:

$$\frac{IA^2}{m_a^2} + \frac{IB^2}{m_b^2} + \frac{IC^2}{m_c^2} \leq \frac{3}{4}.$$

The three medians of any triangle can form the sides of another triangle:

$$m_a < m_b + m_c, \quad m_b < m_c + m_a, \quad m_c < m_a + m_b.$$

Altitudes

The altitudes h_a, etc. each connect a vertex to the opposite side and are perpendicular to that side. They satisfy both

$$h_a + h_b + h_c \leq \frac{\sqrt{3}}{2}(a + b + c)$$

and

$$h_a^2 + h_b^2 + h_c^2 \leq \frac{3}{4}(a^2 + b^2 + c^2).$$

In addition, if $a \geq b \geq c$, then

$$a + h_a \geq b + h_b \geq c + h_c.$$

We also have

$$\frac{h_a^2}{(b^2 + c^2)} \cdot \frac{h_b^2}{(c^2 + a^2)} \cdot \frac{h_c^2}{(a^2 + b^2)} \leq \left(\frac{3}{8}\right)^3.$$

For internal angle bisectors t_a, t_b, t_c from vertices A, B, C and circumcenter R and incenter r, we have

$$\frac{h_a}{t_a} + \frac{h_b}{t_b} + \frac{h_c}{t_c} \geq \frac{R + 4r}{R}.$$

The reciprocals of the altitudes of any triangle can themselves form a triangle:

$$\frac{1}{h_a} < \frac{1}{h_b} + \frac{1}{h_c}, \quad \frac{1}{h_b} < \frac{1}{h_c} + \frac{1}{h_a}, \quad \frac{1}{h_c} < \frac{1}{h_a} + \frac{1}{h_b}.$$

Internal Angle Bisectors and Incenter

The internal angle bisectors are segments in the interior of the triangle reaching from one vertex to the opposite side and bisecting the vertex angle into two equal angles. The angle bisectors t_a etc. satisfy

$$t_a + t_b + t_c \leq \frac{3}{2}(a + b + c)$$

in terms of the sides, and

$$h_a \leq t_a \leq m_a$$

in terms of the altitudes and medians, and likewise for t_b and t_c. Further,

$$\sqrt{m_a} + \sqrt{m_b} + \sqrt{m_c} \geq \sqrt{t_a} + \sqrt{t_b} + \sqrt{t_c}$$

in terms of the medians.

Let T_a, T_b, and T_c be the lengths of the angle bisectors extended to the circumcircle. Then

$$T_a T_b T_c \geq \frac{8\sqrt{3}}{9} abc,$$

with equality only in the equilateral case, and

$$T_a + T_b + T_c \leq 5R + 2r$$

for circumradius R and inradius r, again with equality only in the equilateral case. In addition,.

$$T_a + T_b + T_c \geq \frac{4}{3}(t_a + t_b + t_c).$$

For incenter I (the intersection of the internal angle bisectors),

$$6r \leq AI + BI + CI \leq \sqrt{12(R^2 - Rr + r^2)}.$$

For midpoints L, M, N of the sides,

$$IL^2 + IM^2 + IN^2 \geq r(R + r).$$

For incenter I, centroid G, circumcenter O, nine-point center N, and orthocenter H, we have for non-equilateral triangles the distance inequalities

$$IG < HG,$$

$$IH < HG,$$

$$IG < IO,$$

and

$$IN < \frac{1}{2} IO;$$

and we have the angle inequality

$$\angle IOH < \frac{\pi}{6}.$$

In addition,

$$IG < \frac{1}{3}v,$$

where v is the longest median.

Three triangles with vertex at the incenter, OIH, GIH, and OGI, are obtuse:

$$\angle OIH > \angle GIH > 90° , \ \angle OGI > 90°.$$

Since these triangles have the indicated obtuse angles, we have

$$OI^2 + IH^2 < OH^2, \quad GI^2 + IH^2 < GH^2, \quad OG^2 + GI^2 < OI^2,$$

and in fact the second of these is equivalent to a result stronger than the first, shown by Euler:

$$OI^2 < OH^2 - 2 \cdot IH^2 < 2 \cdot OI^2.$$

The larger of two angles of a triangle has the shorter internal angle bisector:

$$\text{If} \quad A > B \quad \text{then} \quad t_a < t_b.$$

Perpendicular Bisectors of Sides

These inequalities deal with the lengths p_a etc. of the triangle-interior portions of the perpendicular bisectors of sides of the triangle. Denoting the sides so that $a \geq b \geq c$, we have

$$p_a \geq p_b$$

and

$$p_c \geq p_b.$$

Segments from an Arbitrary Point

Consider any point P in the interior of the triangle, with the triangle's vertices denoted A, B, and C and with the lengths of line segments denoted PA etc. We have

$$2(PA + PB + PC) > AB + BC + CA > PA + PB + PC$$

and more strongly

$$PA + PB + PC \leq AC + BC, \quad PA + PB + PC \leq AB + BC, \quad PA + PB + PC \leq AB + AC.$$

We also have Ptolemy's inequality

$$PA{\cdot}BC + PB{\cdot}CA > PC{\cdot}AB$$

and likewise for cyclic permutations of the vertices.

If we draw perpendiculars from P to the sides of the triangle, intersecting the sides at D, E, and F, we have

$$PA \cdot PB \cdot PC \geq (PD + PE)(PE + PF)(PF + PD).$$

Further, the Erdős–Mordell inequality states that

$$\frac{PA + PB + PC}{PD + PE + PF} \geq 2$$

with equality in the equilateral case. More strongly, Barrow's inequality states that if the interior bisectors of the angles at P (namely, of $\angle APB$, $\angle BPC$, and $\angle CPA$) intersect the triangle's sides at U, V, and W, then

$$\frac{PA + PB + PC}{PU + PV + PW} \geq 2.$$

Again with distances PD, PE, PF of the interior point P from the sides we have these three inequalities:

$$\frac{PA^2}{PE \cdot PF} + \frac{PB^2}{PF \cdot PD} + \frac{PC^2}{PD \cdot PE} \geq 12;$$

$$\frac{PA}{\sqrt{PE \cdot PF}} + \frac{PB}{\sqrt{PF \cdot PD}} + \frac{PC}{\sqrt{PD \cdot PE}} \geq 6;$$

$$\frac{PA}{PE + PF} + \frac{PB}{PF + PD} + \frac{PC}{PD + PE} \geq 3.$$

For interior point P with distances PA, PB, PC from the vertices and with triangle area T,

$$(b + c)PA + (c + a)PB + (a + b)PC \geq 8T$$

and

$$\frac{PA}{a} + \frac{PB}{b} + \frac{PC}{c} \geq \sqrt{3}.$$

For an interior point P, centroid G, midpoints L, M, N of the sides, and semiperimeter s,

$$2(PL + PM + PN) \leq 3PG + PA + PB + PC \leq s + 2(PL + PM + PN).$$

Moreover, for positive numbers k_1, k_2, k_3, and t with t less than or equal to 1:

$$(PA)^t + k_2 \cdot (PB)^t + k_3 \cdot (PC)^t \geq 2^t \sqrt{k_1 k_2 k_3} \left(\frac{(PD)^t}{\sqrt{k_1}} + \frac{(PE)^t}{\sqrt{k_2}} + \frac{(PF)^t}{\sqrt{k_3}} \right),$$

while for $t > 1$ we have

$$(PA)^t + k_2 \cdot (PB)^t + k_3 \cdot (PC)^t \ge 2\sqrt{k_1 k_2 k_3}\left(\frac{(PD)^t}{\sqrt{k_1}} + \frac{(PE)^t}{\sqrt{k_2}} + \frac{(PF)^t}{\sqrt{k_3}}\right).$$

There are various inequalities for an arbitrary interior or exterior point in the plane in terms of the radius r of the triangle's inscribed circle. For example,

$$PA + PB + PC \ge 6r.$$

Others include:

$$PA^3 + PB^3 + PC^3 + k \cdot (PA \cdot PB \cdot PC) \ge 8(k+3)r^3$$

for $k = 0, 1, ..., 6$;

$$PA^2 + PB^2 + PC^2 + (PA \cdot PB \cdot PC)^{2/3} \ge 16r^2;$$

$$PA^2 + PB^2 + PC^2 + 2(PA \cdot PB \cdot PC)^{2/3} \ge 20r^2;$$

and

$$PA^4 + PB^4 + PC^4 + k(PA \cdot PB \cdot PC)^{4/3} \ge 16(k+3)r^4$$

for $k = 0, 1, ..., 9$.

Furthermore, for circumradius R,

$$(PA \cdot PB)^{3/2} + (PB \cdot PC)^{3/2} + (PC \cdot PA)^{3/2} \ge 12Rr^2;$$

$$(PA \cdot PB)^2 + (PB \cdot PC)^2 + (PC \cdot PA)^2 \ge 8(R+r)Rr^2;$$

$$(PA \cdot PB)^2 + (PB \cdot PC)^2 + (PC \cdot PA)^2 \ge 48r^4;$$

$$(PA \cdot PB)^2 + (PB \cdot PC)^2 + (PC \cdot PA)^2 \ge 6(7R - 6r)r^3.$$

Inradius, Exradii, and Circumradius

Inradius and Circumradius

The Euler inequality for the circumradius R and the inradius r states that

$$\frac{R}{r} \ge 2,$$

with equality only in the equilateral case.

A stronger version is

$$\frac{R}{r} \geq \frac{abc + a^3 + b^3 + c^3}{2abc} \geq \frac{a}{b} + \frac{b}{c} + \frac{c}{a} - 1 \geq \frac{2}{3}\left(\frac{a}{b} + \frac{b}{c} + \frac{c}{a}\right) \geq 2.$$

By comparison,

$$\frac{r}{R} \geq \frac{4abc - a^3 - b^3 - c^3}{2abc},$$

where the right side could be positive or negative.

Two other refinements of Euler's inequality are

$$\frac{R}{r} \geq \frac{(b+c)}{3a} + \frac{(c+a)}{3b} + \frac{(a+b)}{3c} \geq 2$$

and

$$\left(\frac{R}{r}\right)^3 \geq \left(\frac{a}{b} + \frac{b}{a}\right)\left(\frac{b}{c} + \frac{c}{b}\right)\left(\frac{c}{a} + \frac{a}{c}\right) \geq 8.$$

Moreover,

$$\frac{R}{r} \geq \frac{2(a^2 + b^2 + c^2)}{ab + bc + ca};$$

$$a^3 + b^3 + c^3 \leq 8s(R^2 - r^2)$$

in terms of the semiperimeter s;

$$r(r + 4R) \geq \sqrt{3} \cdot T$$

in terms of the area T;

$$s\sqrt{3} \leq r + 4R$$

and

$$s^2 \geq 16Rr - 5r^2$$

in terms of the semiperimeter s; and

$$2R^2 + 10Rr - r^2 - 2(R - 2r)\sqrt{R^2 - 2Rr} \leq s^2$$

$$\leq 2R^2 + 10Rr - r^2 + 2(R - 2r)\sqrt{R^2 - 2Rr}$$

also in terms of the semiperimeter. Here the expression $\sqrt{R^2 - 2Rr} = d$ where d is the distance between the incenter and the circumcenter. In the latter double inequality, the first part holds with equality if and only if the triangle is isosceles with an apex angle of at least 60°, and the last part holds with equality if and only if the triangle is isosceles with an apex angle of at most 60°. Thus both are equalities if and only if the triangle is equilateral.

We also have for any side a

$$(R-d)^2 - r^2 \le 4R^2 r^2 (\frac{(R+d)^2 - r^2}{(R+d)^4}) \le \frac{a^2}{4} \le Q \le (R+d)^2 - r^2,$$

where $Q = R^2$ if the circumcenter is on or outside of the incircle and $Q = 4R^2 r^2 \left(\dfrac{(R-d)^2 - r^2}{(R-d)^4} \right)$ if the circumcenter is inside the incircle. The circumcenter is inside the incircle if and only if

$$\frac{R}{r} < \sqrt{2} + 1.$$

Further,

$$\frac{9r}{2T} \le \frac{1}{a} + \frac{1}{b} + \frac{1}{c} \le \frac{9R}{4T}.$$

Blundon's inequality states that

$$s \le (3\sqrt{3} - 4)r + 2R.$$

For incircle center I, let AI, BI, and CI extend beyond I to intersect the circumcircle at D, E, and F respectively. Then

$$\frac{AI}{ID} + \frac{BI}{IE} + \frac{CI}{IF} \ge 3.$$

In terms of the vertex angles we have

$$\cos A \cdot \cos B \cdot \cos C \le \left(\frac{r}{R\sqrt{2}} \right)^2.$$

Circumradius and Other Lengths

For the circumradius R we have

$$18R^3 \ge (a^2 + b^2 + c^2)R + abc\sqrt{3}$$

and

$$a^{2/3} + b^{2/3} + c^{2/3} \le 3^{7/4} R^{3/2}.$$

We also have

$$a + b + c \leq 3\sqrt{3} \cdot R,$$

$$9R^2 \geq a^2 + b^2 + c^2,$$

$$h_a + h_b + h_c \leq 3\sqrt{3} \cdot R$$

in terms of the altitudes,

$$m_a^2 + m_b^2 + m_c^2 \leq \frac{27}{4} R^2$$

in terms of the medians, and

$$\frac{ab}{a+b} + \frac{bc}{b+c} + \frac{ca}{c+a} \geq \frac{2T}{R}$$

in terms of the area.

Moreover, for circumcenter O, let lines AO, BO, and CO intersect the opposite sides BC, CA, and AB at U, V, and W respectively. Then

$$OU + OV + OW \geq \frac{3}{2} R.$$

For an acute triangle the distance between the circumcenter O and the orthocenter H satisfies

$$OH < R,$$

with the opposite inequality holding for an obtuse triangle.

The circumradius is at least twice the distance between the first and second Brocard points B_1 and B_2:

$$R \geq 2B_1 B_2.$$

Inradius, Exradii, and Other Lengths

For the inradius r we have

$$\frac{1}{a} + \frac{1}{b} + \frac{1}{c} \leq \frac{\sqrt{3}}{2r},$$

$$9r \leq h_a + h_b + h_c$$

in terms of the altitudes, and

$$\sqrt{r_a^2 + r_b^2 + r_c^2} \geq 6r$$

in terms of the radii of the excircles. We additionally have

$$\sqrt{s}(\sqrt{a} + \sqrt{b} + \sqrt{c}) \leq \sqrt{2}(r_a + r_b + r_c)$$

and

$$\frac{abc}{r} \geq \frac{a^3}{r_a} + \frac{b^3}{r_b} + \frac{c^3}{r_c}.$$

The exradii and medians are related by

$$\frac{r_a r_b}{m_a m_b} + \frac{r_b r_c}{m_b m_c} + \frac{r_c r_a}{m_c m_a} \geq 3.$$

In addition, for an acute triangle the distance between the incircle center I and orthocenter H satisfies

$$IH < r\sqrt{2},$$

with the reverse inequality for an obtuse triangle.

Also, an acute triangle satisfies

$$r^2 + r_a^2 + r_b^2 + r_c^2 < 8R^2,$$

in terms of the circumradius R, again with the reverse inequality holding for an obtuse triangle.

If the internal angle bisectors of angles A, B, C meet the opposite sides at U, V, W then

$$\frac{1}{4} < \frac{AI \cdot BI \cdot CI}{AU \cdot BV \cdot CW} \leq \frac{8}{27}.$$

If the internal angle bisectors through incenter I extend to meet the circumcircle at X, Y and Z then

$$\frac{1}{IX} + \frac{1}{IY} + \frac{1}{IZ} \geq \frac{3}{R}$$

for circumradius R, and

$$0 \leq (IX - IA) + (IY - IB) + (IZ - IC) \leq 2(R - 2r).$$

If the incircle is tangent to the sides at D, E, F, then

$$EF^2 + FD^2 + DE^2 \leq \frac{s^2}{3}$$

for semiperimeter s.

Inscribed Figures

Inscribed Hexagon

If a tangential hexagon is formed by drawing three segments tangent to a triangle's incircle and parallel to a side, so that the hexagon is inscribed in the triangle with its other three sides coinciding with parts of the triangle's sides, then

$$\text{Perimeter of hexagon} \le \frac{2}{3}(\text{Perimeter of triangle}).$$

Inscribed Triangle

If three points D, E, F on the respective sides AB, BC, and CA of a reference triangle ABC are the vertices of an inscribed triangle, which thereby partitions the reference triangle into four triangles, then the area of the inscribed triangle is greater than the area of at least one of the other interior triangles, unless the vertices of the inscribed triangle are at the midpoints of the sides of the reference triangle (in which case the inscribed triangle is the medial triangle and all four interior triangles have equal areas):

$$\text{Area(DEF)} \ge \min(\text{Area(BED)}, \text{Area(CFE)}, \text{Area(ADF)}).$$

Inscribed Squares

An acute triangle has three inscribed squares, each with one side coinciding with part of a side of the triangle and with the square's other two vertices on the remaining two sides of the triangle. (A right triangle has only two distinct inscribed squares.) If one of these squares has side length x_a and another has side length x_b with $x_a < x_b$, then

$$1 \ge \frac{x_a}{x_b} \ge \frac{2\sqrt{2}}{3} \approx 0.94.a$$

Moreover, for any square inscribed in any triangle we have

$$\frac{\text{Area of triangle}}{\text{Area of inscribed square}} \ge 2.$$

Euler Line

A triangle's Euler line goes through its orthocenter, its circumcenter, and its centroid, but does not go through its incenter unless the triangle is isosceles. For all non-isosceles triangles, the distance d from the incenter to the Euler line satisfies the following inequalities in terms of the triangle's longest median v, its longest side u, and its semiperimeter s:

$$\frac{d}{s} < \frac{d}{u} < \frac{d}{v} < \frac{1}{3}.$$

For all of these ratios, the upper bound of 1/3 is the tightest possible.

Right Triangle

In right triangles the legs a and b and the hypotenuse c obey the following, with equality only in the isosceles case:

$$a+b \leq c\sqrt{2}.$$

In terms of the inradius, the hypotenuse obeys

$$2r \leq c(\sqrt{2}-1),$$

and in terms of the altitude from the hypotenuse the legs obey

$$h_c \leq \frac{\sqrt{2}}{4}(a+b).$$

Isosceles Triangle

If the two equal sides of an isosceles triangle have length a and the other side has length c, then the internal angle bisector t from one of the two equal-angled vertices satisfies

$$\frac{2ac}{a+c} > t > \frac{ac\sqrt{2}}{a+c}.$$

Equilateral Triangle

For any point P in the plane of an equilateral triangle ABC, the distances of P from the vertices, PA, PB, and PC, are such that, unless P is on the triangle's circumcircle, they obey the basic triangle inequality and thus can themselves form the sides of a triangle:

$$PA+PB > PC, \quad PB+PC > PA, \quad PC+PA > PB.$$

However, when P is on the circumcircle the sum of the distances from P to the nearest two vertices exactly equals the distance to the farthest vertex.

A triangle is equilateral if and only if, for *every* point P in the plane, with distances PD, PE, and PF to the triangle's sides and distances PA, PB, and PC to its vertices,

$$4(PD^2 + PE^2 + PF^2) \geq PA^2 + PB^2 + PC^2.$$

Two Triangles

Pedoe's inequality for two triangles, one with sides a, b, and c and area T, and the other with sides d, e, and f and area S, states that

$$d^2(b^2+c^2-a^2)+e^2(a^2+c^2-b^2)+f^2(a^2+b^2-c^2) \geq 16TS,$$

with equality if and only if the two triangles are similar.

The hinge theorem or open-mouth theorem states that if two sides of one triangle are congruent to two sides of another triangle, and the included angle of the first is larger than the included angle of the second, then the third side of the first triangle is longer than the third side of the second triangle. That is, in triangles ABC and DEF with sides a, b, c, and d, e, f respectively (with a opposite A etc.), if $a = d$ and $b = e$ and angle $C >$ angle F, then

$$c > f.$$

The converse also holds: if $c > f$, then $C > F$.

The angles in any two triangles ABC and DEF are related in terms of the cotangent function according to

$$\cot A(\cot E + \cot F) + \cot B(\cot F + \cot D) + \cot C(\cot D + \cot E) \geq 2.$$

Non-Euclidean Triangles

In a triangle on the surface of a sphere, as well as in elliptic geometry,

$$\angle A + \angle B + \angle C > 180°.$$

This inequality is reversed for hyperbolic triangles.

References

- Zeidler, Eberhard (2004). Oxford Users' Guide to Mathematics. Oxford University Press. p. 729. ISBN 978-0-19-850763-5.

- Heath, Thomas L. (1956), The Thirteen Books of Euclid's Elements, 1 (2nd ed. [Facsimile. Original publication: Cambridge University Press, 1925] ed.), New York: Dover Publications, ISBN 0-486-60088-2

- Yurii, N. Maltsev and Anna S. Kuzmina, "An improvement of Birsan's inequalities for the sides of a triangle", Forum Geometricorum 16, 2016, pp. 81–84.

- Dao, Thanh Oai (2015). "Equilateral triangles and Kiepert perspectors in complex numbers" (PDF). Forum Geometricorum. 15: 105–114.

- Birsan, Temistocle (2015). "Bounds for elements of a triangle expressed by R, r, and s" (PDF). Forum Geometricorum. 15: 99–103.

- Wladimir G. Boskoff, Laurenţiu Homentcovschi, and Bogdan D. Suceava, "Gossard's Perspector and Projective Consequences", Forum Geometricorum, Volume 13 (2013), 169–184. [1]

Circle: A Comprehensive Study

Circles have a center and are closed in shape. The distance from the center to any point of the circle is known as radius. Some of the basic aspects of a circle are circumference, diameter, arc and radius. The information provided in this section helps the reader to delve deep into the topics related to it.

Circle

A circle is a simple closed shape in Euclidean geometry. It is the set of all points in a plane that are at a given distance from a given point, the centre; equivalently it is the curve traced out by a point that moves so that its distance from a given point is constant. The distance between any of the points and the centre is called the radius.

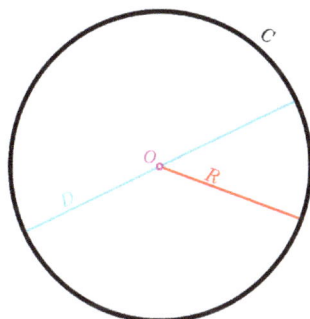

A circle (black) which is measured by its circumference (C), diameter (D) in cyan, and radius (R) in red; its centre (O) is in magenta.

A circle is a simple closed curve which divides the plane into two regions: an interior and an exterior. In everyday use, the term "circle" may be used interchangeably to refer to either the boundary of the figure, or to the whole figure including its interior; in strict technical usage, the circle is only the boundary and the whole figure is called a disc.

A circle may also be defined as a special kind of ellipse in which the two foci are coincident and the eccentricity is 0, or the two-dimensional shape enclosing the most area per unit perimeter squared, using calculus of variations.

A circle is a plane figure bounded by one line, and such that all right lines drawn from a certain point within it to the bounding line, are equal. The bounding line is called its circumference and the point, its centre.

— *Euclid. Elements Book I.*

Terminology

- Annulus: the ring-shaped object, the region bounded by two concentric circles.

- Arc: any connected part of the circle.

- Centre: the point equidistant from the points on the circle.

- Chord: a line segment whose endpoints lie on the circle.

- Circumference: the length of one circuit along the circle, or the distance around the circle.

- Diameter: a line segment whose endpoints lie on the circle and which passes through the centre; or the length of such a line segment, which is the largest distance between any two points on the circle. It is a special case of a chord, namely the longest chord, and it is twice the radius.

- Disc: the region of the plane bounded by a circle.

- Lens: the intersection of two discs.

- Passant: a coplanar straight line that does not touch the circle.

- Radius: a line segment joining the centre of the circle to any point on the circle itself; or the length of such a segment, which is half a diameter.

- Sector: a region bounded by two radii and an arc lying between the radii.

- Segment: a region, not containing the centre, bounded by a chord and an arc lying between the chord's endpoints.

- Secant: an extended chord, a coplanar straight line cutting the circle at two points.

- Semicircle: an arc that extends from one of a diameter's endpoints to the other. In non-technical common usage it may mean the diameter, arc, and its interior, a two dimensional region, that is technically called a half-disc. A half-disc is a special case of a segment, namely the largest one.

- Tangent: a coplanar straight line that touches the circle at a single point.

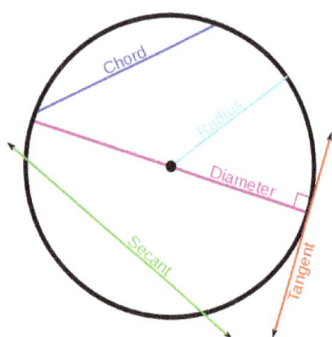

Chord, secant, tangent, radius, and diameter

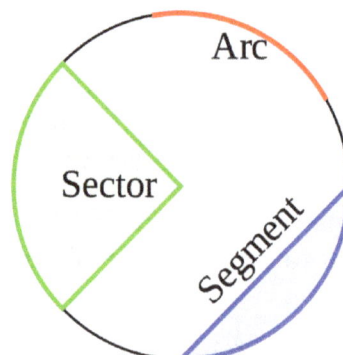

Arc, sector, and segment

History

The compass in this 13th-century manuscript is a symbol of God's act of Creation.
Notice also the circular shape of the halo.

The origins of the words *circus* and *circuit* are closely related.

Circular piece of silk with Mongol images

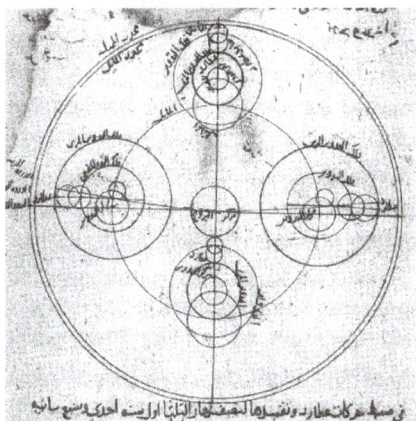

Circles in an old Arabic astronomical drawing.

The circle has been known since before the beginning of recorded history. Natural circles would have been observed, such as the Moon, Sun, and a short plant stalk blowing in the wind on sand, which forms a circle shape in the sand. The circle is the basis for the wheel, which, with related inventions such as gears, makes much of modern machinery possible. In mathematics, the study of the circle has helped inspire the development of geometry, astronomy and calculus.

Early science, particularly geometry and astrology and astronomy, was connected to the divine for most medieval scholars, and many believed that there was something intrinsically "divine" or "perfect" that could be found in circles.

Some highlights in the history of the circle are:

- 1700 BCE – The Rhind papyrus gives a method to find the area of a circular field. The result corresponds to 256/81 (3.16049...) as an approximate value of π.

Tughrul Tower from inside

- 300 BCE – Book 3 of Euclid's *Elements* deals with the properties of circles.

- In Plato's Seventh Letter there is a detailed definition and explanation of the circle. Plato explains the perfect circle, and how it is different from any drawing, words, definition or explanation.

- 1880 CE – Lindemann proves that π is transcendental, effectively settling the millennia-old problem of squaring the circle.

Analytic Results

Length of Circumference

The ratio of a circle's circumference to its diameter is π (pi), an irrational constant approximately equal to 3.141592654. Thus the length of the circumference C is related to the radius r and diameter d by:

$$C = 2\pi r = \pi d.$$

Area Enclosed

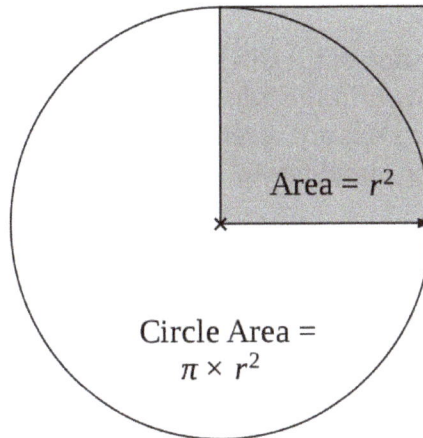

Area enclosed by a circle = π × area of the shaded square

As proved by Archimedes, in his Measurement of a Circle, the area enclosed by a circle is equal to that of a triangle whose base has the length of the circle's circumference and whose height equals the circle's radius, which comes to π multiplied by the radius squared:

$$\text{Area} = \pi r^2.$$

Equivalently, denoting diameter by d,

$$\text{Area} = \frac{\pi d^2}{4} \approx 0.7854 d^2,$$

that is, approximately 79% of the circumscribing square (whose side is of length d).

The circle is the plane curve enclosing the maximum area for a given arc length. This relates the circle to a problem in the calculus of variations, namely the isoperimetric inequality.

Equations

Cartesian Coordinates

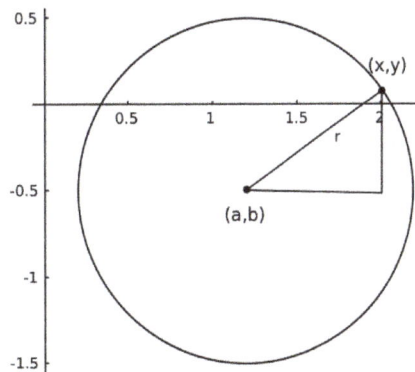

Circle of radius $r = 1$, centre $(a, b) = (1.2, -0.5)$

In an x–y Cartesian coordinate system, the circle with centre coordinates (a, b) and radius r is the set of all points (x, y) such that

$$\left(x-a\right)^2+\left(y-b\right)^2=r^2.$$

This equation, known as the Equation of the Circle, follows from the Pythagorean theorem applied to any point on the circle: as shown in the adjacent diagram, the radius is the hypotenuse of a right-angled triangle whose other sides are of length $|x-a|$ and $|y-b|$. If the circle is centred at the origin (0, 0), then the equation simplifies to

$$x^2+y^2=r^2.$$

The equation can be written in parametric form using the trigonometric functions sine and cosine as

$$x=a+r\cos t,$$

$$y=b+r\sin t$$

where t is a parametric variable in the range 0 to 2π, interpreted geometrically as the angle that the ray from (a, b) to (x, y) makes with the positive x-axis.

An alternative parametrisation of the circle is:

$$x=a+r\frac{1-t^2}{1+t^2}.$$

$$y=b+r\frac{2t}{1+t^2}$$

In this parametrisation, the ratio of t to r can be interpreted geometrically as the stereographic projection of the line passing through the centre parallel to the x-axis. However, this parametrisation works only if t is made to range not only through all reals but also to a point at infinity; otherwise, the bottom-most point of the circle would be omitted.

In homogeneous coordinates each conic section with the equation of a circle has the form

$$x^2+y^2-2axz-2byz+cz^2=0.$$

It can be proven that a conic section is a circle exactly when it contains (when extended to the complex projective plane) the points $I(1: i: 0)$ and $J(1: -i: 0)$. These points are called the circular points at infinity.

Polar Coordinates

In polar coordinates the equation of a circle is:

$$r^2-2rr_0\cos(\theta-\phi)+r_0^2=a^2$$

where a is the radius of the circle, (r,θ) is the polar coordinate of a generic point on the circle, and (r_0,ϕ) is the polar coordinate of the centre of the circle (i.e., r_0 is the distance from the origin to the centre of the circle, and φ is the anticlockwise angle from the positive x-axis to the line connecting the origin to the centre of the circle). For a circle centred at the origin, i.e. $r_0 = 0$, this reduces to simply $r = a$. When $r_0 = a$, or when the origin lies on the circle, the equation becomes

$$r = 2a\cos(\theta - \phi).$$

In the general case, the equation can be solved for r, giving

$$r = r_0 \cos(\theta - \phi) \pm \sqrt{a^2 - r_0^2 \sin^2(\theta - \phi)},$$

Note that without the $\pm$ sign, the equation would in some cases describe only half a circle.

Complex Plane

In the complex plane, a circle with a centre at c and radius (r) has the equation $|z - c| = r$. In parametric form this can be written $z = re^{it} + c$.

The slightly generalised equation $pz\bar{z} + gz + \overline{gz} = q$ for real p, q and complex g is sometimes called a generalised circle. This becomes the above equation for a circle with $p = 1$, $g = -\bar{c}$, $q = r^2 - |c|^2$, since $|z - c|^2 = z\bar{z} - \bar{c}z - c\bar{z} + c\bar{c}$. Not all generalised circles are actually circles: a generalised circle is either a (true) circle or a line.

Tangent Lines

The tangent line through a point P on the circle is perpendicular to the diameter passing through P. If $P = (x_1, y_1)$ and the circle has centre (a, b) and radius r, then the tangent line is perpendicular to the line from (a, b) to (x_1, y_1), so it has the form $(x_1 - a)x + (y_1 - b)y = c$. Evaluating at (x_1, y_1) determines the value of c and the result is that the equation of the tangent is

$$(x_1 - a)x + (y_1 - b)y = (x_1 - a)x_1 + (y_1 - b)y_1$$

or

$$(x_1 - a)(x - a) + (y_1 - b)(y - b) = r^2.$$

If $y_1 \neq b$ then the slope of this line is

$$\frac{dy}{dx} = -\frac{x_1 - a}{y_1 - b}.$$

This can also be found using implicit differentiation.

When the centre of the circle is at the origin then the equation of the tangent line becomes

$$x_1 x + y_1 y = r^2,$$

and its slope is

$$\frac{dy}{dx} = -\frac{x_1}{y_1}.$$

Properties

- The circle is the shape with the largest area for a given length of perimeter.

- The circle is a highly symmetric shape: every line through the centre forms a line of reflection symmetry and it has rotational symmetry around the centre for every angle. Its symmetry group is the orthogonal group O(2,R). The group of rotations alone is the circle group T.

- All circles are similar.

  - A circle's circumference and radius are proportional.

  - The area enclosed and the square of its radius are proportional.

  - The constants of proportionality are 2π and π, respectively.

- The circle which is centred at the origin with radius 1 is called the unit circle.

  - Thought of as a great circle of the unit sphere, it becomes the Riemannian circle.

- Through any three points, not all on the same line, there lies a unique circle. In Cartesian coordinates, it is possible to give explicit formulae for the coordinates of the centre of the circle and the radius in terms of the coordinates of the three given points.

Chord

- Chords are equidistant from the centre of a circle if and only if they are equal in length.

- The perpendicular bisector of a chord passes through the centre of a circle; equivalent statements stemming from the uniqueness of the perpendicular bisector are:

  - A perpendicular line from the centre of a circle bisects the chord.

  - The line segment through the centre bisecting a chord is perpendicular to the chord.

- If a central angle and an inscribed angle of a circle are subtended by the same chord and on the same side of the chord, then the central angle is twice the inscribed angle.

- If two angles are inscribed on the same chord and on the same side of the chord, then they are equal.

- If two angles are inscribed on the same chord and on opposite sides of the chord, then they are supplementary.

  - For a cyclic quadrilateral, the exterior angle is equal to the interior opposite angle.

- An inscribed angle subtended by a diameter is a right angle.

- The diameter is the longest chord of the circle.

- If the intersection of any two chords divides one chord into lengths a and b and divides the other chord into lengths c and d, then $ab = cd$.

- If the intersection of any two perpendicular chords divides one chord into lengths a and b and divides the other chord into lengths c and d, then $a^2 + b^2 + c^2 + d^2$ equals the square of the diameter.

- The sum of the squared lengths of any two chords intersecting at right angles at a given point is the same as that of any other two perpendicular chords intersecting at the same point, and is given by $8r^2 - 4p^2$ (where r is the circle's radius and p is the distance from the centre point to the point of intersection).

- The distance from a point on the circle to a given chord times the diameter of the circle equals the product of the distances from the point to the ends of the chord.[p.71]

Tangent

- A line drawn perpendicular to a radius through the end point of the radius lying on the circle is a tangent to the circle.

- A line drawn perpendicular to a tangent through the point of contact with a circle passes through the centre of the circle.

- Two tangents can always be drawn to a circle from any point outside the circle, and these tangents are equal in length.

- If a tangent at A and a tangent at B intersect at the exterior point P, then denoting the centre as O, the angles $\angle BOA$ and $\angle BPA$ are supplementary.

- If AD is tangent to the circle at A and if AQ is a chord of the circle, then $\angle DAQ = 1/2\,\mathrm{arc}(AQ)$.

Theorems

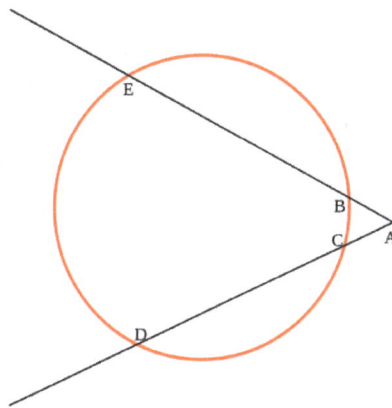

Secant-secant theorem

- The chord theorem states that if two chords, CD and EB, intersect at A, then $AC \times AD = AB \times AE$.

- If two secants, *AE* and *AD*, also cut the circle at *B* and *C* respectively, then $AC \times AD = AB \times AE$. (Corollary of the chord theorem.)

- A tangent can be considered a limiting case of a secant whose ends are coincident. If a tangent from an external point *A* meets the circle at *F* and a secant from the external point *A* meets the circle at *C* and *D* respectively, then $AF^2 = AC \times AD$. (Tangent-secant theorem.)

- The angle between a chord and the tangent at one of its endpoints is equal to one half the angle subtended at the centre of the circle, on the opposite side of the chord (Tangent Chord Angle).

- If the angle subtended by the chord at the centre is 90 degrees then $\ell = r \sqrt{2}$, where ℓ is the length of the chord and *r* is the radius of the circle.

- If two secants are inscribed in the circle as shown at right, then the measurement of angle *A* is equal to one half the difference of the measurements of the enclosed arcs (*DE* and *BC*). I.e. $2\angle CAB = \angle DOE - \angle BOC$, where *O* is the centre of the circle. This is the secant-secant theorem.

Inscribed Angles

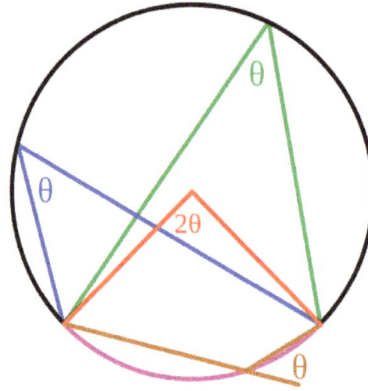

Inscribed angle theorem

An inscribed angle (examples are the blue and green angles in the figure) is exactly half the corresponding central angle (red). Hence, all inscribed angles that subtend the same arc (pink) are equal. Angles inscribed on the arc (brown) are supplementary. In particular, every inscribed angle that subtends a diameter is a right angle (since the central angle is 180 degrees).

Sagitta

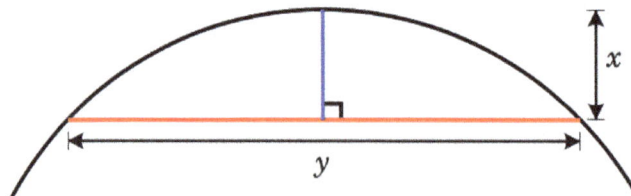

The sagitta is the vertical segment.

- The sagitta (also known as the versine) is a line segment drawn perpendicular to a chord, between the midpoint of that chord and the arc of the circle.

- Given the length y of a chord, and the length x of the sagitta, the Pythagorean theorem can be used to calculate the radius of the unique circle which will fit around the two lines:

$$r = \frac{y^2}{8x} + \frac{x}{2}.$$

Another proof of this result which relies only on two chord properties given above is as follows. Given a chord of length y and with sagitta of length x, since the sagitta intersects the midpoint of the chord, we know it is part of a diameter of the circle. Since the diameter is twice the radius, the "missing" part of the diameter is $(2r - x)$ in length. Using the fact that one part of one chord times the other part is equal to the same product taken along a chord intersecting the first chord, we find that $(2r - x)x = (y / 2)^2$. Solving for r, we find the required result.

Compass and Straightedge Constructions

There are many compass-and-straightedge constructions resulting in circles.

The simplest and most basic is the construction given the centre of the circle and a point on the circle. Place the fixed leg of the compass on the centre point, the movable leg on the point on the circle and rotate the compass.

Construct a Circle with a Given Diameter

- Construct the midpoint M of the diameter.

- Construct the circle with centre M passing through one of the endpoints of the diameter (it will also pass through the other endpoint).

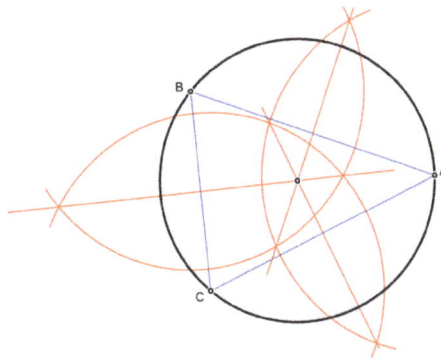

Construct a circle through points A, B and C by finding the perpendicular bisectors (red) of the sides of the triangle (blue). Only two of the three bisectors are needed to find the centre.

Construct a Circle Through 3 Noncollinear Points

- Name the points P, Q and R,

- Construct the perpendicular bisector of the segment PQ.

- Construct the perpendicular bisector of the segment PR.

- Label the point of intersection of these two perpendicular bisectors M. (They meet because the points are not collinear).

- Construct the circle with centre M passing through one of the points P, Q or R (it will also pass through the other two points).

Circle of Apollonius

Apollonius of Perga showed that a circle may also be defined as the set of points in a plane having a constant *ratio* (other than 1) of distances to two fixed foci, A and B. (The set of points where the distances are equal is the perpendicular bisector of A and B, a line.) That circle is sometimes said to be drawn *about* two points.

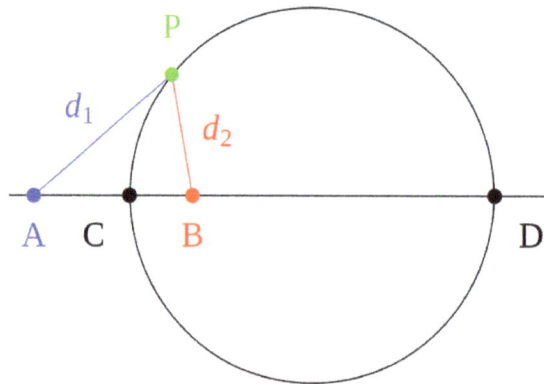

Apollonius' definition of a circle: d_1/d_2 constant

The proof is in two parts. First, one must prove that, given two foci A and B and a ratio of distances, any point P satisfying the ratio of distances must fall on a particular circle. Let C be another point, also satisfying the ratio and lying on segment AB. By the angle bisector theorem the line segment PC will bisect the interior angle APB, since the segments are similar:

$$\frac{AP}{BP} = \frac{AC}{BC}.$$

Analogously, a line segment PD through some point D on AB extended bisects the corresponding exterior angle BPQ where Q is on AP extended. Since the interior and exterior angles sum to 180 degrees, the angle CPD is exactly 90 degrees, i.e., a right angle. The set of points P such that angle CPD is a right angle forms a circle, of which CD is a diameter.

Cross-ratios

A closely related property of circles involves the geometry of the cross-ratio of points in the complex plane. If A, B, and C are as above, then the circle of Apollonius for these three points is the collection of points P for which the absolute value of the cross-ratio is equal to one:

$$|[A,B;C,P]|=1.$$

Stated another way, P is a point on the circle of Apollonius if and only if the cross-ratio $[A,B;C,P]$ is on the unit circle in the complex plane.

Generalised Circles

If C is the midpoint of the segment AB, then the collection of points P satisfying the Apollonius condition

$$\frac{|AP|}{|BP|} = \frac{|AC|}{|BC|}$$

is not a circle, but rather a line.

Thus, if A, B, and C are given distinct points in the plane, then the locus of points P satisfying the above equation is called a "generalised circle." It may either be a true circle or a line. In this sense a line is a generalised circle of infinite radius.

Circles Inscribed in or Circumscribed about Other Figures

In every triangle a unique circle, called the incircle, can be inscribed such that it is tangent to each of the three sides of the triangle.

About every triangle a unique circle, called the circumcircle, can be circumscribed such that it goes through each of the triangle's three vertices.

A tangential polygon, such as a tangential quadrilateral, is any convex polygon within which a circle can be inscribed that is tangent to each side of the polygon.

A cyclic polygon is any convex polygon about which a circle can be circumscribed, passing through each vertex. A well-studied example is the cyclic quadrilateral.

A hypocycloid is a curve that is inscribed in a given circle by tracing a fixed point on a smaller circle that rolls within and tangent to the given circle.

Circle as Limiting Case of Other Figures

The circle can be viewed as a limiting case of each of various other figures:

- A Cartesian oval is a set of points such that a weighted sum of the distances from any of its points to two fixed points (foci) is a constant. An ellipse is the case in which the weights are equal. A circle is an ellipse with an eccentricity of zero, meaning that the two foci coincide with each other as the centre of the circle. A circle is also a different special case of a Cartesian oval in which one of the weights is zero.

- A superellipse has an equation of the form $\left|\frac{x}{a}\right|^n + \left|\frac{y}{b}\right|^n = 1$ for positive a, b, and n. A supercircle has $b = a$. A circle is the special case of a supercircle in which $n = 2$.

- A Cassini oval is a set of points such that the product of the distances from any of its points to two fixed points is a constant. When the two fixed points coincide, a circle results.

- A curve of constant width is a figure whose width, defined as the perpendicular distance between two distinct parallel lines each intersecting its boundary in a single point, is the same regardless of the direction of those two parallel lines. The circle is the simplest example of this type of figure.

Squaring the Circle

Squaring the circle is the problem, proposed by ancient geometers, of constructing a square with the same area as a given circle by using only a finite number of steps with compass and straightedge.

In 1882, the task was proven to be impossible, as a consequence of the Lindemann–Weierstrass theorem which proves that pi (π) is a transcendental number, rather than an algebraic irrational number; that is, it is not the root of any polynomial with rational coefficients.

Arc (Geometry)

In Euclidean geometry, an arc (symbol: ⌒) is a closed segment of a differentiable curve. A common example in the plane (a two-dimensional manifold), is a segment of a circle called a circular arc. In space, if the arc is part of a great circle (or great ellipse), it is called a great arc.

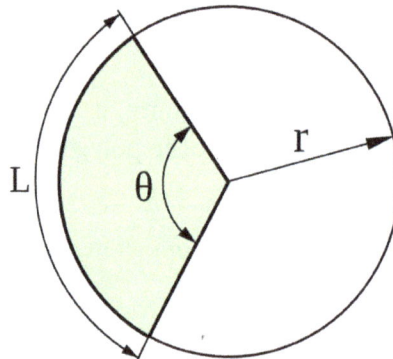

A circular sector is shaded in green. Its curved boundary of length L is a circular arc.

Every pair of distinct points on a circle determines two arcs. If the two points are not directly opposite each other, one of these arcs, the minor arc, will subtend an angle at the centre of the circle that is less than π radians (180 degrees), and the other arc, the major arc, will subtend an angle greater than π radians.

Circular Arcs

Length of an Arc of a Circle

The length (more precisely, arc length), L, of an arc of a circle with radius r and subtending an angle θ (measured in radians) with the circle center — i.e., the central angle — equals θr. This is because

$$\frac{L}{\text{circumference}} = \frac{\theta}{2\pi}.$$

Substituting in the circumference

$$\frac{L}{2\pi r} = \frac{\theta}{2\pi},$$

and, with α being the same angle measured in degrees, since $\theta = \alpha/180\pi$, the arc length equals

$$L = \frac{\alpha\pi r}{180}.$$

A practical way to determine the length of an arc in a circle is to plot two lines from the arc's endpoints to the center of the circle, measure the angle where the two lines meet the center, then solve for L by cross-multiplying the statement:

measure of angle in degrees/360° = L/circumference.

For example, if the measure of the angle is 60 degrees and the circumference is 24 inches, then

$$\frac{60}{360} = \frac{L}{24}$$
$$360L = 1440$$
$$L = 4.$$

This is so because the circumference of a circle and the degrees of a circle, of which there are always 360, are directly proportional.

Arc Sector Area

The area of the sector formed by an arc and the center of a circle (bounded by the arc and the two radii drawn to its endpoints) is

$$A = \tfrac{1}{2}r^2\theta.$$

The area A has the same proportion to the circle area as the angle θ to a full circle:

$$\frac{A}{\pi r^2} = \frac{\theta}{2\pi}.$$

We can cancel π on both sides:

$$\frac{A}{r^2} = \frac{\theta}{2}.$$

By multiplying both sides by r^2, we get the final result:

$$A = \frac{1}{2}r^2\theta.$$

Using the conversion described above, we find that the area of the sector for a central angle measured in degrees is

$$A = \frac{\alpha}{360}\pi r^2.$$

Arc Segment Area

The area of the shape bounded by the arc and the straight line between its two end points is

$$\frac{1}{2}r^2\left(\theta - \sin\theta\right).$$

To get the area of the arc segment, we need to subtract the area of the triangle, determined by the circle's center and the two end points of the arc, from the area A.

Arc Radius

Using the intersecting chords theorem (also known as power of a point or secant tangent theorem) it is possible to calculate the radius r of a circle given the height H and the width W of an arc:

Consider the chord with the same endpoints as the arc. Its perpendicular bisector is another chord, which is a diameter of the circle. The length of the first chord is W, and it is divided by the bisector into two equal halves, each with length $W/2$. The total length of the diameter is $2r$, and it is divided into two parts by the first chord. The length of one part is the height of the arc, H, and the other part is the remainder of the diameter, with length $2r - H$. Applying the intersecting chords theorem to these two chords produces

$$H(2r - H) = \left(\frac{W}{2}\right)^2,$$

whence

$$2r - H = \frac{W^2}{4H},$$

so

$$r = \frac{W^2}{8H} + \frac{H}{2}.$$

Tangent Lines to Circles

In Euclidean plane geometry, a tangent line to a circle is a line that touches the circle at exactly one point, never entering the circle's interior. Roughly speaking, it is a line through a pair of infinitely close points on the circle. Tangent lines to circles form the subject of several theorems, and play an important role in many geometrical constructions and proofs. Since the tangent line to a circle at a point P is perpendicular to the radius to that point, theorems involving tangent lines often involve radial lines and orthogonal circles.

Tangent Lines to One Circle

A tangent line t to a circle C intersects the circle at a single point T. For comparison, secant lines intersect a circle at two points, whereas another line may not intersect a circle at all. This property of tangent lines is preserved under many geometrical transformations, such as scalings, rotation, translations, inversions, and map projections. In technical language, these transformations do not change the incidence structure of the tangent line and circle, even though the line and circle may be deformed.

The radius of a circle is perpendicular to the tangent line through its endpoint on the circle's circumference. Conversely, the perpendicular to a radius through the same endpoint is a tangent line. The resulting geometrical figure of circle and tangent line has a reflection symmetry about the axis of the radius.

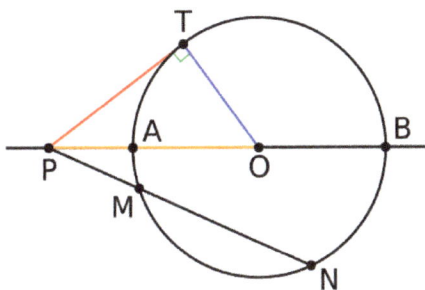

By the power-of-a-point theorem, the product of lengths PM·PN for any ray PMN equals to the square of PT, the length of the tangent line segment (red).

No tangent line can be drawn through a point within a circle, since any such line must be a secant line. However, *two* tangent lines can be drawn to a circle from a point P outside of the circle. The geometrical figure of a circle and both tangent lines likewise has a reflection symmetry about the radial axis joining P to the center point O of the circle. Thus the lengths of the segments from P to the two tangent points are equal. By the secant-tangent theorem, the square of this tangent length equals the power of the point P in the circle C. This power equals the product of distances from P to any two intersection points of the circle with a secant line passing through P.

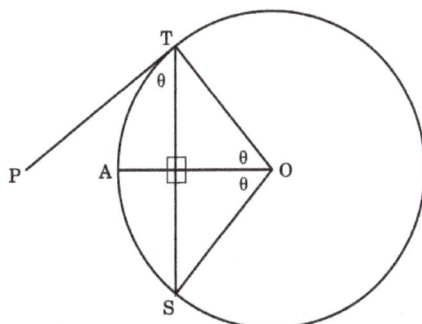

The angle θ between a chord and a tangent is half the arc belonging to the chord.

The tangent line t and the tangent point T have a conjugate relationship to one another, which has been generalized into the idea of pole points and polar lines. The same reciprocal relation exists between a point P outside the circle and the secant line joining its two points of tangency.

If a point P is exterior to a circle with center O, and if the tangent lines from P touch the circle at points T and S, then $\angle$TPS and $\angle$TOS are supplementary (sum to 180°).

If a chord TM is drawn from the tangency point T of exterior point P and $\angle$PTM $\leq$ 90° then $\angle$PTM $= (1/2)\angle$TOM.

Compass and Straightedge Constructions

It is relatively straightforward to construct a line t tangent to a circle at a point T on the circumference of the circle:

- A line a is drawn from O, the center of the circle, through the radial point T;

- The line t is the perpendicular line to a.

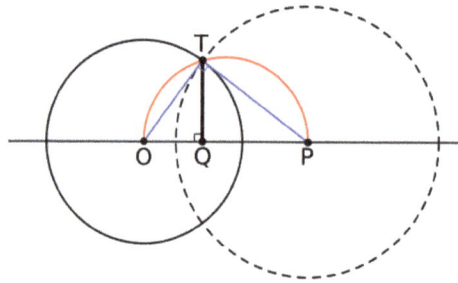

Construction of a tangent to a given circle (black) from a given exterior point (P).

Thales' theorem may be used to construct the tangent lines to a point P external to the circle C:

- A circle is drawn centered on the midpoint of the line segment OP, where O is again the center of the circle C.

- The intersection points T_1 and T_2 are the tangent points for lines passing through P, by the following argument.

The line segments OT_1 and OT_2 are radii of the circle C; since both are inscribed in a semicircle, they are perpendicular to the line segments PT_1 and PT_2, respectively. But only a tangent line is perpendicular to the radial line. Hence, the two lines from P and passing through T_1 and T_2 are tangent to the circle C.

Another method to construct the tangent lines to a point P external to the circle using only a straightedge:

- Draw any three different lines through the given point P that intersect the circle twice.

- Let $A_1, A_2, B_1, B_2, C_1, C_2$ be the six intersection points, with the same letter corresponding to the same line and the index 1 corresponding to the point closer to P.

- Let D be the point where the lines A_1B_2 and A_2B_1 intersect,

- Similarly E for the lines B_1C_2 and B_2C_1.

- Draw a line through D and E.

- This line meets the circle at two points, F and G.

- The tangents are the lines PF and PG.

Tangent quadrilateral theorem and inscribed circles

A tangential quadrilateral ABCD is a closed figure of four straight sides that are tangent to a given circle C. Equivalently, the circle C is inscribed in the quadrilateral ABCD. By the Pitot theorem, the sums of opposite sides of any such quadrilateral are equal, i.e.,

$$\overline{AB} + \overline{CD} = \overline{BC} + \overline{DA}.$$

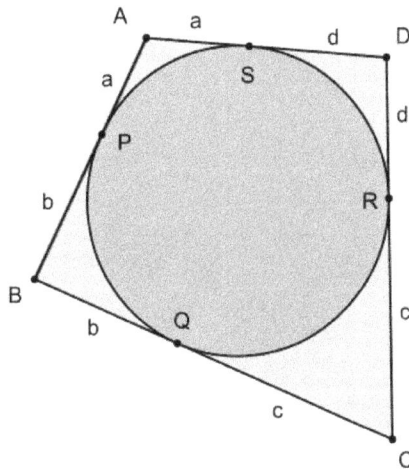

Tangential quadrilateral

This conclusion follows from the equality of the tangent segments from the four vertices of the quadrilateral. Let the tangent points be denoted as P (on segment AB), Q (on segment BC), R (on segment CD) and S (on segment DA). The symmetric tangent segments about each point of ABCD are equal, e.g., BP=BQ=b, CQ=CR=c, DR=DS=d, and AS=AP=a. But each side of the quadrilateral is composed of two such tangent segments

$$\overline{AB} + \overline{CD} = (a+b)+(c+d) = \overline{BC} + \overline{DA} = (b+c)+(d+a)$$

proving the theorem.

The converse is also true: a circle can be inscribed into every quadrilateral in which the lengths of opposite sides sum to the same value.

This theorem and its converse have various uses. For example, they show immediately that no rectangle can have an inscribed circle unless it is a square, and that every rhombus has an inscribed circle, whereas a general parallelogram does not.

Tangent Lines to Two Circles

For two circles, there are generally four distinct lines that are tangent to both (bitangent) – if the two circles are outside each other – but in degenerate cases there may be any number between zero and four bitangent lines; these are addressed below. For two of these, the external tangent lines, the circles fall on the same side of the line; for the two others, the internal tangent lines, the circles fall on opposite sides of the line. The external tangent lines intersect in the external homothetic center, whereas the internal tangent lines intersect at the internal homothetic center. Both the external and internal homothetic centers lie on the line of centers (the line connecting the centers of the two circles), closer to the center of the smaller circle: the internal center is in the segment between the two circles, while the external center is not between the points, but rather outside, on the side of the center of the smaller circle. If the two circles have equal radius, there are still four bitangents, but the external tangent lines are parallel and there is no external center in the affine plane; in the projective plane, the external homothetic center lies at the point at infinity corresponding to the slope of these lines.

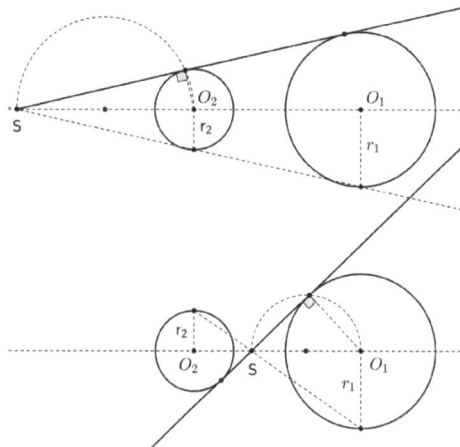

The external (above) and internal (below) homothetic center S of the two circles.

Outer Tangent

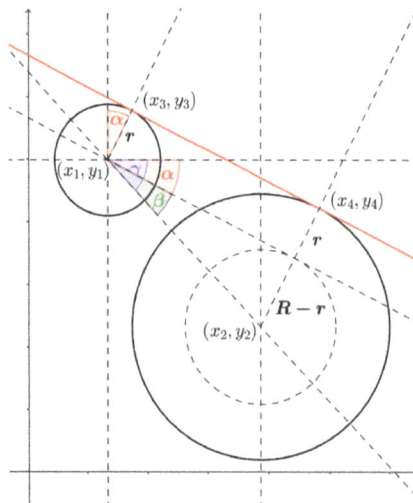

Finding outer tangent. Two circles' outer tangents.

The red line joining the points (x_3, y_3) and (x_4, y_4) is the outer tangent between the two circles. Given points (x_1, y_1), (x_2, y_2) the points (x_3, y_3), (x_3, y_3) can easily be calculated with help of the angle α :

$$x_3 = x_1 + r \cdot \cos(\tfrac{\pi}{2} - \alpha)$$
$$y_3 = y_1 + r \cdot \sin(\tfrac{\pi}{2} - \alpha)$$
$$x_4 = x_2 + R \cdot \cos(\tfrac{\pi}{2} - \alpha)$$
$$y_4 = y_2 + R \cdot \sin(\tfrac{\pi}{2} - \alpha)$$

Here R and r notate the radii of the two circles and the angle α can be computed using basic trigonometry. You have $\alpha = \gamma - \beta$ with $\gamma = \arctan\left(\frac{y_1 - y_2}{x_2 - x_1}\right)$ and $\beta = \arcsin\left(\frac{R - r}{\sqrt{(x_2 - x_1)^2 + (y_2 - y_1)^2}}\right)$..

Inner Tangent

An inner tangent is a tangent that intersects the segment joining two circles' centers. Note that the inner tangent will not be defined for cases when the two circles overlap.

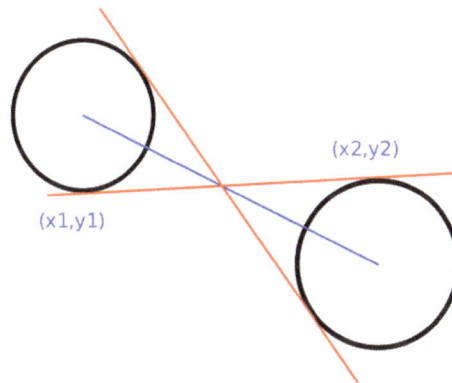

Inner tangent. The external tangent lines pass through the internal homothetic center.

Construction

The bitangent lines can be constructed either by constructing the homothetic centers, as described at that article, and then constructing the tangent lines through the homothetic center that is tangent to one circle, by one of the methods described above. The resulting line will then be tangent to the other circle as well. Alternatively, the tangent lines and tangent points can be constructed more directly, as detailed below. Note that in degenerate cases these constructions break down; to simplify exposition this is not discussed in this section, but a form of the construction can work in limit cases (e.g., two circles tangent at one point).

Synthetic Geometry

Let O_1 and O_2 be the centers of the two circles, C_1 and C_2 and let r_1 and r_2 be their radii, with $r_1 > r_2$; in other words, circle C_1 is defined as the larger of the two circles. Two different methods may be used to construct the external and internal tangent lines.

External tangents

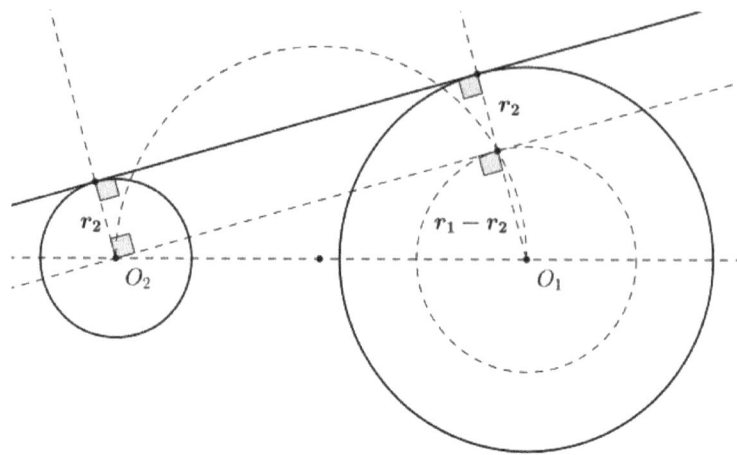

Construction of the outer tangent

A new circle C_3 of radius $r_1 - r_2$ is drawn centered on O_1. Using the method above, two lines are drawn from O_2 that are tangent to this new circle. These lines are parallel to the desired tangent lines, because the situation corresponds to shrinking both circles C_1 and C_2 by a constant amount, r_2, which shrinks C_2 to a point. Two radial lines may be drawn from the center O_1 through the tangent points on C_3; these intersect C_1 at the desired tangent points. The desired external tangent lines are the lines perpendicular to these radial lines at those tangent points, which may be constructed as described above.

Internal tangents

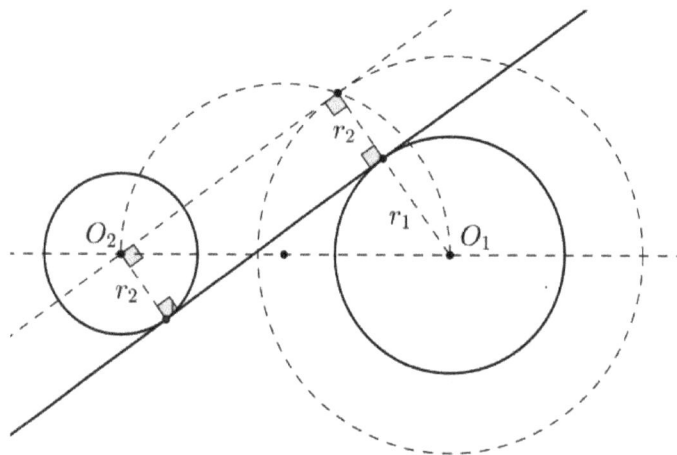

Construction of the immer tangent

A new circle C_3 of radius $r_1 + r_2$ is drawn centered on O_1. Using the method above, two lines are drawn from O_2 that are tangent to this new circle. These lines are parallel to the desired tangent lines, because the situation corresponds to shrinking C_2 to a point while expanding C_1 by a constant amount, r_2. Two radial lines may be drawn from the center O_1 through the tangent points on C_3; these intersect C_1 at the desired tangent points. The desired internal tangent lines are the lines perpendicular to these radial lines at those tangent points, which may be constructed as described above.

Analytic Geometry

Let the circles have centres $c_1 = (x_1, y_1)$ and $c_2 = (x_2, y_2)$ with radius r_1 and r_2 respectively. Expressing a line by the equation $ax + by + c = 0$, with the normalization $a^2 + b^2 = 1$, then a bitangent line satisfies:

$$ax_1 + by_1 + c = r_1 \text{ and}$$

$$ax_2 + by_2 + c = r_2.$$

Solving for (a, b, c) by subtracting the first from the second yields

$$a\Delta x + b\Delta y = \Delta r$$

where $\Delta x = x_2 - x_1$, $\Delta y = y_2 - y_1$ and $\Delta r = r_2 - r_1$.

If $d = \sqrt{(\Delta x)^2 + (\Delta y)^2}$ is the distance from c_1 to c_2 we can normalize by $X = \Delta x / d$, $Y = \Delta y / d$ and $R = \Delta r / d$ to simplify equations, yielding the equations $aX + bY = R$ and $a^2 + b^2 = 1$, solve these to get two solutions ($k = \pm 1$) for the two external tangent lines:

$$a = RX - kY\sqrt{(1 - R^2)}$$

$$b = RY + kX\sqrt{(1 - R^2)}$$

$$c = r_1 - (ax_1 + by_1)$$

Geometrically this corresponds to computing the angle formed by the tangent lines and the line of centers, and then using that to rotate the equation for the line of centers to yield an equation for the tangent line. The angle is computed by computing the trigonometric functions of a right triangle whose vertices are the (external) homothetic center, a center of a circle, and a tangent point; the hypotenuse lies on the tangent line, the radius is opposite the angle, and the adjacent side lies on the line of centers.

(X, Y) is the unit vector pointing from c_1 to c_2, while R is $\cos\theta$ where θ is the angle between the line of centers and a tangent line. $\sin\theta$ is then $\pm\sqrt{1 - R^2}$ (depending on the sign of θ, equivalently the direction of rotation), and the above equations are rotation of (X, Y) by $\pm\theta$, using the rotation matrix:

$$\begin{pmatrix} R & \mp\sqrt{1 - R^2} \\ \pm\sqrt{1 - R^2} & R \end{pmatrix}$$

$k = 1$ is the tangent line to the right of the circles looking from c_1 to c_2.

$k = -1$ is the tangent line to the right of the circles looking from c_2 to c_1.

The above assumes each circle has positive radius. If r_1 is positive and r_2 negative then c_1 will lie to the left of each line and c_2 to the right, and the two tangent lines will cross. In this way all four solutions are obtained. Switching signs of both radii switches $k = 1$ and $k = -1$.

Vectors

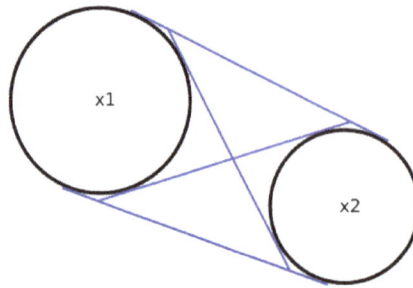

Finding outer tangent. Circle tangents.

In general the points of tangency t_1 and t_2 for the four lines tangent to two circles with centers v_1 and v_2 and radii r_1 and r_2 are given by solving the simultaneous equations:

$$
\begin{aligned}
(t_2 - v_2) \cdot (t_2 - t_1) &= 0 \\
(t_1 - v_1) \cdot (t_2 - t_1) &= 0 \\
(t_1 - v_1) \cdot (t_1 - v_1) &= r_1^2 \\
(t_2 - v_2) \cdot (t_2 - v_2) &= r_2^2
\end{aligned}
$$

These equations express that the tangent line, which is parallel to $t_2 - t_1$, is perpendicular to the radii, and that the tangent points lie on their respective circles.

These are four quadratic equations in two two-dimensional vector variables, and in general position will have four pairs of solutions.

Degenerate Cases

Two distinct circles may have between zero and four bitangent lines, depending on configuration; these can be classified in terms of the distance between the centers and the radii. If counted with multiplicity (counting a common tangent twice) there are zero, two, or four bitangent lines. Bitangent lines can also be generalized to circles with negative or zero radius. The degenerate cases and the multiplicities can also be understood in terms of limits of other configurations – e.g., a limit of two circles that almost touch, and moving one so that they touch, or a circle with small radius shrinking to a circle of zero radius.

- If the circles are outside each other ($d > r_1 + r_2$), which is general position, there are four bitangents.

- If they touch externally at one point ($d = r_1 + r_2$) – have one point of external tangency – then they have two external bitangents and one internal bitangent, namely the common tangent line. This common tangent line has multiplicity two, as it separates the circles (one on the left, one on the right) for either orientation (direction).

- If the circles intersect in two points ($|r_1 - r_2| < d < r_1 + r_2$), then they have no internal bitangents and two external bitangents (they cannot be separated, because they intersect, hence no internal bitangents).

- If the circles touch internally at one point ($d = |r_1 - r_2|$) – have one point of internal tangency – then they have no internal bitangents and one external bitangent, namely the common tangent line, which has multiplicity two, as above.

- If one circle is completely inside the other ($d < |r_1 - r_2|$) then they have no bitangents, as a tangent line to the outer circle does not intersect the inner circle, or conversely a tangent line to the inner circle is a secant line to the outer circle.

Finally, if the two circles are identical, any tangent to the circle is a common tangent and hence (external) bitangent, so there is a circle's worth of bitangents.

Further, the notion of bitangent lines can be extended to circles with negative radius (the same locus of points, $x^2 + y^2 = (-r)^2$, but considered "inside out"), in which case if the radii have opposite sign (one circle has negative radius and the other has positive radius) the external and internal homothetic centers and external and internal bitangents are switched, while if the radii have the same sign (both positive radii or both negative radii) "external" and "internal" have the same usual sense (switching one sign switches them, so switching both switches them back).

Bitangent lines can also be defined when one or both of the circles has radius zero. In this case the circle with radius zero is a double point, and thus any line passing through it intersects the point with multiplicity two, hence is "tangent". If one circle has radius zero, a bitangent line is simply a line tangent to the circle and passing through the point, and is counted with multiplicity two. If both circles have radius zero, then the bitangent line is the line they define, and is counted with multiplicity four.

Note that in these degenerate cases the external and internal homothetic center do generally still exist (the external center is at infinity if the radii are equal), except if the circles coincide, in which case the external center is not defined, or if both circles have radius zero, in which case the internal center is not defined.

Applications

Belt Problem

The internal and external tangent lines are useful in solving the *belt problem*, which is to calculate the length of a belt or rope needed to fit snugly over two pulleys. If the belt is considered to be a mathematical line of negligible thickness, and if both pulleys are assumed to lie in exactly the same plane, the problem devolves to summing the lengths of the relevant tangent line segments with the lengths of circular arcs subtended by the belt. If the belt is wrapped about the wheels so as to cross, the interior tangent line segments are relevant. Conversely, if the belt is wrapped exteriorly around the pulleys, the exterior tangent line segments are relevant; this case is sometimes called the *pulley problem*.

Tangent Lines to Three Circles: Monge's Theorem

For three circles denoted by C_1, C_2, and C_3, there are three pairs of circles (C_1C_2, C_2C_3, and C_1C_3). Since each pair of circles has two homothetic centers, there are six homothetic centers altogether. Gaspard Monge showed in the early 19th century that these six points lie on four lines, each line having three collinear points.

Problem of Apollonius

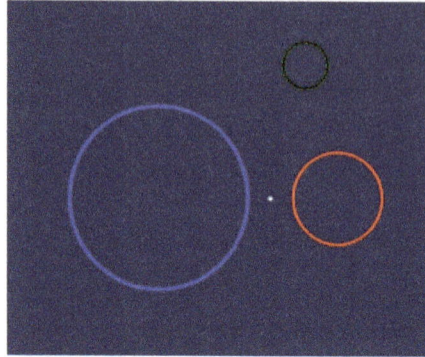

Animation showing the inversive transformation of an Apollonius problem. The blue and red circles swell to tangency, and are inverted in the grey circle, producing two straight lines. The yellow solutions are found by sliding a circle between them until it touches the transformed green circle from within or without.

Many special cases of Apollonius's problem involve finding a circle that is tangent to one or more lines. The simplest of these is to construct circles that are tangent to three given lines (the LLL problem). To solve this problem, the center of any such circle must lie on an angle bisector of any pair of the lines; there are two angle-bisecting lines for every intersection of two lines. The intersections of these angle bisectors give the centers of solution circles. There are four such circles in general, the inscribed circle of the triangle formed by the intersection of the three lines, and the three exscribed circles.

A general Apollonius problem can be transformed into the simpler problem of circle tangent to one circle and two parallel lines (itself a special case of the LLC special case). To accomplish this, it suffices to scale two of the three given circles until they just touch, i.e., are tangent. An inversion in their tangent point with respect to a circle of appropriate radius transforms the two touching given circles into two parallel lines, and the third given circle into another circle. Thus, the solutions may be found by sliding a circle of constant radius between two parallel lines until it contacts the transformed third circle. Re-inversion produces the corresponding solutions to the original problem.

Generalizations

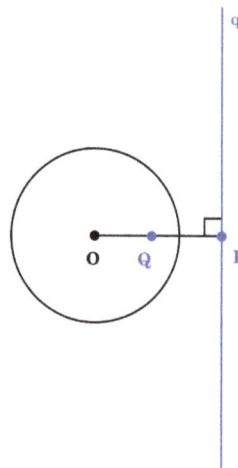

The concept of a tangent line and tangent point can be generalized to a pole point **Q** and its corresponding polar line q. The points **P** and **Q** are inverses of each other with respect to the circle.

The concept of a tangent line to one or more circles can be generalized in several ways. First, the conjugate relationship between tangent points and tangent lines can be generalized to pole points and polar lines, in which the pole points may be anywhere, not only on the circumference of the circle. Second, the union of two circles is a special (reducible) case of a quartic plane curve, and the external and internal tangent lines are the bitangents to this quartic curve. A generic quartic curve has 28 bitangents.

A third generalization considers tangent circles, rather than tangent lines; a tangent line can be considered as a tangent circle of infinite radius. In particular, the external tangent lines to two circles are limiting cases of a family of circles which are internally or externally tangent to both circles, while the internal tangent lines are limiting cases of a family of circles which are internally tangent to one and externally tangent to the other of the two circles.

In Möbius or inversive geometry, lines are viewed as circles through a point "at infinity" and for any line and any circle, there is a Möbius transformation which maps one to the other. In Möbius geometry, tangency between a line and a circle becomes a special case of tangency between two circles. This equivalence is extended further in Lie sphere geometry.

References

- Katz, Victor J. (1998), A History of Mathematics / An Introduction (2nd ed.), Addison Wesley Longman, p. 108, ISBN 978-0-321-01618-8

An Integrated Study of Quadrilateral

A quadrilateral is any figure that has four sides and four corners. They can be simple or complex and simple quadrilaterals can further be divided into convex or concave. This section will provide an integrated study of quadrilateral.

Quadrilateral

In Euclidean plane geometry, a quadrilateral is a polygon with four edges (or sides) and four vertices or corners. Sometimes, the term quadrangle is used, by analogy with triangle, and sometimes tetragon for consistency with pentagon (5-sided), hexagon (6-sided) and so on.

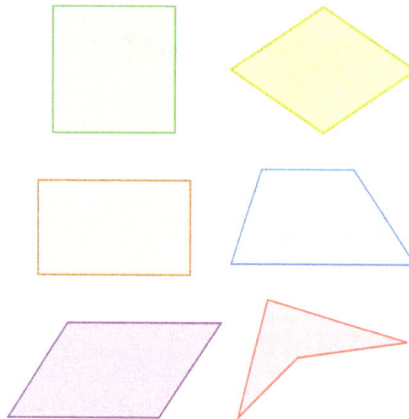

Some types of quadrilaterals

The origin of the word "quadrilateral" is the two Latin words *quadri*, a variant of four, and *latus*, meaning "side".

Quadrilaterals are simple (not self-intersecting) or complex (self-intersecting), also called crossed. Simple quadrilaterals are either convex or concave.

The interior angles of a simple (and planar) quadrilateral *ABCD* add up to 360 degrees of arc, that is

$$\angle A + \angle B + \angle C + \angle D = 360^{\circ}.$$

This is a special case of the *n*-gon interior angle sum formula $(n - 2) \times 180°$.

All non-self-crossing quadrilaterals tile the plane by repeated rotation around the midpoints of their edges.

Simple Quadrilaterals

Any quadrilateral that is not self-intersecting is a simple quadrilateral.

Convex Quadrilaterals

In a convex quadrilateral, all interior angles are less than 180° and the two diagonals both lie inside the quadrilateral.

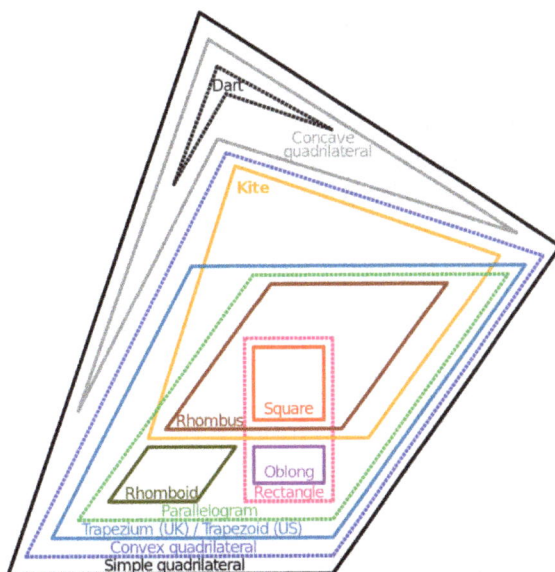

Euler diagram of some types of simple quadrilaterals. (UK) denotes British English
and (US) denotes American English.

- Irregular quadrilateral (British English) or trapezium (North American English): no sides are parallel. (In British English this was once called a *trapezoid*.)

- Trapezium (UK) or trapezoid (US): at least one pair of opposite sides are parallel.

- Isosceles trapezium (UK) or isosceles trapezoid (US): one pair of opposite sides are parallel and the base angles are equal in measure. Alternative definitions are a quadrilateral with an axis of symmetry bisecting one pair of opposite sides, or a trapezoid with diagonals of equal length.

- Parallelogram: a quadrilateral with two pairs of parallel sides. Equivalent conditions are that opposite sides are of equal length; that opposite angles are equal; or that the diagonals bisect each other. Parallelograms also include the square, rectangle, rhombus and rhomboid.

- Rhombus or rhomb: all four sides are of equal length. An equivalent condition is that the diagonals perpendicularly bisect each other. Informally: "a pushed-over square" (but strictly including a square too).

- Rhomboid: a parallelogram in which adjacent sides are of unequal lengths and angles are oblique (not right angles). A parallelogram which is not a rhombus. Informally: "a pushed-over oblong" (but strictly including an oblong too).

- Rectangle: all four angles are right angles. An equivalent condition is that the diagonals bisect each other and are equal in length. Informally: "a box or oblong" (including a square).

- Square (regular quadrilateral): all four sides are of equal length (equilateral), and all four angles are right angles. An equivalent condition is that opposite sides are parallel (a square is a parallelogram), that the diagonals perpendicularly bisect each other, and are of equal length. A quadrilateral is a square if and only if it is both a rhombus and a rectangle (four equal sides and four equal angles).

- Oblong: a term sometimes used to denote a rectangle which has unequal adjacent sides (i.e. a rectangle that is not a square).

- Kite: two pairs of adjacent sides are of equal length. This implies that one diagonal divides the kite into congruent triangles, and so the angles between the two pairs of equal sides are equal in measure. It also implies that the diagonals are perpendicular.

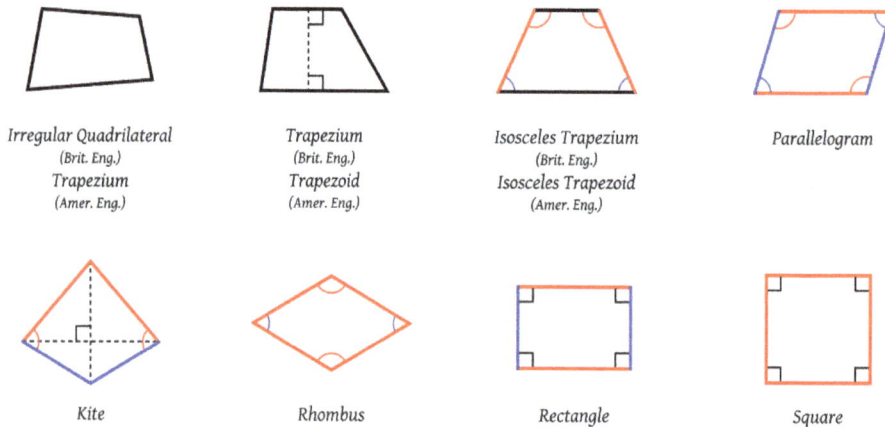

| Irregular Quadrilateral (Brit. Eng.) Trapezium (Amer. Eng.) | Trapezium (Brit. Eng.) Trapezoid (Amer. Eng.) | Isosceles Trapezium (Brit. Eng.) Isosceles Trapezoid (Amer. Eng.) | Parallelogram |

| Kite | Rhombus | Rectangle | Square |

- Tangential quadrilateral: the four sides are tangents to an inscribed circle. A convex quadrilateral is tangential if and only if opposite sides have equal sums.

- Tangential trapezoid: a trapezoid where the four sides are tangents to an inscribed circle.

- Cyclic quadrilateral: the four vertices lie on a circumscribed circle. A convex quadrilateral is cyclic if and only if opposite angles sum to 180°.

- Right kite: a kite with two opposite right angles. It is a type of cyclic quadrilateral.

- Bicentric quadrilateral: it is both tangential and cyclic.

- Orthodiagonal quadrilateral: the diagonals cross at right angles.

- Equidiagonal quadrilateral: the diagonals are of equal length.

- Ex-tangential quadrilateral: the four extensions of the sides are tangent to an excircle.

- An *equilic quadrilateral* has two opposite equal sides that, when extended, meet at 60°.

- A *Watt quadrilateral* is a quadrilateral with a pair of opposite sides of equal length.

- A *quadric quadrilateral* is a convex quadrilateral whose four vertices all lie on the perimeter of a square.

- A *diametric quadrilateral* is a cyclic quadrilateral having one of its sides as a diameter of the circumcircle.

Concave Quadrilaterals

In a concave quadrilateral, one interior angle is bigger than 180° and one of the two diagonals lies outside the quadrilateral.

- A *dart* (or arrowhead) is a concave quadrilateral with bilateral symmetry like a kite, but one interior angle is reflex.

Complex Quadrilaterals

A self-intersecting quadrilateral is called variously a cross-quadrilateral, crossed quadrilateral, butterfly quadrilateral or bow-tie quadrilateral. In a crossed quadrilateral, the four "interior" angles on either side of the crossing (two acute and two reflex, all on the left or all on the right as the figure is traced out) add up to 720°.

An antiparallelogram

- Antiparallelogram: a crossed quadrilaterals in which (like a parallelogram) each pair of nonadjacent sides have equal lengths.

- Crossed rectangle: an antiparallelogram whose sides are two opposite sides and the two diagonals of a rectangle, hence having one pair of opposite sides parallel.

- Crossed square: a special case of a crossed rectangle where two of the sides intersect at right angles.

Special Line Segments

The two diagonals of a convex quadrilateral are the line segments that connect opposite vertices.

The two bimedians of a convex quadrilateral are the line segments that connect the midpoints of opposite sides. They intersect at the "vertex centroid" of the quadrilateral.

The four maltitudes of a convex quadrilateral are the perpendiculars to a side through the midpoint of the opposite side.

Area of a Convex Quadrilateral

There are various general formulas for the area K of a convex quadrilateral $ABCD$ with sides $a = AB$, $b = BC$, $c = CD$ and $d = DA$.

Trigonometric Formulas

The area can be expressed in trigonometric terms as

$$K = \tfrac{1}{2}\, pq \cdot \sin\theta,$$

where the lengths of the diagonals are p and q and the angle between them is θ. In the case of an orthodiagonal quadrilateral (e.g. rhombus, square, and kite), this formula reduces to $K = \tfrac{1}{2} pq$ since θ is 90°.

The area can be also expressed in terms of bimedians as

$$K = mn \cdot \sin\varphi,$$

where the lengths of the bimedians are m and n and the angle between them is φ.

Bretschneider's formula expresses the area in terms of the sides and two opposite angles:

$$K \qquad = \sqrt{(s-a)(s-b)(s-c)(s-d) - \frac{1}{2}abcd\,[1 + \cos(A+C)]}$$

$$= \sqrt{(s-a)(s-b)(s-c)(s-d) - abcd[\cos^2(\frac{A+C}{2})]}$$

where the sides in sequence are a, b, c, d, where s is the semiperimeter, and A and C are two (in fact, any two) opposite angles. This reduces to Brahmagupta's formula for the area of a cyclic quadrilateral when $A+C = 180°$.

Another area formula in terms of the sides and angles, with angle C being between sides b and c, and A being between sides a and d, is

$$K = \tfrac{1}{2}ad \cdot \sin A + \tfrac{1}{2}bc \cdot \sin C.$$

In the case of a cyclic quadrilateral, the latter formula becomes $K = \dfrac{1}{2}(ad + bc)\sin A$.

In a parallelogram, where both pairs of opposite sides and angles are equal, this formula reduces to $K = ab \cdot \sin A$.

Alternatively, we can write the area in terms of the sides and the intersection angle θ of the diagonals, so long as this angle is not 90°:

$$K = \frac{|\tan\theta|}{4} \cdot |a^2 + c^2 - b^2 - d^2|.$$

In the case of a parallelogram, the latter formula becomes $K = \dfrac{1}{2}|\tan\theta| \cdot |a^2 - b^2|$.

Another area formula including the sides a, b, c, d is

$$K = \tfrac{1}{4}\sqrt{(2(a^2+c^2)-4x^2)(2(b^2+d^2)-4x^2)}\sin\varphi$$

where x is the distance between the midpoints of the diagonals and φ is the angle between the bimedians.

The last trigonometric area formula including the sides a, b, c, d and the angle α between a and b is:

$$K = \frac{1}{2}ab\cdot\sin\alpha + \frac{1}{4}\sqrt{4c^2d^2 - (c^2+d^2-a^2-b^2+2ab\cdot\cos\alpha)^2},$$

which can also be used for the area of a concave quadrilateral (having the concave part opposite to angle α) just changing the first sign + to - .

Non-trigonometric Formulas

The following two formulas express the area in terms of the sides a, b, c, d, the semiperimeter s, and the diagonals p, q:

$$K = \sqrt{(s-a)(s-b)(s-c)(s-d) - \tfrac{1}{4}(ac+bd+pq)(ac+bd-pq)},$$

$$K = \tfrac{1}{4}\sqrt{4p^2q^2 - \left(a^2+c^2-b^2-d^2\right)^2}.$$

The first reduces to Brahmagupta's formula in the cyclic quadrilateral case, since then $pq = ac + bd$.

The area can also be expressed in terms of the bimedians m, n and the diagonals p, q:

$$K = \tfrac{1}{2}\sqrt{(m+n+p)(m+n-p)(m+n+q)(m+n-q)},$$

$$K = \tfrac{1}{2}\sqrt{p^2q^2 - (m^2-n^2)^2}.$$

In fact, any three of the four values m, n, p, and q suffice for determination of the area, since in any quadrilateral the four values are related by $p^2 + q^2 = 2(m^2+n^2)$. The corresponding expressions are:

$$K = \tfrac{1}{2}\sqrt{[(m+n)^2 - p^2]\cdot[p^2 - (m-n)^2]},$$

if the lengths of two bimedians and one diagonal are given, and

$$K = \tfrac{1}{2}\sqrt{[(m+n)^2 - p^2]\cdot[p^2 - (m-n)^2]},$$

if the lengths of two diagonals and one bimedian are given.

Vector Formulas

The area of a quadrilateral $ABCD$ can be calculated using vectors. Let vectors AC and BD form the diagonals from A to C and from B to D. The area of the quadrilateral is then

$$K = \tfrac{1}{2} | \mathbf{AC} \times \mathbf{BD} |,$$

which is half the magnitude of the cross product of vectors AC and BD. In two-dimensional Euclidean space, expressing vector AC as a free vector in Cartesian space equal to $(\boldsymbol{x}_1, \boldsymbol{y}_1)$ and BD as $(\boldsymbol{x}_2, \boldsymbol{y}_2)$, this can be rewritten as:

$$K = \tfrac{1}{2} | x_1 y_2 - x_2 y_1 |.$$

Diagonals

Properties of the Diagonals in Some Quadrilaterals

In the following table it is listed if the diagonals in some of the most basic quadrilaterals bisect each other, if their diagonals are perpendicular, and if their diagonals have equal length. The list applies to the most general cases, and excludes named subsets.

Quadrilateral	Bisecting diagonals	Perpendicular diagonals	Equal diagonals
Trapezoid	No	*See note 1*	No
Isosceles trapezoid	No	*See note 1*	Yes
Parallelogram	Yes	No	No
Kite	*See note 2*	Yes	*See note 2*
Rectangle	Yes	No	Yes
Rhombus	Yes	Yes	No
Square	Yes	Yes	Yes

Note 1: The most general trapezoids and isosceles trapezoids do not have perpendicular diagonals, but there are infinite numbers of (non-similar) trapezoids and isosceles trapezoids that do have perpendicular diagonals and are not any other named quadrilateral.

Note 2: In a kite, one diagonal bisects the other. The most general kite has unequal diagonals, but there is an infinite number of (non-similar) kites in which the diagonals are equal in length (and the kites are not any other named quadrilateral).

Lengths of the Diagonals

The lengths of the diagonals in a convex quadrilateral $ABCD$ can be calculated using the law of cosines on each triangle formed by one diagonal and two sides of the quadrilateral. Thus

$$p = \sqrt{a^2 + b^2 - 2ab \cos B} = \sqrt{c^2 + d^2 - 2cd \cos D}$$

and

$$q = \sqrt{a^2 + d^2 - 2ad \cos A} = \sqrt{b^2 + c^2 - 2bc \cos C}.$$

Other, more symmetric formulas for the lengths of the diagonals, are

$$p = \sqrt{\frac{(ac+bd)(ad+bc)-2abcd(\cos B+\cos D)}{ab+cd}}$$

and

$$q = \sqrt{\frac{(ab+cd)(ac+bd)-2abcd(\cos A+\cos C)}{ad+bc}}.$$

Generalizations of the Parallelogram Law and Ptolemy's Theorem

In any convex quadrilateral $ABCD$, the sum of the squares of the four sides is equal to the sum of the squares of the two diagonals plus four times the square of the line segment connecting the midpoints of the diagonals. Thus

$$a^2+b^2+c^2+d^2 = p^2+q^2+4x^2$$

where x is the distance between the midpoints of the diagonals. This is sometimes known as *Euler's quadrilateral theorem* and is a generalization of the parallelogram law.

The German mathematician Carl Anton Bretschneider derived in 1842 the following generalization of Ptolemy's theorem, regarding the product of the diagonals in a convex quadrilateral

$$p^2q^2 = a^2c^2+b^2d^2-2abcd\cos(A+C).$$

This relation can be considered to be a law of cosines for a quadrilateral. In a cyclic quadrilateral, where $A + C = 180°$, it reduces to $pq = ac + bd$. Since $\cos(A + C) \geq -1$, it also gives a proof of Ptolemy's inequality.

Other Metric Relations

If X and Y are the feet of the normals from B and D to the diagonal $AC = p$ in a convex quadrilateral $ABCD$ with sides $a = AB$, $b = BC$, $c = CD$, $d = DA$, then

$$XY = \frac{|a^2+c^2-b^2-d^2|}{2p}.$$

In a convex quadrilateral $ABCD$ with sides $a = AB$, $b = BC$, $c = CD$, $d = DA$, and where the diagonals intersect at E,

$$efgh(a+c+b+d)(a+c-b-d) = (agh+cef+beh+dfg)(agh+cef-beh-dfg)$$

where $e = AE$, $f = BE$, $g = CE$, and $h = DE$.

The shape and size of a convex quadrilateral are fully determined by the lengths of its sides in sequence and of one diagonal between two specified vertices. The two diagonals p, q and the four side lengths a, b, c, d of a quadrilateral are related by the Cayley-Menger determinant, as follows:

$$\det \begin{bmatrix} 0 & a^2 & p^2 & d^2 & 1 \\ a^2 & 0 & b^2 & q^2 & 1 \\ p^2 & b^2 & 0 & c^2 & 1 \\ d^2 & q^2 & c^2 & 0 & 1 \\ 1 & 1 & 1 & 1 & 0 \end{bmatrix} = 0.$$

Angle Bisectors

The internal angle bisectors of a convex quadrilateral either form a cyclic quadrilateral (that is, the four intersection points of adjacent angle bisectors are concyclic) or they are concurrent. In the latter case the quadrilateral is a tangential quadrilateral.

In quadrilateral $ABCD$, if the angle bisectors of A and C meet on diagonal BD, then the angle bisectors of B and D meet on diagonal AC.

Bimedians

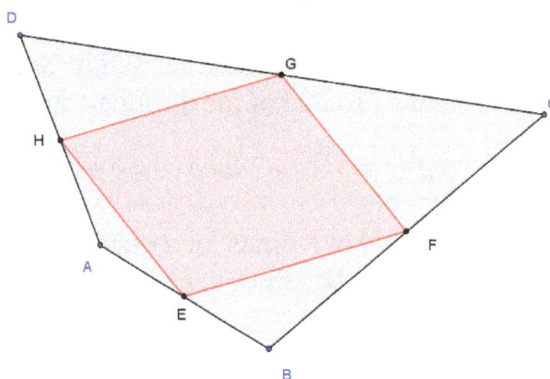

The Varignon parallelogram $EFGH$

The bimedians of a quadrilateral are the line segments connecting the midpoints of the opposite sides. The intersection of the bimedians is the centroid of the vertices of the quadrilateral.

The midpoints of the sides of any quadrilateral (convex, concave or crossed) are the vertices of a parallelogram called the Varignon parallelogram. It has the following properties:

- Each pair of opposite sides of the Varignon parallelogram are parallel to a diagonal in the original quadrilateral.

- A side of the Varignon parallelogram is half as long as the diagonal in the original quadrilateral it is parallel to.

- The area of the Varignon parallelogram equals half the area of the original quadrilateral. This is true in convex, concave and crossed quadrilaterals provided the area of the latter is defined to be the difference of the areas of the two triangles it is composed of.

- The perimeter of the Varignon parallelogram equals the sum of the diagonals of the original quadrilateral.

- The diagonals of the Varignon parallelogram are the bimedians of the original quadrilateral.

The two bimedians in a quadrilateral and the line segment joining the midpoints of the diagonals in that quadrilateral are concurrent and are all bisected by their point of intersection.

In a convex quadrilateral with sides a, b, c and d, the length of the bimedian that connects the midpoints of the sides a and c is

$$m = \tfrac{1}{2}\sqrt{-a^2 + b^2 - c^2 + d^2 + p^2 + q^2}$$

where p and q are the length of the diagonals. The length of the bimedian that connects the midpoints of the sides b and d is

$$n = \tfrac{1}{2}\sqrt{a^2 - b^2 + c^2 - d^2 + p^2 + q^2}.$$

Hence

$$p^2 + q^2 = 2(m^2 + n^2).$$

This is also a corollary to the parallelogram law applied in the Varignon parallelogram.

The lengths of the bimedians can also be expressed in terms of two opposite sides and the distance x between the midpoints of the diagonals. This is possible when using Euler's quadrilateral theorem in the above formulas. Whence

$$m = \tfrac{1}{2}\sqrt{2(b^2 + d^2) - 4x^2}$$

and

$$n = \tfrac{1}{2}\sqrt{2(a^2 + c^2) - 4x^2}.$$

Note that the two opposite sides in these formulas are not the two that the bimedian connects.

In a convex quadrilateral, there is the following dual connection between the bimedians and the diagonals:

- The two bimedians have equal length if and only if the two diagonals are perpendicular.

- The two bimedians are perpendicular if and only if the two diagonals have equal length.

Trigonometric identities

The four angles of a simple quadrilateral $ABCD$ satisfy the following identities:

$$A + \sin B + \sin C + \sin D = 4\sin\frac{A+B}{2}\sin\frac{A+C}{2}\sin\frac{A+D}{2}$$

and

$$\frac{\tan A \tan B - \tan C \tan D}{\tan A \tan C - \tan B \tan D} = \frac{\tan(A+C)}{\tan(A+B)}.$$

Also,

$$\frac{A + \tan B + \tan C + \tan D}{\cot A + \cot B + \cot C + \cot D} = \tan A \tan B \tan C \tan D.$$

In the last two formulas, no angle is allowed to be a right angle, since $\tan 90°$ is not defined.

Inequalities

Area

If a convex quadrilateral has the consecutive sides a, b, c, d and the diagonals p, q, then its area K satisfies

$K \le \frac{1}{4}(a+c)(b+d)$ with equality only for a rectangle.

$K \le \frac{1}{4}(a^2 + b^2 + c^2 + d^2)$ with equality only for a square.

$K \le \frac{1}{4}(p^2 + q^2)$ with equality only if the diagonals are perpendicular and equal.

$K \le \frac{1}{2}\sqrt{(a^2 + c^2)(b^2 + d^2)}$ with equality only for a rectangle.

From Bretschneider's formula it directly follows that the area of a quadrilateral satisfies

$$K \le \sqrt{(s-a)(s-b)(s-c)(s-d)}$$

with equality if and only if the quadrilateral is cyclic or degenerate such that one side is equal to the sum of the other three (it has collapsed into a line segment, so the area is zero).

The area of any quadrilateral also satisfies the inequality

$$K \le \frac{1}{2}\sqrt[3]{(ab+cd)(ac+bd)(ad+bc)}.$$

Denoting the perimeter as L, we have

$$K \le \frac{1}{16}L^2,$$

with equality only in the case of a square.

The area of a convex quadrilateral also satisfies

$$K \le \frac{1}{2}pq$$

for diagonal lengths p and q, with equality if and only if the diagonals are perpendicular.

Diagonals and Bimedians

A corollary to Euler's quadrilateral theorem is the inequality

$$a^2 + b^2 + c^2 + d^2 \geq p^2 + q^2$$

where equality holds if and only if the quadrilateral is a parallelogram.

Euler also generalized Ptolemy's theorem, which is an equality in a cyclic quadrilateral, into an inequality for a convex quadrilateral. It states that

$$pq \leq ac + bd$$

where there is equality if and only if the quadrilateral is cyclic. This is often called Ptolemy's inequality.

In any convex quadrilateral the bimedians m, n and the diagonals p, q are related by the inequality

$$pq \leq m^2 + n^2,$$

with equality holding if and only if the diagonals are equal. This follows directly from the quadrilateral identity $m^2 + n^2 = \frac{1}{2}(p^2 + q^2)$.

Sides

The sides a, b, c, and d of any quadrilateral satisfy

$$a^2 + b^2 + c^2 > \frac{d^2}{3}$$

and

$$a^4 + b^4 + c^4 \geq \frac{d^4}{27}.$$

Maximum and Minimum Properties

Among all quadrilaterals with a given perimeter, the one with the largest area is the square. This is called the *isoperimetric theorem for quadrilaterals*. It is a direct consequence of the area inequality

$$K \leq \tfrac{1}{16} L^2$$

where K is the area of a convex quadrilateral with perimeter L. Equality holds if and only if the quadrilateral is a square. The dual theorem states that of all quadrilaterals with a given area, the square has the shortest perimeter.

The quadrilateral with given side lengths that has the maximum area is the cyclic quadrilateral.

Of all convex quadrilaterals with given diagonals, the orthodiagonal quadrilateral has the largest area. This is a direct consequence of the fact that the area of a convex quadrilateral satisfies

$$K = \tfrac{1}{2} pq \sin \theta \le \tfrac{1}{2} pq,$$

where θ is the angle between the diagonals p and q. Equality holds if and only if $\theta = 90°$.

If P is an interior point in a convex quadrilateral $ABCD$, then

$$AP + BP + CP + DP \ge AC + BD.$$

From this inequality it follows that the point inside a quadrilateral that minimizes the sum of distances to the vertices is the intersection of the diagonals. Hence that point is the Fermat point of a convex quadrilateral.

Remarkable Points and Lines in a Convex Quadrilateral

The centre of a quadrilateral can be defined in several different ways. The "vertex centroid" comes from considering the quadrilateral as being empty but having equal masses at its vertices. The "side centroid" comes from considering the sides to have constant mass per unit length. The usual centre, called just centroid (centre of area) comes from considering the surface of the quadrilateral as having constant density. These three points are in general not all the same point.

The "vertex centroid" is the intersection of the two bimedians. As with any polygon, the x and y coordinates of the vertex centroid are the arithmetic means of the x and y coordinates of the vertices.

The "area centroid" of quadrilateral $ABCD$ can be constructed in the following way. Let G_a, G_b, G_c, G_d be the centroids of triangles BCD, ACD, ABD, ABC respectively. Then the "area centroid" is the intersection of the lines G_aG_c and G_bG_d.

In a general convex quadrilateral $ABCD$, there are no natural analogies to the circumcenter and orthocenter of a triangle. But two such points can be constructed in the following way. Let O_a, O_b, O_c, O_d be the circumcenters of triangles BCD, ACD, ABD, ABC respectively; and denote by H_a, H_b, H_c, H_d the orthocenters in the same triangles. Then the intersection of the lines O_aO_c and O_bO_d is called the quasicircumcenter, and the intersection of the lines H_aH_c and H_bH_d is called the *quasiorthocenter* of the convex quadrilateral. These points can be used to define an Euler line of a quadrilateral. In a convex quadrilateral, the quasiorthocenter H, the "area centroid" G, and the quasicircumcenter O are collinear in this order, and $HG = 2GO$.

There can also be defined a *quasinine-point center* E as the intersection of the lines E_aE_c and E_bE_d, where E_a, E_b, E_c, E_d are the nine-point centers of triangles BCD, ACD, ABD, ABC respectively. Then E is the midpoint of OH.

Another remarkable line in a convex non-parallelogram quadrilateral is the Newton line, which connects the midpoints of the diagonals, and which contains the intersection of the bimedians.

Other Properties of Convex Quadrilaterals

- Let exterior squares be drawn on all sides of a quadrilateral. The segments connecting the centers of opposite squares are (a) equal in length, and (b) perpendicular. Thus these centers are the vertices of an orthodiagonal quadrilateral. This is called Van Aubel's theorem.

- For any simple quadrilateral with given edge lengths, there is a cyclic quadrilateral with the same edge lengths.

- The four smaller triangles formed by the diagonals and sides of a convex quadrilateral have the property that the product of the areas of two opposite triangles equals the product of the areas of the other two triangles.

Taxonomy

A hierarchical taxonomy of quadrilaterals is illustrated by the figure to the right. Lower classes are special cases of higher classes they are connected to. Note that "trapezoid" here is referring to the North American definition (the British equivalent is a trapezium). Inclusive definitions are used throughout.

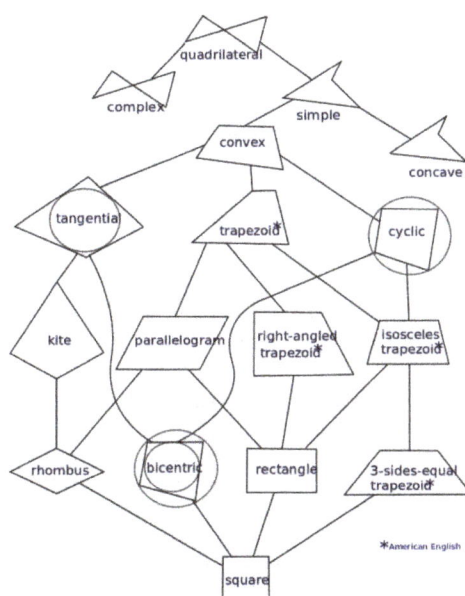

A taxonomy of quadrilaterals.

Skew Quadrilaterals

A non-planar quadrilateral is called a skew quadrilateral. Formulas to compute its dihedral angles from the edge lengths and the angle between two adjacent edges were derived for work on the properties of molecules such as cyclobutane that contain a "puckered" ring of four atoms. Skew polygon for more. Historically the term gauche quadrilateral was also used to mean a skew quadrilateral. A skew quadrilateral together with its diagonals form a (possibly non-regular) tetrahedron, and conversely every skew quadrilateral comes from a tetrahedron where a pair of opposite edges is removed.

Rectangle

In Euclidean plane geometry, a rectangle is a quadrilateral with four right angles. It can also be defined as an equiangular quadrilateral, since equiangular means that all of its angles are equal ($360°/4 = 90°$). It can also be defined as a parallelogram containing a right angle. A rectangle with four sides of equal length is a square. The term oblong is occasionally used to refer to a non-square rectangle. A rectangle with vertices *ABCD* would be denoted as □ *ABCD*.

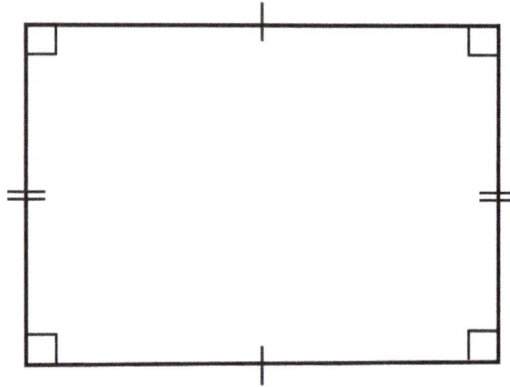

Rectangle

The word rectangle comes from the Latin *rectangulus*, which is a combination of *rectus* (right) and *angulus* (angle).

A crossed rectangle is a crossed (self-intersecting) quadrilateral which consists of two opposite sides of a rectangle along with the two diagonals. It is a special case of an antiparallelogram, and its angles are not right angles. Other geometries, such as spherical, elliptic, and hyperbolic, have so-called rectangles with opposite sides equal in length and equal angles that are not right angles.

Rectangles are involved in many tiling problems, such as tiling the plane by rectangles or tiling a rectangle by polygons.

Characterizations

A convex quadrilateral is a rectangle if and only if it is any one of the following:

- an equiangular quadrilateral

- a quadrilateral with four right angles

- a parallelogram with at least one right angle

- a parallelogram with diagonals of equal length

- a parallelogram *ABCD* where triangles *ABD* and *DCA* are congruent

- a convex quadrilateral with successive sides a, b, c, d whose area is $\frac{1}{4}(a+c)(b+d)$.

- a convex quadrilateral with successive sides a, b, c, d whose area is $\frac{1}{2}\sqrt{(a^2+c^2)(b^2+d^2)}$.

Classification

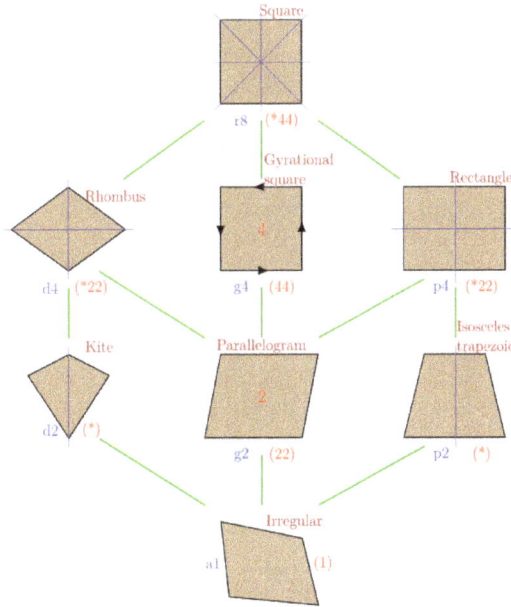

A rectangle is a special case of both parallelogram and trapezoid. A square is a special case of a rectangle.

Traditional Hierarchy

A rectangle is a special case of a parallelogram in which each pair of adjacent sides is perpendicular.

A parallelogram is a special case of a trapezium (known as a trapezoid in North America) in which *both* pairs of opposite sides are parallel and equal in length.

A trapezium is a convex quadrilateral which has at least one pair of parallel opposite sides.

A convex quadrilateral is

- Simple: The boundary does not cross itself.

- Star-shaped: The whole interior is visible from a single point, without crossing any edge.

Alternative Hierarchy

De Villiers defines a rectangle more generally as any quadrilateral with axes of symmetry through each pair of opposite sides. This definition includes both right-angled rectangles and crossed rectangles. Each has an axis of symmetry parallel to and equidistant from a pair of opposite sides, and another which is the perpendicular bisector of those sides, but, in the case of the crossed rectangle, the first axis is not an axis of symmetry for either side that it bisects.

Quadrilaterals with two axes of symmetry, each through a pair of opposite sides, belong to the larger class of quadrilaterals with at least one axis of symmetry through a pair of opposite sides. These quadrilaterals comprise isosceles trapezia and crossed isosceles trapezia (crossed quadrilaterals with the same vertex arrangement as isosceles trapezia).

Properties

Symmetry

A rectangle is cyclic: all corners lie on a single circle.

It is equiangular: all its corner angles are equal (each of 90 degrees).

It is isogonal or vertex-transitive: all corners lie within the same symmetry orbit.

It has two lines of reflectional symmetry and rotational symmetry of order 2 (through 180°).

Rectangle-rhombus Duality

The dual polygon of a rectangle is a rhombus, as shown in the table below.

Rectangle	Rhombus
All *angles* are equal.	All *sides* are equal.
Alternate *sides* are equal.	Alternate *angles* are equal.
Its centre is equidistant from its *vertices*, hence it has a *circumcircle*.	Its centre is equidistant from its *sides*, hence it has an *incircle*.
Its axes of symmetry bisect opposite *sides*.	Its axes of symmetry bisect opposite *angles*.
Diagonals are equal in *length*.	Diagonals intersect at equal *angles*.

- The figure formed by joining, in order, the midpoints of the sides of a rectangle is a rhombus and vice versa.

Miscellaneous

The two diagonals are equal in length and bisect each other. Every quadrilateral with both these properties is a rectangle.

A rectangle is rectilinear: its sides meet at right angles.

A rectangle in the plane can be defined by five independent degrees of freedom consisting, for example, of three for position (comprising two of translation and one of rotation), one for shape (aspect ratio), and one for overall size (area).

Two rectangles, neither of which will fit inside the other, are said to be incomparable.

Formulae

$$P = 2\ell + 2w$$

The formula for the perimeter of a rectangle.

If a rectangle has length ℓ and width w

- it has area $A = \ell w$,

- it has perimeter $P = 2\ell + 2w = 2(\ell + w)$,

- each diagonal has length $d = \sqrt{\ell^2 + w^2}$,

- and when $\ell = w$, the rectangle is a square.

Theorems

The isoperimetric theorem for rectangles states that among all rectangles of a given perimeter, the square has the largest area.

The midpoints of the sides of any quadrilateral with perpendicular diagonals form a rectangle.

A parallelogram with equal diagonals is a rectangle.

The Japanese theorem for cyclic quadrilaterals states that the incentres of the four triangles determined by the vertices of a cyclic quadrilateral taken three at a time form a rectangle.

The British flag theorem states that with vertices denoted A, B, C, and D, for any point P on the same plane of a rectangle:

$$(AP)^2 + (CP)^2 = (BP)^2 + (DP)^2.$$

For every convex body C in the plane, we can inscribe a rectangle r in C such that a homothetic copy R of r is circumscribed about C and the positive homothety ratio is at most 2 and $0.5 \times \text{Area}(R) \leq \text{Area}(C) \leq 2 \times \text{Area}(r)..$

Crossed Rectangles

A crossed (self-intersecting) quadrilateral consists of two opposite sides of a non-self-intersecting quadrilateral along with the two diagonals. Similarly, a crossed rectangle is a crossed quadrilateral which consists of two opposite sides of a rectangle along with the two diagonals. It has the same vertex arrangement as the rectangle. It appears as two identical triangles with a common vertex, but the geometric intersection is not considered a vertex.

A crossed quadrilateral is sometimes likened to a bow tie or butterfly. A three-dimensional rectangular wire frame that is twisted can take the shape of a bow tie. A crossed rectangle is sometimes called an "angular eight".

The interior of a crossed rectangle can have a polygon density of ±1 in each triangle, dependent upon the winding orientation as clockwise or counterclockwise.

A crossed rectangle is not equiangular. The sum of its interior angles (two acute and two reflex), as with any crossed quadrilateral, is 720°.

A rectangle and a crossed rectangle are quadrilaterals with the following properties in common:

- Opposite sides are equal in length.

- The two diagonals are equal in length.

- It has two lines of reflectional symmetry and rotational symmetry of order 2 (through 180°).

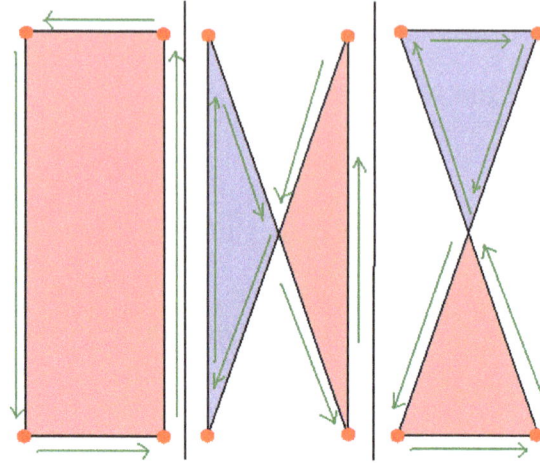

Other Rectangles

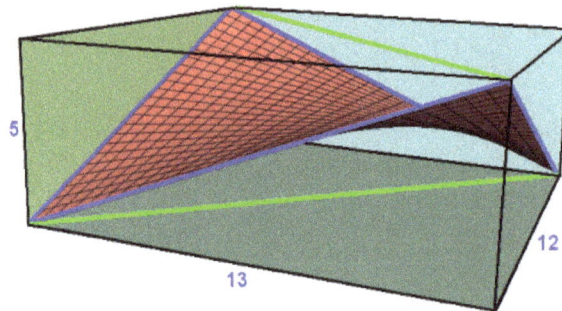

A saddle rectangle has 4 nonplanar vertices, alternated from vertices of a cuboid, with a unique minimal surface interior defined as a linear combination of the four vertices, creating a saddle surface. This example shows 4 blue edges of the rectangle, and two green diagonals, all being diagonal of the cuboid rectangular faces.

In spherical geometry, a spherical rectangle is a figure whose four edges are great circle arcs which meet at equal angles greater than 90°. Opposite arcs are equal in length. The surface of a sphere in Euclidean solid geometry is a non-Euclidean surface in the sense of elliptic geometry. Spherical geometry is the simplest form of elliptic geometry.

In elliptic geometry, an elliptic rectangle is a figure in the elliptic plane whose four edges are elliptic arcs which meet at equal angles greater than 90°. Opposite arcs are equal in length.

In hyperbolic geometry, a hyperbolic rectangle is a figure in the hyperbolic plane whose four edges are hyperbolic arcs which meet at equal angles less than 90°. Opposite arcs are equal in length.

Tessellations

The rectangle is used in many periodic tessellation patterns, in brickwork, for example, these tilings:

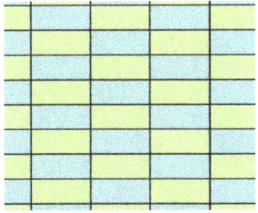    

Stacked bond   Running bond   Basket weave   Basket weave   Herringbone pattern

Squared, Perfect, and Other Tiled Rectangles

A rectangle tiled by squares, rectangles, or triangles is said to be a "squared", "rectangled", or "triangulated" (or "triangled") rectangle respectively. The tiled rectangle is *perfect* if the tiles are similar and finite in number and no two tiles are the same size. If two such tiles are the same size, the tiling is *imperfect*. In a perfect (or imperfect) triangled rectangle the triangles must be right triangles.

A rectangle has commensurable sides if and only if it is tileable by a finite number of unequal squares. The same is true if the tiles are unequal isosceles right triangles.

The tilings of rectangles by other tiles which have attracted the most attention are those by congruent non-rectangular polyominoes, allowing all rotations and reflections. There are also tilings by congruent polyaboloes.

Square

In geometry, a square is a regular quadrilateral, which means that it has four equal sides and four equal angles (90-degree angles, or right angles). It can also be defined as a rectangle in which two adjacent sides have equal length. A square with vertices *ABCD* would be denoted □*ABCD*.

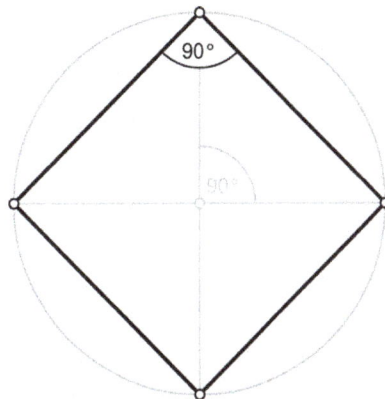

A regular quadrilateral (tetragon)

Properties

A square is a special case of a rhombus (equal sides, opposite equal angles), a kite (two pairs of adjacent equal sides), a parallelogram (opposite sides parallel), a quadrilateral or tetragon (four-sided polygon), and a rectangle (opposite sides equal, right-angles) and therefore has all the properties of all these shapes, namely:

- The diagonals of a square bisect each other and meet at 90°

- The diagonals of a square bisect its angles.

- Opposite sides of a square are both parallel and equal in length.

- All four angles of a square are equal. (Each is 360°/4 = 90°, so every angle of a square is a right angle.)

- All four sides of a square are equal.

- The diagonals of a square are equal.

- The square is the n=2 case of the families of n-hypercubes and n-orthoplexes.

- A square has Schläfli symbol {4}. A truncated square, t{4}, is an octagon, {8}. An alternated square, h{4}, is a digon, {2}.

Perimeter and Area

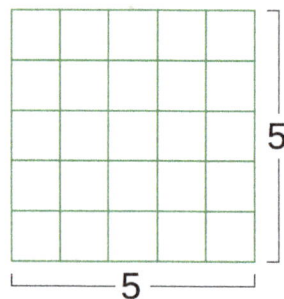

$$5 \times 5 = 25$$
$$5^2 = 25$$

The area of a square is the product of the length of its sides.

The perimeter of a square whose four sides have length ℓ is

$$P = 4\ell$$

and the area A is

In classical times, the second power was described in terms of the area of a square, as in the above formula. This led to the use of the term *square* to mean raising to the second power.

The area can also be calculated using the diagonal d according to

$$A = \frac{d^2}{2}.$$

In terms of the circumradius R, the area of a square is

$$A = 2R^2;$$

since the area of the circle is πR^2, the square fills approximately 0.6366 of its circumscribed circle.

In terms of the inradius r, the area of the square is

$$A = 4r^2.$$

Because it is a regular polygon, a square is the quadrilateral of least perimeter enclosing a given area. Dually, a square is the quadrilateral containing the largest area within a given perimeter. Indeed, if A and P are the area and perimeter enclosed by a quadrilateral, then the following isoperimetric inequality holds:

$$16A \le P^2$$

with equality if and only if the quadrilateral is a square.

A convex quadrilateral with successive sides a, b, c, d is a square if and only if

$$A = \frac{1}{2}(a^2 + c^2) = \frac{1}{2}(b^2 + d^2).$$

Other Facts

- The diagonals of a square are $\sqrt{2}$ (about 1.414) times the length of a side of the square. This value, known as the square root of 2 or Pythagoras' constant, was the first number proven to be irrational.

- A square can also be defined as a parallelogram with equal diagonals that bisect the angles.

- If a figure is both a rectangle (right angles) and a rhombus (equal edge lengths), then it is a square.

- If a circle is circumscribed around a square, the area of the circle is $\pi/2$ (about 1.5708) times the area of the square.

- If a circle is inscribed in the square, the area of the circle is $\pi/4$ (about 0.7854) times the area of the square.

- A square has a larger area than any other quadrilateral with the same perimeter.

- A square tiling is one of three regular tilings of the plane (the others are the equilateral triangle and the regular hexagon).

- The square is in two families of polytopes in two dimensions: hypercube and the cross polytope. The Schläfli symbol for the square is {4}.

- The square is a highly symmetric object. There are four lines of reflectional symmetry and it has rotational symmetry of order 4 (through 90°, 180° and 270°). Its symmetry group is the dihedral group D_4.

- If the inscribed circle of a square $ABCD$ has tangency points E on AB, F on BC, G on CD, and H on DA, then for any point P on the inscribed circle,

$$2(PH^2 - PE^2) = PD^2 - PB^2.$$

- If d_i is the distance from an arbitrary point in the plane to the i-th vertex of a square and R is the circumradius of the square, then

$$\tfrac{1}{4}(d_1^4 + d_2^4 + d_3^4 + d_4^4) + 3R^4 = \left(\frac{d_1^2 + d_2^2 + d_3^2 + d_4^2}{4} + R^2 \right)^2.$$

Coordinates and equations

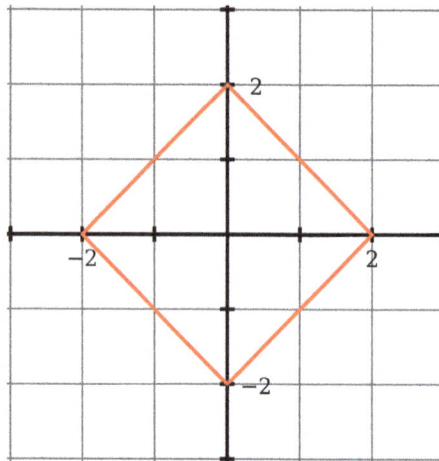

plotted on *Cartesian coordinates*.

The coordinates for the vertices of a square with vertical and horizontal sides, centered at the origin and with side length 2 are $(\pm 1, \pm 1)$, while the interior of this square consists of all points (x_i, y_i) with $-1 < x_i < 1$ and $-1 < y_i < 1$. The equation

$$\max(x^2, y^2) = 1$$

specifies the boundary of this square. This equation means "x^2 or y^2, whichever is larger, equals 1." The circumradius of this square (the radius of a circle drawn through the square's vertices) is half the square's diagonal, and equals $\sqrt{2}$. Then the circumcircle has the equation

$$x^2 + y^2 = 2.$$

Alternatively the equation

$$|x - a| + |y - b| = r.$$

can also be used to describe the boundary of a square with center coordinates (a, b) and a horizontal or vertical radius of r.

Construction

The following animations show how to construct a square using a compass and straightedge. This is possible as $4 = 2^2$, a power of two.

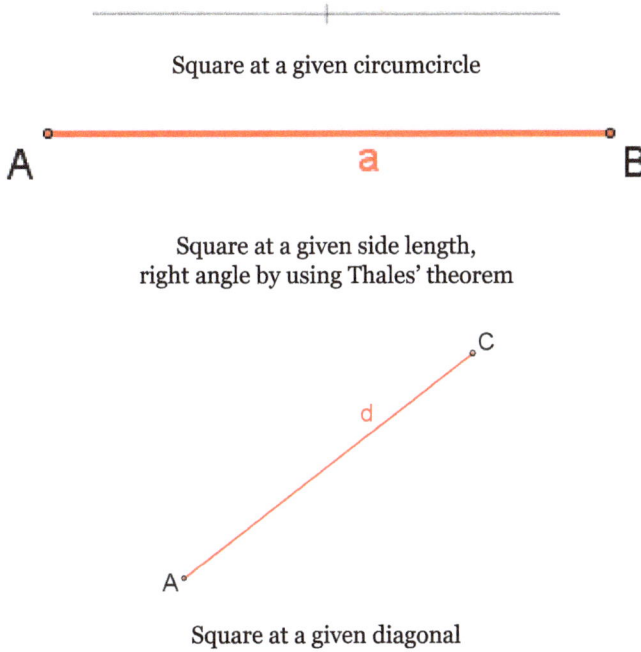

Square at a given circumcircle

A ——————————— a ——————————— B

Square at a given side length,
right angle by using Thales' theorem

.C

d

A°

Square at a given diagonal

Symmetry

The *square* has Dih_4 symmetry, order 8. There are 2 dihedral subgroups: Dih_2, Dih_1, and 3 cyclic subgroups: Z_4, Z_2, and Z_1.

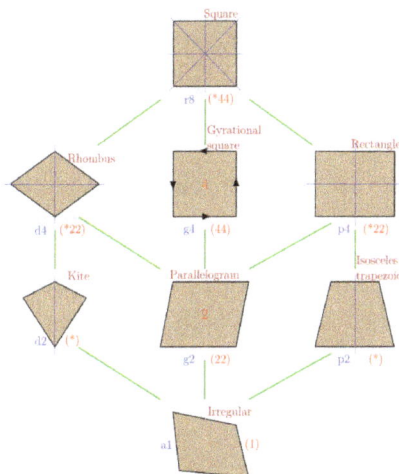

The dihedral symmetries are divided depending on whether they pass through vertices (**d** for diagonal) or edges (**p** for perpendiculars) Cyclic symmetries in the middle column are labeled as **g** for their central gyration orders. Full symmetry of the square is **r12** and no symmetry is labeled **a1**.

A square is a special case of many lower symmetry convex quadrilaterals:

- a rectangle with two adjacent equal sides

- a quadrilateral with four equal sides and four right angles

- a parallelogram with one right angle and two adjacent equal sides

- a rhombus with a right angle

- a rhombus with all angles equal

- a quadrilateral where the diagonals are equal and are the perpendicular bisectors of each other, i.e. a rhombus with equal diagonals

These 6 symmetries express 8 distinct symmetries on a square. John Conway labels these by a letter and group order.

Each subgroup symmetry allows one or more degrees of freedom for irregular quadrilaterals. r8 is full symmetry of the square, and a1 is no symmetry. d4, is the symmetry of a rectangle and p4, is the symmetry of a rhombus. These two forms are duals of each other and have half the symmetry order of the square. d2 is the symmetry of an isosceles trapezoid, and p2 is the symmetry of a kite. g2 defines the geometry of a parallelogram.

Only the g4 subgroup has no degrees of freedom but can seen as a square with directed edges.

Squares Inscribed in Triangles

Every acute triangle has three inscribed squares (squares in its interior such that all four of a square's vertices lie on a side of the triangle, so two of them lie on the same side and hence one side of the square coincides with part of a side of the triangle). In a right triangle two of the squares coincide and have a vertex at the triangle's right angle, so a right triangle has only two *distinct* inscribed squares. An obtuse triangle has only one inscribed square, with a side coinciding with part of the triangle's longest side.

The fraction of the triangle's area that is filled by the square is no more than 1/2.

Squaring the Circle

Squaring the circle is the problem, proposed by ancient geometers, of constructing a square with the same area as a given circle by using only a finite number of steps with compass and straightedge.

In 1882, the task was proven to be impossible, as a consequence of the Lindemann–Weierstrass theorem which proves that pi (π) is a transcendental number, rather than an algebraic irrational number; that is, it is not the root of any polynomial with rational coefficients.

Non-Euclidean Geometry

In non-Euclidean geometry, squares are more generally polygons with 4 equal sides and equal angles.

In spherical geometry, a square is a polygon whose edges are great circle arcs of equal distance, which meet at equal angles. Unlike the square of plane geometry, the angles of such a square are larger than a right angle. Larger spherical squares have larger angles.

In hyperbolic geometry, squares with right angles do not exist. Rather, squares in hyperbolic geometry have angles of less than right angles. Larger hyperbolic squares have smaller angles.

Examples:

Two squares can tile the sphere with 2 squares around each vertex and 180-degree internal angles. Each square covers an entire hemisphere and their vertices lie along a great circle. This is called a spherical square dihedron. The Schläfli symbol is {4,2}.	Six squares can tile the sphere with 3 squares around each vertex and 120-degree internal angles. This is called a spherical cube. The Schläfli symbol is {4,3}.	Squares can tile the Euclidean plane with 4 around each vertex, with each square having an internal angle of 90°. The Schläfli symbol is {4,4}.	Squares can tile the hyperbolic plane with 5 around each vertex, with each square having 72-degree internal angles. The Schläfli symbol is {4,5}. In fact, for any n ≥ 5 there is a hyperbolic tiling with n squares about each vertex.

Crossed Square

A crossed square is a faceting of the square, a self-intersecting polygon created by removing two opposite edges of a square and reconnecting by its two diagonals. It has half the symmetry of the square, Dih_2, order 4. It has the same vertex arrangement as the square, and is vertex-transitive. It appears as two 45-45-90 triangle with a common vertex, but the geometric intersection is not considered a vertex.

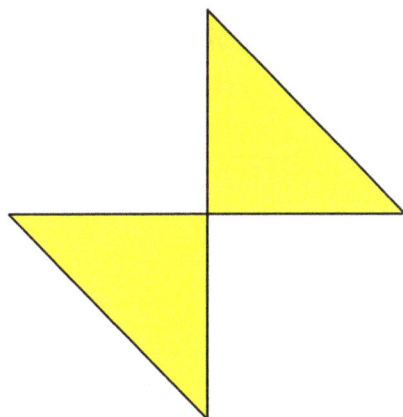

Crossed-square

A crossed square is sometimes likened to a bow tie or butterfly. the crossed rectangle is related, as a faceting of the rectangle, both special cases of crossed quadrilaterals.

The interior of a crossed square can have a polygon density of ±1 in each triangle, dependent upon the winding orientation as clockwise or counterclockwise.

A square and a crossed square have the following properties in common:

- Opposite sides are equal in length.

- The two diagonals are equal in length.

- It has two lines of reflectional symmetry and rotational symmetry of order 2 (through 180°).

It exists in the vertex figure of a uniform star polyhedra, the tetrahemihexahedron.

Graphs

The K_4 complete graph is often drawn as a square with all 6 possible edges connected, hence appearing as a square with both diagonals drawn. This graph also represents an orthographic projection of the 4 vertices and 6 edges of the regular 3-simplex (tetrahedron).

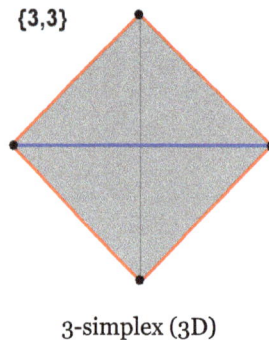

{3,3}

3-simplex (3D)

Trapezoid

Trapezoid (AmE)
Trapezium (BrE)

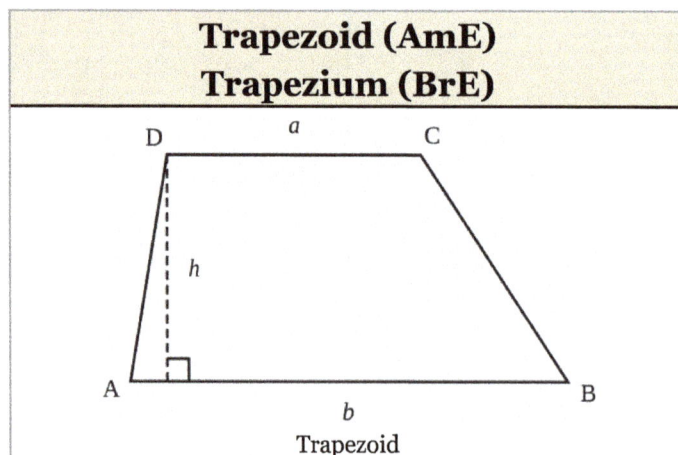

Trapezoid

In Euclidean geometry, a convex quadrilateral with at least one pair of parallel sides is referred to as a trapezoid in American and Canadian English but as a trapezium in English outside North America. The parallel sides are called the *bases* of the trapezoid and the other two sides are called the *legs* or the lateral sides (if they are not parallel; otherwise there are two pairs of bases). A *scalene trapezoid* is a trapezoid with no sides of equal measure, in contrast to the special cases below.

Etymology

The term *trapezium* has been in use in English since 1570, from Late Latin *trapezium*, from Greek τραπέζιον (*trapézion*), literally "a little table", a diminutive of τράπεζα (*trápeza*), "a table", itself from τετράς (*tetrás*), "four" + πέζα (*péza*), "a foot, an edge". The first recorded use of the Greek word translated *trapezoid* (τραπεζοειδή, *trapezoeidé*, "table-like") was by Marinus Proclus (412 to 485 AD) in his Commentary on the first book of Euclid's Elements.

This article uses the term *trapezoid* in the sense that is current in the United States and Canada. In many other languages using a word derived from the Greek for this figure, the form closest to *trapezium* (e.g. French *trapèze*, Italian *trapezio*, Spanish *trapecio*, German *Trapez*, Russian *трапеция*) is used.

Special cases

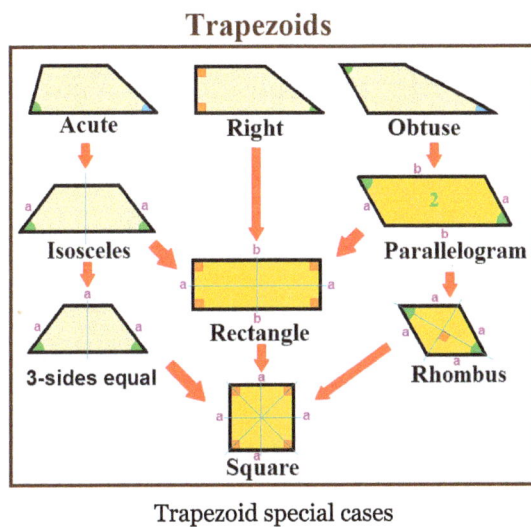

Trapezoid special cases

A right trapezoid (also called *right-angled trapezoid*) has two adjacent right angles. Right trapezoids are used in the trapezoidal rule for estimating areas under a curve.

An acute trapezoid has two adjacent acute angles on its longer *base* edge, while an obtuse trapezoid has one acute and one obtuse angle on each *base*.

An *acute trapezoid* is also an isosceles trapezoid, if its sides (legs) have the same length, and the base angles have the same measure. It has reflection symmetry.

An *obtuse trapezoid* with two pairs of parallel sides is a parallelogram. A parallelogram has central 2-fold rotational symmetry (or point reflection symmetry).

A Saccheri quadrilateral is similar to a trapezoid in the hyperbolic plane, with two adjacent right angles, while it is a rectangle in the Euclidean plane. A Lambert quadrilateral in the hyperbolic plane has 3 right angles.

A tangential trapezoid is a trapezoid that has an incircle.

Inclusive vs Exclusive Definition

There is some disagreement whether parallelograms, which have two pairs of parallel sides, should be regarded as trapezoids. Some define a trapezoid as a quadrilateral having *only* one pair of parallel sides (the exclusive definition), thereby excluding parallelograms. Others define a trapezoid as a quadrilateral with *at least* one pair of parallel sides (the inclusive definition), making the parallelogram a special type of trapezoid. The latter definition is consistent with its uses in higher mathematics such as calculus. The former definition would make such concepts as the trapezoidal approximation to a definite integral ill-defined. This article uses the inclusive definition and considers parallelograms as special cases of a trapezoid. This is also advocated in the taxonomy of quadrilaterals.

Under the inclusive definition, all parallelograms (including rhombuses, rectangles and squares) are trapezoids. Rectangles have mirror symmetry on mid-edges; rhombuses have mirror symmetry on vertices, while squares have mirror symmetry on both mid-edges and vertices.

Condition of Existence

Four lengths a, c, b, d can constitute the consecutive sides of a non-parallelogram trapezoid with a and b parallel only when

$$| d - c | < | b - a | < d + c.$$

The quadrilateral is a parallelogram when $d - c = b - a = 0$, but it is an ex-tangential quadrilateral (which is not a trapezoid) when $| d - c | = | b - a | \neq 0$.

Characterizations

Given a convex quadrilateral, the following properties are equivalent, and each implies that the quadrilateral is a trapezoid:

- It has two adjacent angles that are supplementary, that is, they add up to 180 degrees.

- The angle between a side and a diagonal is equal to the angle between the opposite side and the same diagonal.

- The diagonals cut each other in mutually the same ratio (this ratio is the same as that between the lengths of the parallel sides).

- The diagonals cut the quadrilateral into four triangles of which one opposite pair are similar.

- The diagonals cut the quadrilateral into four triangles of which one opposite pair have equal areas.

- The product of the areas of the two triangles formed by one diagonal equals the product of the areas of the two triangles formed by the other diagonal.

- The areas S and T of some two opposite triangles of the four triangles formed by the diagonals satisfy the equation

$$\sqrt{K} = \sqrt{S} + \sqrt{T},$$

  where K is the area of the quadrilateral.

- The midpoints of two opposite sides and the intersection of the diagonals are collinear.

- $\sin A \sin C = \sin B \sin D$.

- The cosines of two adjacent angles sum to 0, as do the cosines of the other two angles.

- The cotangents of two adjacent angles sum to 0, as do the cotangents of the other two adjacent angles.

- One bimedian divides the quadrilateral into two quadrilaterals of equal areas.

- Twice the length of the bimedian connecting the midpoints of two opposite sides equals the sum of the lengths of the other sides.

Additionally, the following properties are equivalent, and each implies that opposite sides a and b are parallel:

- The consecutive sides a, c, b, d and the diagonals p, q satisfy the equation

$$p^2 + q^2 = c^2 + d^2 + 2ab.$$

- The distance v between the midpoints of the diagonals satisfies the equation

$$v = \frac{|a-b|}{2}.$$

Midsegment and Height

The *midsegment* (also called the median or midline) of a trapezoid is the segment that joins the midpoints of the legs. It is parallel to the bases. Its length m is equal to the average of the lengths of the bases a and b of the trapezoid,

$$m = \frac{a+b}{2}.$$

The midsegment of a trapezoid is one of the two bimedians (the other bimedian divides the trapezoid into equal areas).

The *height* (or altitude) is the perpendicular distance between the bases. In the case that the two bases have different lengths ($a \neq b$), the height of a trapezoid h can be determined by the length of its four sides using the formula

$$h = \frac{\sqrt{(-a+b+c+d)(a-b+c+d)(a-b+c-d)(a-b-c+d)}}{2|b-a|}$$

where c and d are the lengths of the legs.

Area

The area K of a trapezoid is given by

$$K = \frac{a+b}{2} \cdot h = mh$$

where a and b are the lengths of the parallel sides, h is the height (the perpendicular distance between these sides), and m is the arithmetic mean of the lengths of the two parallel sides. In 499 AD Aryabhata, a great mathematician-astronomer from the classical age of Indian mathematics and Indian astronomy, used this method in the *Aryabhatiya* (section 2.8). This yields as a special case the well-known formula for the area of a triangle, by considering a triangle as a degenerate trapezoid in which one of the parallel sides has shrunk to a point.

The 7th-century Indian mathematician Bhāskara I derived the following formula for the area of a trapezoid with consecutive sides a, c, b, d:

$$K = \frac{1}{2}(a+b)\sqrt{c^2 - \frac{1}{4}\left((b-a) + \frac{c^2-d^2}{b-a}\right)^2}$$

where a and b are parallel and $b > a$. This formula can be factored into a more symmetric version

$$K = \frac{a+b}{4|b-a|}\sqrt{(-a+b+c+d)(a-b+c+d)(a-b+c-d)(a-b-c+d)}.$$

When one of the parallel sides has shrunk to a point (say $a = 0$), this formula reduces to Heron's formula for the area of a triangle.

Another equivalent formula for the area, which more closely resembles Heron's formula, is

$$K = \frac{a+b}{|b-a|}\sqrt{(s-b)(s-a)(s-b-c)(s-b-d)},$$

where $s = \frac{1}{2}(a+b+c+d)$ is the semiperimeter of the trapezoid. (This formula is similar to Brahmagupta's formula, but it differs from it, in that a trapezoid might not be cyclic (inscribed in a circle). The formula is also a special case of Bretschneider's formula for a general quadrilateral).

From Bretschneider's formula, it follows that

$$K = \sqrt{\frac{(ab^2 - a^2 b - ad^2 + bc^2)(ab^2 - a^2 b - ac^2 + bd^2)}{(2(b-a))^2} - (\frac{b^2 + d^2 - a^2 - c^2}{4})^2}.$$

The line that joins the midpoints of the parallel sides, bisects the area.

Diagonals

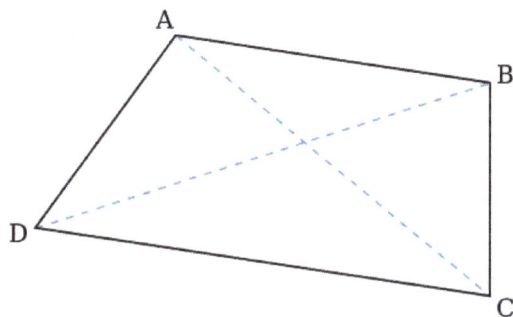

The lengths of the diagonals are

$$p = \sqrt{\frac{ab^2 - a^2 b - ac^2 + bd^2}{b-a}},$$

$$q = \sqrt{\frac{ab^2 - a^2 b - ad^2 + bc^2}{b-a}}$$

where a and b are the bases, c and d are the other two sides, and $a < b$.

If the trapezoid is divided into four triangles by its diagonals AC and BD (as shown on the right), intersecting at O, then the area of $\triangle AOD$ is equal to that of $\triangle BOC$, and the product of the areas of $\triangle AOD$ and $\triangle BOC$ is equal to that of $\triangle AOB$ and $\triangle COD$. The ratio of the areas of each pair of adjacent triangles is the same as that between the lengths of the parallel sides.

Let the trapezoid have vertices A, B, C, and D in sequence and have parallel sides AB and DC. Let E be the intersection of the diagonals, and let F be on side DA and G be on side BC such that FEG is parallel to AB and CD. Then FG is the harmonic mean of AB and DC:

$$\frac{1}{FG} = \frac{1}{2}\left(\frac{1}{AB} + \frac{1}{DC}\right).$$

The line that goes through both the intersection point of the extended nonparallel sides and the intersection point of the diagonals, bisects each base.

Other Properties

The center of area (center of mass for a uniform lamina) lies along the line segment joining the midpoints of the parallel sides, at a perpendicular distance x from the longer side b given by

$$x = \frac{h}{3}\left(\frac{2a+b}{a+b}\right).$$

The center of area divides this segment in the ratio (when taken from the short to the long side)

$$\frac{a+2b}{2a+b}.$$

If the angle bisectors to angles A and B intersect at P, and the angle bisectors to angles C and D intersect at Q, then

$$PQ = \frac{|AD + BC - AB - CD|}{2}.$$

More on Terminology

The term *trapezoid* was once defined as a quadrilateral without any parallel sides in Britain and elsewhere. (The Oxford English Dictionary says "Often called by English writers in the 19th century".) According to the *Oxford English Dictionary*, the sense of a figure with no sides parallel is the meaning for which Proclus introduced the term "trapezoid". This is retained in the French *trapézoïde* (), German *Trapezoid*, and in other languages. However, this particular sense is considered obsolete.

A *trapezium* in Proclus' sense is a quadrilateral having one pair of its opposite sides parallel. This was the specific sense in England in 17th and 18th centuries, and again the prevalent one in recent use outside North America. A trapezium as any quadrilateral more general than a parallelogram is the sense of the term in Euclid.

Confusingly, the word *trapezium* was sometimes used in England from c. 1800 to c. 1875, to denote an irregular quadrilateral having no sides parallel. This is now obsolete in England, but continues in North America. However this shape is more usually (and less confusingly) just called an irregular quadrilateral.

Application in Geometry

The crossed ladders problem is the problem of finding the distance between the parallel sides of a right trapezoid, given the diagonal lengths and the distance from the perpendicular leg to the diagonal intersection.

Architecture

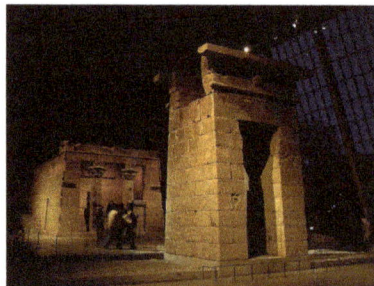

The Temple of Dendur in the Metropolitan Museum of Art in New York City

In architecture the word is used to refer to symmetrical doors, windows, and buildings built wider at the base, tapering toward the top, in Egyptian style. If these have straight sides and sharp angular corners, their shapes are usually isosceles trapezoids. This was the standard style for the doors and windows of the Inca.

Application in Biology

In morphology, taxonomy and other descriptive disciplines in which a term for such shapes is necessary, terms such as *trapezoidal* or *trapeziform* commonly are useful in descriptions of particular organs or forms.

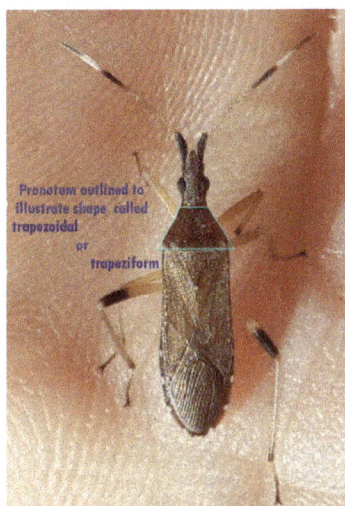

Example of a trapeziform pronotum outlined on a spurge bug

References

- John L. Capinera (11 August 2008). Encyclopedia of Entomology. Springer Science & Business Media. pp. 386, 1062, 1247. ISBN 978-1-4020-6242-1.

- John H. Conway, Heidi Burgiel, Chaim Goodman-Strauss, (2008) The Symmetries of Things, ISBN 978-1-56881-220-5 (Chapter 20, Generalized Schaefli symbols, Types of symmetry of a polygon pp. 275-278)

- Zalman Usiskin and Jennifer Griffin, "The Classification of Quadrilaterals. A Study of Definition", Information Age Publishing, 2008, p. 59, ISBN 1-59311-695-0.

- Zalman Usiskin and Jennifer Griffin, "The Classification of Quadrilaterals. A Study of Definition", Information Age Publishing, 2008, pp. 34–36 ISBN 1-59311-695-0.

- Owen Byer; Felix Lazebnik; Deirdre L. Smeltzer (19 August 2010). Methods for Euclidean Geometry. MAA. pp. 53–. ISBN 978-0-88385-763-2. Retrieved 2011-11-13.

- Park, Poo-Sung. "Regular polytope distances", Forum Geometricorum 16, 2016, 227-232. http://forumgeom.fau.edu/FG2016volume16/FG201627.pdf

- Tom M. Apostol and Mamikon A. Mnatsakanian (December 2004). "Figures Circumscribing Circles" (PDF). American Mathematical Monthly: 853–863. Retrieved 2016-04-06.

Fundamental Study of Polyhedron Cone and Sphere

Polyhedrons have three dimensions with straight edges and vertices. Some of the examples of polyhedrons are pyramids, prisms, cylinders, cones and spheres. The chapter strategically encompasses and incorporates the major components and key concepts of polyhedron, providing a complete understanding.

Polyhedron

In elementary geometry, a polyhedron (plural polyhedra or polyhedrons) is a solid in three dimensions with flat polygonal faces, straight edges and sharp corners or vertices.

Some Polyhedra

Regular tetrahedron   Small stellated dodecahedron   Icosidodecahedron

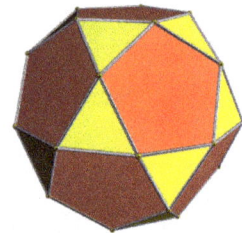

Great cubicuboctahedron   Rhombic triacontahedron   Octagonal prism

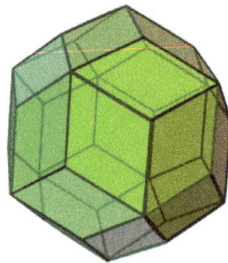

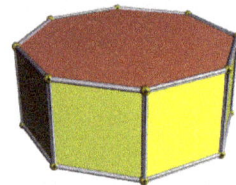

Cubes and pyramids are examples of polyhedra.

A polyhedron is said to be convex if its surface (comprising its faces, edges and vertices) does not intersect itself and the line segment joining any two points of the polyhedron is contained in the interior or surface.

A polyhedron is a 3-dimensional example of the more general polytope in any number of dimensions.

Basis for Definition

In elementary geometry, the faces are polygons – regions of planes – meeting in pairs along their edges which are straight-line segments, and with the edges meeting in vertex points. Treating a polyhedron as a solid bounded by flat faces and straight edges is not very precise; for example it is difficult to reconcile with star polyhedra. Grünbaum (1994, p. 43) observed, "The Original Sin in the theory of polyhedra goes back to Euclid, and through Kepler, Poinsot, Cauchy and many others … [in that] at each stage … the writers failed to define what are the 'polyhedra' …." Many definitions of "polyhedron" have been given within particular contexts, some more rigorous than others. For example, definitions based on the idea of a bounding surface rather than a solid are common. However such definitions are not always compatible in other mathematical contexts.

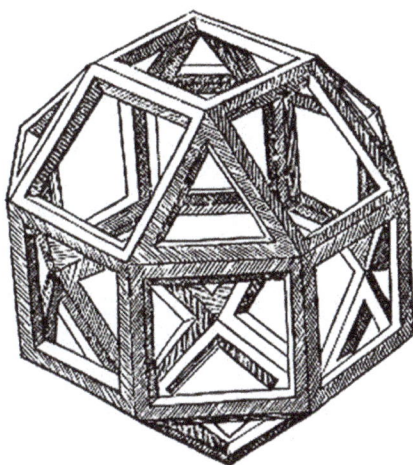

A skeletal polyhedron (specifically, a rhombicuboctahedron) drawn by Leonardo da Vinci to illustrate a book by Luca Pacioli

One modern approach treats a geometric polyhedron as an injection into real space, a *realisation*, of some abstract polyhedron. Any such polyhedron can be built up from different kinds of element or entity, each associated with a different number of dimensions:

- 3 dimensions: The interior is the volume bounded by the faces. It might or might not be realised as a solid body.

- 2 dimensions: A face is a *polygon* bounded by a circuit of edges, and usually also realises the flat (plane) region inside the boundary. These polygonal faces together make up the polyhedral surface.

- 1 dimension: An edge joins one vertex to another and one face to another, and is usually a line segment. The edges together make up the polyhedral skeleton.

- 0 dimensions: A vertex (plural vertices) is a corner point.

Different approaches - and definitions - may require different realisations. Sometimes the interior volume is considered to be part of the polyhedron, sometimes only the surface is considered, and occasionally only the skeleton of edges or even just the set of vertices.

In such elementary geometric and set-based definitions, a polyhedron is typically understood as a three-dimensional example of the more general polytope in any number of dimensions. For example, a polygon has a two-dimensional body and no faces, while a 4-polytope has a four-dimensional body and an additional set of three-dimensional "cells".

In other mathematical disciplines, the term "polyhedron" may be used to refer to a variety of specialised constructs, some geometric and others purely algebraic or abstract. In such contexts definition of the term "polyhedron" may not be consistent with a polytope but rather in contrast to it.

Characteristics

Polyhedral Surface

A defining characteristic of almost all kinds of polyhedra is that just two faces join along any common edge. Likewise any edge meets just two vertices, one at each end. These two characteristics are dual to each other and they ensure that the polyhedral surface is continuously connected and does not end abruptly or split off in different directions.

For similar reasons, the surface may not be divisible into two parts such that each part is a valid polyhedron. This rules out both self-intersecting compound polyhedra or figures joined only by a vertex or an edge, such as two tetrahedra joined at a common apex.

Every simple (non-self-intersecting) polyhedron has at least two faces with the same number of edges.

Number of Faces

Polyhedra may be classified and are often named according to the number of faces. The naming system is based on Classical Greek, for example tetrahedron (4), pentahedron (5), hexahedron (6), triacontahedron (30), and so on.

For a complete list of the Greek numeral prefixes Numeral prefixes>Table of number prefixes in English>Greek>Quantitative

Topological Characteristics

The topological class of a polyhedron is defined by its Euler characteristic and orientability.

From this perspective, any polyhedral surface may be classed as certain kind of topological manifold. For example, the surface of a convex or indeed any simply connected polyhedron is a topological sphere.

Euler Characteristic

The Euler characteristic χ relates the number of vertices V, edges E, and faces F of a polyhedron:

$$= V - E + F$$

This is equal to the topological Euler characteristic of its surface. For a convex polyhedron or more generally any simply connected polyhedron (i.e. with surface a topological sphere), $\chi = 2$.

For more complicated shapes, the Euler characteristic relates to the number of toroidal holes, handles and/or cross-caps in the surface and will be less than 2.

Leonhard Euler's discovery of the characteristic which bears his name marked the beginning of the modern discipline of topology.

Orientability

Some polyhedra have two distinct sides to their surface, for example the inside and outside of a convex polyhedron paper model can each be given a different colour (although the inside colour will be hidden from view). We say that the figure is orientable. Some non-convex orientable polyhedra have regions turned "inside out" so that both colours appear on the outside in different places.

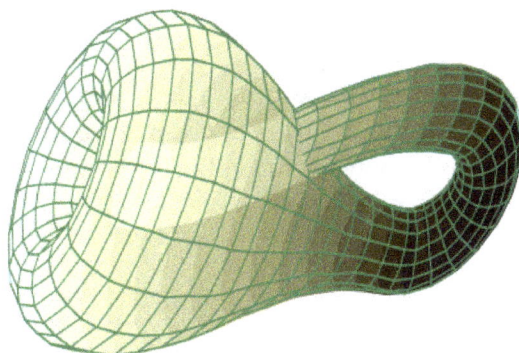

Self-intersecting Klein bottle approximated as quadrilateral polyhedron

But for some polyhedra, such as the tetrahemihexahedron, this is not possible and the surface is said to be one-sided. Such a polyhedron is non-orientable.

All polyhedra with odd-numbered Euler characteristic χ are non-orientable. A given figure with even $\chi < 2$ may or may not be orientable. For example, the one-holed toroid and the Klein bottle both have $\chi = 0$, with the first being orientable and the other not.

Duality

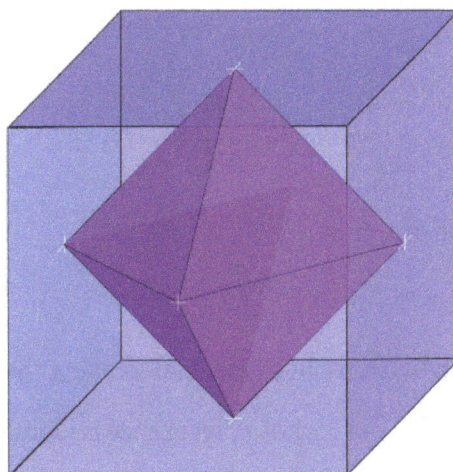

The octahedron is dual to the cube

For every polyhedron there exists a dual polyhedron having:

- faces in place of the original's vertices and vice versa,

- the same number of edges

- the same Euler characteristic and orientability

The dual of a convex polyhedron and of many other polyhedra can be obtained by the process of polar reciprocation.

Dual polyhedra exist in pairs. The dual of a dual is just the original polyhedron again. Some polyhedra are self-dual, meaning that the dual of the polyhedron is congruent to the original polyhedron.

Vertex Figures

For every vertex one can define a vertex figure, which describes the local structure of the polyhedron around the vertex. Precise definitions vary, but a vertex figure can be thought of as the polygon exposed where a slice through the polyhedron cuts off a corner. If the vertex figure is a regular polygon, then the vertex itself is said to be regular.

Volume

Regular polyhedra

Any regular polyhedron can be divided up into congruent pyramids, with each pyramid having a face of the polyhedron as its base and the centre of the polyhedron as its apex. The height of a pyramid is equal to the inradius of the polyhedron. If the area of a face is A and the in-radius is r then the volume of the pyramid is one-third of the base times the height, or $Ar/3$. For a regular polyhedron with n faces, its volume is then simply

$$\text{volume} = nAr/3 .$$

For instance, a cube with edges of length L has six faces, each face being a square with area $A = L^2$. The inradius from the center of the face to the center of the cube is $r = L/2$. Then the volume is given by

$$\text{volume} = \frac{6 \cdot L^2 \cdot \dfrac{L}{2}}{3} = L^3 ,$$

the usual formula for the volume of a cube.

Orientable Polyhedra

The volume of any orientable polyhedron can be calculated using the divergence theorem. Consider the vector field $\vec{F}(\vec{x}) = \frac{1}{3}\vec{x} = \left(\frac{x_1}{3}, \frac{x_2}{3}, \frac{x_3}{3} \right)$, whose divergence is identically 1. The divergence theorem implies that the volume is equal to a surface integral of $F(x)$:

$$\text{volume}(\Omega) = \int_{\Omega} \nabla \cdot \vec{F} d\Omega = \oint_{S} \vec{F} \cdot \hat{n} dS.$$

When Ω is the region enclosed by a polyhedron, since the faces of a polyhedron are planar and have piecewise constant normal vectors, this simplifies to

$$\text{volume} = \frac{1}{3} \sum_{\text{face } i} \vec{x}_i \cdot \hat{n}_i A_i$$

where $\vec{x}_i$ is the ith face's barycenter, $\hat{n}_i$ is its normal vector, and A_i is its area. Once the faces are decomposed in a set of non-overlapping triangles with surface normals pointing away from the volume, the volume is one sixth of the sum over the triple products of the nine Cartesian vertex coordinates of the triangles.

Since it may be difficult to enumerate the faces, volume computation may be challenging, and hence there exist specialized algorithms to determine the volume (many of these generalize to convex polytopes in higher dimensions).

Convex Polyhedra

A polyhedron is said to be convex if its surface (comprising its faces, edges and vertices) does not intersect itself and the line segment joining any two points of the polyhedron is contained in the interior or surface. A convex polyhedron is sometimes defined as a convex set of points in space, the intersection of a set of half-spaces, or the convex hull of a set of points. However many such definitions cannot easily be extended to include self-intersecting figures such as star polyhedra.

Convex polyhedron blocks on display at the Universum museum in Mexico City

Important classes of convex polyhedra include the highly symmetrical Platonic solids, Archimedean solids and Archimedean duals or Catalan solids, and the regular-faced deltahedra and Johnson solids.

Convex polyhedra, and especially triangular pyramids or 3-simplexes, are important in many areas of mathematics, especially those relating to topology.

Symmetries

Many of the most studied polyhedra are highly symmetrical.

A symmetrical polyhedron can be rotated and superimposed on its original position such that its faces and so on have changed position. All the elements which can be superimposed on each other in this way are said to lie in a given "symmetry orbit". For example, all the faces of a cube lie in one orbit, while all the edges lie in another. If all the elements of a given dimension, say all the faces, lie in the same orbit, the figure is said to be "transitive" on that orbit. For example, a cube has one kind of face so it is face-transitive, while a truncated cube has two kinds of face and is not.

Such polyhedra can be distorted so that they are no longer symmetrical. But where a polyhedral name is given, such as icosidodecahedron, the most symmetrical geometry is almost always implied, unless otherwise stated.

There are several types of highly symmetric polyhedron, classified by which kind of element - faces, edges and/or vertices - belong to a single symmetry orbit:

- Regular if it is vertex-transitive, edge-transitive and face-transitive (this implies that every face is the same regular polygon; it also implies that every vertex is regular).

- Quasi-regular if it is vertex-transitive and edge-transitive (and hence has regular faces) but not face-transitive. A quasi-regular dual is face-transitive and edge-transitive (and hence every vertex is regular) but not vertex-transitive.

- Semi-regular if it is vertex-transitive but not edge-transitive, and every face is a regular polygon. (This is one of several definitions of the term, depending on author. Some definitions overlap with the quasi-regular class). These polyhedra include the semiregular prisms and antiprisms. A semi-regular dual is face-transitive but not vertex-transitive, and every vertex is regular.

- Uniform if it is vertex-transitive and every face is a regular polygon, i.e. it is regular, quasi-regular or semi-regular. A uniform dual is face-transitive and has regular vertices, but is not necessarily vertex-transitive).

- Isogonal or vertex-transitive if all vertices are the same, in the sense that for any two vertices there exists a symmetry of the polyhedron mapping the first isometrically onto the second.

- Isotoxal or edge-transitive if all edges are the same, in the sense that for any two edges there exists a symmetry of the polyhedron mapping the first isometrically onto the second.

- Isohedral or face-transitive if all faces are the same, in the sense that for any two faces there exists a symmetry of the polyhedron mapping the first isometrically onto the second.

- Noble if it is face-transitive and vertex-transitive (but not necessarily edge-transitive). The regular polyhedra are also noble; they are the only noble uniform polyhedra.

A polyhedron can belong to the same overall symmetry group as one of higher symmetry, but will be of lower symmetry if it has several groups of elements in different symmetry orbits. For example, the truncated cube has its triangles and octagons in different orbits.

Some classes of polyhedron have only a single main axis of symmetry. These include the pyramids as well as the semiregular prisms and antiprisms.

Regular Polyhedra

Regular polyhedra are the most highly symmetrical. Altogether there are nine regular polyhedra.

The five convex examples have been known since antiquity and are called the Platonic solids. These are the triangular pyramid or tetrahedron, cube (regular hexahedron), octahedron, dodecahedron and icosahedron:

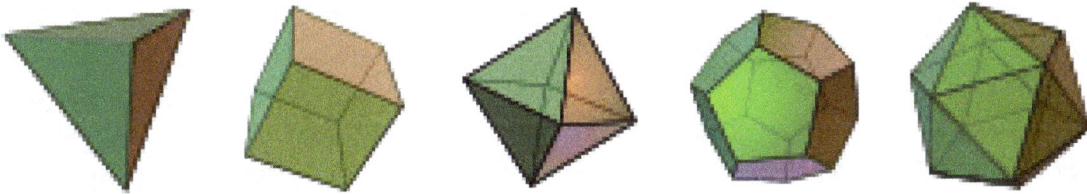

There are also four regular star polyhedra, known as the Kepler-Poinsot polyhedra after their discoverers.

The dual of a regular polyhedron is also regular.

Uniform Polyhedra and their Duals

Uniform polyhedra are *vertex-transitive* and every face is a regular polygon. They may be subdivided into the regular, quasi-regular, or semi-regular, and may be convex or starry.

The uniform duals have irregular faces but are face-transitive and every vertex figure is a regular polygon. A uniform polyhedron has the same symmetry orbits as its dual, with the faces and vertices simply swapped over. The duals of the convex Archimedean polyhedra are sometimes called the Catalan solids.

The uniform polyhedra and their duals are traditionally classified according to their degree of symmetry, and whether they are convex or not.

	Convex uniform	**Convex uniform dual**	**Star uniform**	**Star uniform dual**
Regular	Platonic solids		Kepler–Poinsot polyhedra	
Quasiregular	Archimedean solids	Catalan solids	(no special name)	(no special name)
			(no special name)	(no special name)
Semiregular	Prisms	Bipyramids	Star prisms	Star Bipyramids
	Antiprisms	Trapezohedra	Star antiprisms	Star trapezohedra

Pyramids

Symmetrical pyramids include some of the most time-honoured and famous of all polyhedra, such as the four-sided Egyptian pyramids.

Noble Polyhedra

A noble polyhedron is both isohedral (equal-faced) and isogonal (equal-cornered), but not necessarily equal-sided. Besides the regular polyhedra, there are many other examples.

The dual of a noble polyhedron is also noble.

Isohedra

An isohedron is a polyhedron with symmetries acting transitively on its faces. Their topology can be represented by a face configuration. All 5 Platonic solids and 13 Catalan solids are isohedra, as well as the infinite families of trapezohedra and bipyramids. Some isohedra allow geometric variations including concave and self-intersecting forms.

Symmetry Groups

Many of the symmetries or Point groups in three dimensions are named after polyhedra having the associated symmetry. These include:

- T - chiral tetrahedral symmetry; the rotation group for a regular tetrahedron; order 12.

- T_d - full tetrahedral symmetry; the symmetry group for a regular tetrahedron; order 24.

- T_h - pyritohedral symmetry; the symmetry of a pyritohedron; order 24.

- O - chiral octahedral symmetry; the rotation group of the cube and octahedron; order 24.

- O_h - full octahedral symmetry; the symmetry group of the cube and octahedron; order 48.

- I - chiral icosahedral symmetry; the rotation group of the icosahedron and the dodecahedron; order 60.

- I_h - full icosahedral symmetry; the symmetry group of the icosahedron and the dodecahedron; order 120.

- C_{nv} - n-fold pyramidal symmetry

- D_{nh} - n-fold prismatic symmetry

- D_{nv} - n-fold antiprismatic symmetry.

Those with chiral symmetry do not have reflection symmetry and hence have two enantiomorphous forms which are reflections of each other. Examples include the snub cuboctahedron and snub icosidodecahedron.

Polyhedra with Regular Faces

Besides the regular and uniform polyhedra, there are some other classes which have regular faces but lower overall symmetry.

Equal Regular Faces

Convex polyhedra where every face is the same kind of regular polygon may be found among three families:

- Triangles: These polyhedra are called deltahedra. There are eight convex deltahedra, comprising three of the regular (Platonic) polyhedra and five non-uniform examples.

- Squares: The cube is the only convex example. Others can be obtained by joining cubes together, although care must be taken if coplanar faces are to be avoided.

- Pentagons: The regular dodecahedron is the only convex example.

Polyhedra with congruent regular faces of six or more sides are all non-convex, because the vertex of three regular hexagons defines a plane.

The total number of convex polyhedra with equal regular faces is thus ten, comprising the five Platonic solids and the five non-uniform deltahedra.

There are infinitely many non-convex examples. Infinite sponge-like examples called infinite skew polyhedra exist in some of these families.

Johnson Solids

Norman Johnson sought which convex non-uniform polyhedra had regular faces, although not necessarily all alike. In 1966, he published a list of 92 such solids, gave them names and numbers, and conjectured that there were no others. Victor Zalgaller proved in 1969 that the list of these Johnson solids was complete.

Other Important Families of Polyhedra

Stellations and Facettings

Stellation of a polyhedron is the process of extending the faces (within their planes) so that they meet to form a new polyhedron.

It is the exact reciprocal to the process of facetting which is the process of removing parts of a polyhedron without creating any new vertices.

Zonohedra

A zonohedron is a convex polyhedron where every face is a polygon with inversion symmetry or, equivalently, symmetry under rotations through 180°.

Toroidal Polyhedra

A toroidal polyhedron is a polyhedron with an Euler characteristic of 0 or smaller, equivalent to a genus of 1 or greater, representing a torus surface having one or more holes through the middle.

Spacefilling Polyhedra

A spacefilling polyhedron packs with copies of itself to fill space. Such a close-packing or spacefilling is often called a tessellation of space or a honeycomb. Some honeycombs involve more than one kind of polyhedron.

Compounds

A polyhedral compound is made of two or more polyhedra sharing a common centre.

Symmetrical compounds often share the same vertices as other well-known polyhedra and may often also be formed by stellation. Some are listed in the list of Wenninger polyhedron models.

Orthogonal Polyhedra

An orthogonal polyhedron is one all of whose faces meet at right angles, and all of whose edges are parallel to axes of a Cartesian coordinate system. Aside from a rectangular box, orthogonal polyhedra are nonconvex. They are the 3D analogs of 2D orthogonal polygons, also known as rectilinear polygons. Orthogonal polyhedra are used in computational geometry, where their constrained structure has enabled advances on problems unsolved for arbitrary polyhedra, for example, unfolding the surface of a polyhedron to a polygonal net.

Generalisations of Polyhedra

The name 'polyhedron' has come to be used for a variety of objects having similar structural properties to traditional polyhedra.

Apeirohedra

A classical polyhedral surface has a finite number of faces, joined in pairs along edges. If the number of faces extends indefinitely it is called an apeirohedron. Examples include:

- Tilings or tessellations of the plane.

- Sponge-like structures called infinite skew polyhedra.

Complex Polyhedra

A complex polyhedron is one which is constructed in complex Hilbert 3-space. This space has six dimensions: three real ones corresponding to ordinary space, with each accompanied by an imaginary dimension. A complex polyhedron is mathematically more closely related to configurations than to real polyhedra.

Curved Polyhedra

Some fields of study allow polyhedra to have curved faces and edges.

Spherical Polyhedra

The surface of a sphere may be divided by line segments into bounded regions, to form a spherical polyhedron. Much of the theory of symmetrical polyhedra is most conveniently derived in this way.

The first known man-made polyhedra are spherical polyhedra carved in stone, Poinsot used spherical polyhedra to discover the four regular star polyhedra and Coxeter used them to enumerate all but one of the uniform polyhedra.

Some polyhedra, such as hosohedra and dihedra, exist only as spherical polyhedra and have no flat-faced analogue.

Curved Spacefilling Polyhedra

If faces are allowed to be concave as well as convex, adjacent faces may be made to meet together with no gap. Some of these curved polyhedra can pack together to fill space. Two important types are:

- Bubbles in froths and foams, such as Weaire-Phelan bubbles.

- Spacefilling forms used in architecture.

Hollow-faced or Skeletal Polyhedra

It is not necessary to fill in the face of a figure before we can call it a polyhedron. For example, Leonardo da Vinci devised frame models of the regular solids, which he drew for Pacioli's book *Divina Proportione*. In modern times, Branko Grünbaum (1994) made a special study of this class of polyhedra, in which he developed an early idea of abstract polyhedra. He defined a face as a cyclically ordered set of vertices, and allowed faces to be skew as well as planar.

Alternative Usages

From the latter half of the twentieth century, various mathematical constructs have been found to have properties also present in traditional polyhedra. Rather than confining the term "polyhedron" to describe a three-dimensional polytope, it has been adopted to describe various related but distinct kinds of structure.

General Polyhedra

A polyhedron has been defined as a set of points in real affine (or Euclidean) space of any dimension n that has flat sides. It may alternatively be defined as the union of a finite number of convex polyhedra, where a *convex polyhedron* is any point set that is the intersection of a finite number of half-spaces. Unlike an elementary polyhedron, it may be bounded or unbounded. In this meaning, a polytope is a bounded polyhedron.

Analytically, such a convex polyhedron is expressed as the solution set for a system of linear inequalities. Defining polyhedra in this way provides a geometric perspective for problems in linear programming.

Many traditional polyhedral forms are general polyhedra. Other examples include:

- A quadrant in the plane. For instance, the region of the cartesian plane consisting of all points above the horizontal axis and to the right of the vertical axis: $\{ (x, y) : x \geq 0, y \geq 0 \}$. Its sides are the two positive axes, and it is otherwise unbounded.

- An octant in Euclidean 3-space, $\{ (x, y, z) : x \geq 0, y \geq 0, z \geq 0 \}$.

- A prism of infinite extent. For instance a doubly infinite square prism in 3-space, consisting of a square in the xy-plane swept along the z-axis: $\{ (x, y, z) : 0 \leq x \leq 1, 0 \leq y \leq 1 \}$.

- Each cell in a Voronoi tessellation is a convex polyhedron. In the Voronoi tessellation of a set S, the cell A corresponding to a point $c \in S$ is bounded (hence a traditional polyhedron) when c lies in the interior of the convex hull of S, and otherwise (when c lies on the boundary of the convex hull of S) A is unbounded.

Topological Polyhedra

A topological polytope is a topological space given along with a specific decomposition into shapes that are topologically equivalent to convex polytopes and that are attached to each other in a regular way.

Such a figure is called *simplicial* if each of its regions is a simplex, i.e. in an n-dimensional space each region has $n+1$ vertices. The dual of a simplicial polytope is called *simple*. Similarly, a widely studied class of polytopes (polyhedra) is that of cubical polyhedra, when the basic building block is an n-dimensional cube.

Abstract Polyhedra

An abstract polytope is a partially ordered set (poset) of elements whose partial ordering obeys certain rules of incidence (connectivity) and ranking. The elements of the set correspond to the vertices, edges, faces and so on of the polytope: vertices have rank 0, edges rank 1, etc. with the partially ordered ranking corresponding to the dimensionality of the geometric elements. The empty set, required by set theory, has a rank of -1 and is sometimes said to correspond to the null polytope. An abstract polyhedron is an abstract polytope having the following ranking:

- rank 3: The maximal element, sometimes identified with the body.

- rank 2: The polygonal faces.

- rank 1: The edges.

- rank 0: the vertices.

- rank -1: The empty set, sometimes identified with the *null polytope*.

Any geometric polyhedron is then said to be a "realization" in real space of the abstract poset.

Polyhedra as Graphs

Any polyhedron gives rise to a graph, or skeleton, with corresponding vertices and edges. Thus graph terminology and properties can be applied to polyhedra. For example:

- Due to Steinitz theorem convex polyhedra are in one-to-one correspondence with 3-connected planar graphs.

- The tetrahedron gives rise to a complete graph (K_4). It is the only polyhedron to do so.

- The octahedron gives rise to a strongly regular graph, because adjacent vertices always have two common neighbors, and non-adjacent vertices have four.

- The Archimedean solids give rise to regular graphs: 7 of the Archimedean solids are of degree 3, 4 of degree 4, and the remaining 2 are chiral pairs of degree 5.

History

Prehistory

Polyhedra appeared in early architectural forms such as cubes and cuboids, with the earliest four-sided pyramids of ancient Egypt also dating from the Stone Age.

The Etruscans preceded the Greeks in their awareness of at least some of the regular polyhedra, as evidenced by the discovery near Padua (in Northern Italy) in the late 19th century of a dodecahedron made of soapstone, and dating back more than 2,500 years (Lindemann, 1987).

Greek Civilisation

The earliest known *written* records of these shapes come from Classical Greek authors, who also gave the first known mathematical description of them. The earlier Greeks were interested primarily in the convex regular polyhedra, which came to be known as the Platonic solids. Pythagoras knew at least three of them, and Theaetetus (circa 417 B. C.) described all five. Eventually, Euclid described their construction in his *Elements*. Later, Archimedes expanded his study to the convex uniform polyhedra which now bear his name. His original work is lost and his solids come down to us through Pappus.

China

Cubical gaming dice in China have been dated back as early as 600 B.C.

By 236 AD, Liu Hui was describing the dissection of the cube into its characteristic tetrahedron (orthoscheme) and related solids, using assemblages of these solids as the basis for calculating volumes of earth to be moved during engineering excavations.

Islamic Civilisation

After the end of the Classical era, scholars in the Islamic civilisation continued to take the Greek knowledge forward.

The 9th century scholar Thabit ibn Qurra gave formulae for calculating the volumes of polyhedra such as truncated pyramids.

Then in the 10th century Abu'l Wafa described the convex regular and quasiregular spherical polyhedra.

Renaissance

As with other areas of Greek thought maintained and enhanced by Islamic scholars, Western interest in polyhedra revived during the Italian Renaissance. Artists constructed skeletal polyhedra, depicting them from life as a part of their investigations into perspective. Several appear in marquetry panels of the period. Piero della Francesca gave the first written description of direct geometrical construction of such perspective views of polyhedra. Leonardo da Vinci made skeletal models of several polyhedra and drew illustrations of them for a book by Pacioli. A painting by an anonymous artist of Pacioli and a pupil depicts a glass rhombicuboctahedron half-filled with water.

As the Renaissance spread beyond Italy, later artists such as Wenzel Jamnitzer, Dürer and others also depicted polyhedra of various kinds, many of them novel, in imaginative etchings.

Star Polyhedra

For almost 2,000 years, the concept of a polyhedron as a convex solid had remained as developed by the ancient Greek mathematicians.

During the Renaissance star forms were discovered. A marble tarsia in the floor of St. Mark's Basilica, Venice, depicts a stellated dodecahedron. Artists such as Wenzel Jamnitzer delighted in depicting novel star-like forms of increasing complexity.

Johannes Kepler (1571 - 1630) used star polygons, typically pentagrams, to build star polyhedra. Some of these figures may have been discovered before Kepler's time, but he was the first to recognize that they could be considered "regular" if one removed the restriction that regular polytopes must be convex. Later, Louis Poinsot realised that star vertex figures (circuits around each corner) can also be used, and discovered the remaining two regular star polyhedra. Cauchy proved Poinsot's list complete, and Cayley gave them their accepted English names: (Kepler's) the small stellated dodecahedron and great stellated dodecahedron, and (Poinsot's) the great icosahedron and great dodecahedron. Collectively they are called the Kepler-Poinsot polyhedra.

The Kepler-Poinsot polyhedra may be constructed from the Platonic solids by a process called stellation. Most stellations are not regular. The study of stellations of the Platonic solids was given a big push by H. S. M. Coxeter and others in 1938, with the now famous paper *The 59 icosahedra*.

The reciprocal process to stellation is called facetting (or faceting). Every stellation of one polytope is dual, or reciprocal, to some facetting of the dual polytope. The regular star polyhedra can also be obtained by facetting the Platonic solids. Bridge 1974 listed the simpler facettings of the dodecahedron, and reciprocated them to discover a stellation of the icosahedron that was missing from the set of "59". More have been discovered since, and the story is not yet ended.

Euler's Formula and Topology

Two other modern mathematical developments had a profound effect on polyhedron theory.

In 1750 the German Leonhard Euler for the first time considered the edges of a polyhedron, allowing him to discover his polyhedron formula relating the number of vertices, edges and faces. This signalled the birth of topology, sometimes referred to as "rubber sheet geometry", and the Frenchman Henri Poincaré developed its core ideas around the end of the nineteenth century. This allowed many longstanding issues over what was or was not a polyhedron to be resolved.

Max Brückner summarised work on polyhedra to date, including many findings of his own, in his book "Vielecke und Vielflache: Theorie und Geschichte" (Polygons and polyhedra: Theory and History). Published in German in 1900, it remained little known.

Meanwhile, the discovery of higher dimensions led to the idea of a polyhedron as a three-dimensional example of the more general polytope.

The Twentieth Century Revival

By the early years of the twentieth century, mathematicians had moved on and geometry was little studied. Coxeter's analysis in *The Fifty-Nine Icosahedra* introduced modern ideas from graph theory and combinatorics into the study of polyhedra, signalling a rebirth of interest in geometry.

Coxeter himself went on to enumerate the star uniform polyhedra for the first time, to treat tilings of the plane as polyhedra, to discover the regular skew polyhedra and to develop the theory of complex polyhedra first discovered by Shephard in 1952, as well as making fundamental contributions to many other areas of geometry.

In the second part of the twentieth century, Grünbaum published important works in two areas. One was in convex polytopes, where he noted a tendency among mathematicians to define a "polyhedron" in different and sometimes incompatible ways to suit the needs of the moment. The other was a series of papers broadening the accepted definition of a polyhedron, for example discovering many new Regular polyhedra. At the close of the 20th century these latter ideas merged with other work on incidence complexes to create the modern idea of an abstract polyhedron (as an abstract 3-polytope), notably presented by McMullen and Schulte.

Polyhedra in Nature

Irregular polyhedra appear in nature as crystals.

Cone

A cone is a three-dimensional geometric shape that tapers smoothly from a flat base (frequently, though not necessarily, circular) to a point called the apex or vertex.

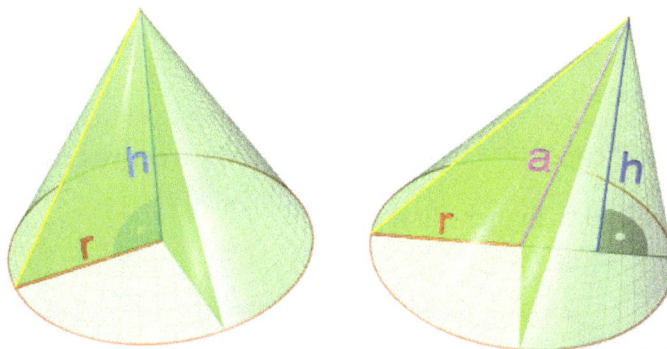

A right circular cone and an oblique circular cone

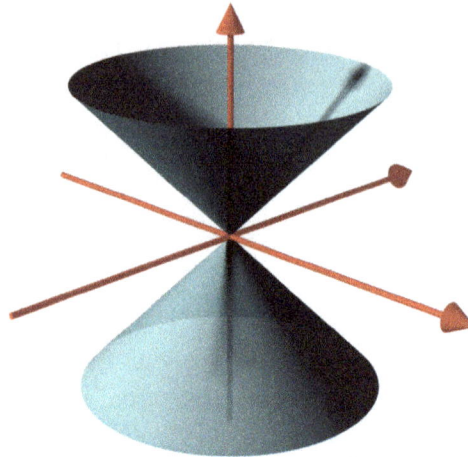

A double cone (not shown infinitely extended)

A cone is formed by a set of line segments, half-lines, or lines connecting a common point, the apex, to all of the points on a base that is in a plane that does not contain the apex. Depending on the author, the base may be restricted to be a circle, any one-dimensional quadratic form in the plane, any closed one-dimensional figure, or any of the above plus all the enclosed points. If the enclosed points are included in the base, the cone is a solid object; otherwise it is a two-dimensional object in three-dimensional space. In the case of a solid object, the boundary formed by these lines or partial lines is called the *lateral surface*; if the lateral surface is unbounded, it is a conical surface.

In the case of line segments, the cone does not extend beyond the base, while in the case of half-lines, it extends infinitely far. In the case of lines, the cone extends infinitely far in both directions from the apex, in which case it is sometimes called a *double cone*. Either half of a double cone on one side of the apex is called a *nappe*.

The axis of a cone is the straight line (if any), passing through the apex, about which the base (and the whole cone) has a circular symmetry.

In common usage in elementary geometry, cones are assumed to be right circular, where *circular* means that the base is a circle and *right* means that the axis passes through the centre of the base at right angles to its plane. If the base is right circular the intersection of a plane with this surface is a conic section. In general, however, the base may be any shape and the apex may lie anywhere (though it is usually assumed that the base is bounded and therefore has finite area, and that the apex lies outside the plane of the base). Contrasted with right cones are oblique cones, in which the axis passes through the centre of the base non-perpendicularly.

A cone with a polygonal base is called a pyramid.

Depending on the context, "cone" may also mean specifically a convex cone or a projective cone.

Cones can also be generalized to higher dimensions.

Further Terminology

The perimeter of the base of a cone is called the "directrix", and each of the line segments between the directrix and apex is a "generatrix" or "generating line" of the lateral surface. (For the connection between this sense of the term "directrix" and the directrix of a conic section, Dandelin spheres.)

The "base radius" of a circular cone is the radius of its base; often this is simply called the radius of the cone. The aperture of a right circular cone is the maximum angle between two generatrix lines; if the generatrix makes an angle θ to the axis, the aperture is 2θ.

A cone with a region including its apex cut off by a plane is called a "truncated cone"; if the truncation plane is parallel to the cone's base, it is called a frustum. An "elliptical cone" is a cone with an elliptical base. A "generalized cone" is the surface created by the set of lines passing through a vertex and every point on a boundary.

Measurements and Equations

Volume

The volume V of any conic solid is one third of the product of the area of the base A_B and the height h

$$V = \frac{1}{3} A_B h.$$

In modern mathematics, this formula can easily be computed using calculus – it is, up to scaling, the integral $\int x^2 dx = \frac{1}{3} x^3$. Without using calculus, the formula can be proven by comparing the cone to a pyramid and applying Cavalieri's principle – specifically, comparing the cone to a (vertically scaled) right square pyramid, which forms one third of a cube. This formula cannot be proven without using such infinitesimal arguments – unlike the 2-dimensional formulae for polyhedral area, though similar to the area of the circle – and hence admitted less rigorous proofs before the advent of calculus, with the ancient Greeks using the method of exhaustion. This is essentially the content of Hilbert's third problem – more precisely, not all polyhedral pyramids are *scissors congruent* (can be cut apart into finite pieces and rearranged into the other), and thus volume cannot be computed purely by using a decomposition argument.

Center of Mass

The center of mass of a conic solid of uniform density lies one-quarter of the way from the center of the base to the vertex, on the straight line joining the two.

Right Circular Cone

Volume

For a circular cone with radius R and height H, the formula for volume becomes

$$V = \int_0^H r^2 \pi \, dh$$

where r is the radius of the cone at height h measured from the apex:

$$r = R\frac{h}{H}.$$

Thus

$$V = \int_0^H \left[R\frac{h}{H} \right]^2 \pi\, dh$$

and so

$$V = \frac{1}{3}\pi R^2 H.$$

Surface Area

The lateral surface area of a right circular cone is $LSA = \pi r l$ where r is the radius of the circle at the bottom of the cone and l is the lateral height (the length of a line segment from the apex of the cone along its side to its base) of the cone (given by the Pythagorean theorem $l = \sqrt{r^2 + h^2}$ where h is the height of the cone). The surface area of the bottom circle of a cone is the same as for any circle, πr^2. Thus, the total surface area of a right circular cone can be expressed as each of the following:

- Radius and height

$$\pi r^2 + \pi r\sqrt{r^2 + h^2}$$

(the area of the base plus the area of the lateral surface; the term $\sqrt{r^2 + h^2}$ is the slant height)

$$\pi r\left(r + \sqrt{r^2 + h^2} \right)$$

where r is the radius and h is the height.

- Radius and lateral height

$$\pi r^2 + \pi r l$$

$$\pi r(r + l)$$

where r is the radius and l is the lateral height.

- Circumference and lateral height

$$\frac{c^2}{4\pi} + \frac{cl}{2}$$

$$\left(\frac{c}{2}\right)\left(\frac{c}{2\pi}+l\right)$$

where c is the circumference and l is the lateral height.

- Apex angle and height

$$\pi h^2 \tan\frac{\dot{\text{E}}}{2}\left(\tan\frac{\dot{\text{E}}}{2}+\sec\frac{\dot{\text{E}}}{2}\right)$$

where Θ is the apex angle and h is the height.

Equation Form

A right solid circular cone with height h and aperture 2θ, whose axis is the z coordinate axis and whose apex is the origin, is described parametrically as

$$F(s,t,u) = \left(u\tan s\cos t, u\tan s\sin t, u\right)$$

where s,t,u range over $[0,\theta)$, $[0,2\pi)$, and $[0,h]$, respectively.

In implicit form, the same solid is defined by the inequalities

$$\{F(x,y,z)\le 0, z\ge 0, z\le h\},$$

where

$$F(x,y,z)=(x^2+y^2)(\cos\theta)^2-z^2(\sin\theta)^2.$$

More generally, a right circular cone with vertex at the origin, axis parallel to the vector d, and aperture 2θ, is given by the implicit vector equation $F(u)=0$ where

$$F(u)=(u{\cdot}d)^2-(d{\cdot}d)(u{\cdot}u)(\cos\theta)^2 \quad\text{or}\quad F(u)=u{\cdot}d-|d\,||\,u|\cos\theta$$

where $u=(x,y,z)$, and $u{\cdot}d$ denotes the dot product.

Elliptic Cone

In the Cartesian coordinate system, an *elliptic cone* is the locus of an equation of the form

$$\frac{x^2}{a^2}+\frac{y^2}{b^2}=\frac{z^2}{c^2}.$$

Projective Geometry

In projective geometry, a cylinder is simply a cone whose apex is at infinity. Intuitively, if one keeps the base fixed and takes the limit as the apex goes to infinity, one obtains a cylinder, the angle of the side increasing as arctan, in the limit forming a right angle. This is useful in the definition of degenerate conics, which require considering the cylindrical conics.

In projective geometry, a cylinder is simply a cone whose apex is at infinity, which corresponds visually to a cylinder in perspective appearing to be a cone towards the sky.

Higher Dimensions

The definition of a cone may be extended to higher dimensions. In this case, one says that a convex set C in the real vector space R^n is a cone (with apex at the origin) if for every vector x in C and every nonnegative real number a, the vector ax is in C. In this context, the analogues of circular cones are not usually special; in fact one is often interested in *polyhedral cones*.

Sphere

A sphere is a perfectly round geometrical object in three-dimensional space that is the surface of a completely round ball, (viz., analogous to a circular object in two dimensions).

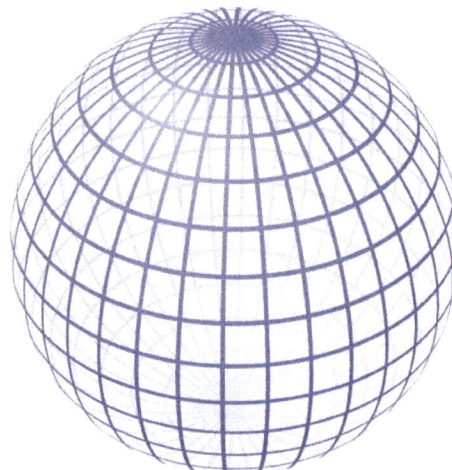

A two-dimensional perspective projection of a sphere

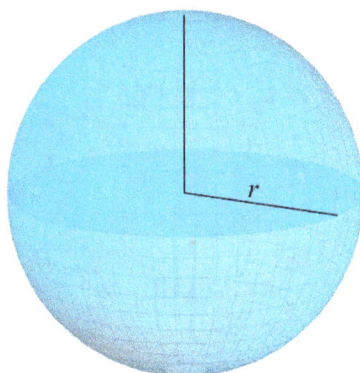

r – radius of the sphere

Like a circle, which geometrically is a two-dimensional object, a sphere is defined mathematically as the set of points that are all at the same distance r from a given point, but in three-dimensional space. This distance r is the radius of the ball, and the given point is the center of the mathematical ball. The longest straight line through the ball, connecting two points of the sphere, passes through the center and its length is thus twice the radius; it is a diameter of the ball.

While outside mathematics the terms "sphere" and "ball" are sometimes used interchangeably, in mathematics a distinction is made between the sphere (a two-dimensional closed surface embedded in three-dimensional Euclidean space) and the ball (a three-dimensional shape that includes the sphere as well as everything inside the sphere). The ball and the sphere share the same radius, diameter, and center.

Surface Area

The surface area of a sphere is:

$$A = 4\pi r^2.$$

Archimedes first derived this formula from the fact that the projection to the lateral surface of a circumscribed cylinder (for example, the Lambert cylindrical equal-area projection) is area-preserving; it equals the derivative of the formula for the volume with respect to r because the total volume inside a sphere of radius r can be thought of as the summation of the surface area of an infinite number of spherical shells of infinitesimal thickness concentrically stacked inside one another from radius 0 to radius r. At infinitesimal thickness the discrepancy between the inner and outer surface area of any given shell is infinitesimal, and the elemental volume at radius r is simply the product of the surface area at radius r and the infinitesimal thickness.

At any given radius r, the incremental volume (δV) equals the product of the surface area at radius r ($A(r)$) and the thickness of a shell (δr):

$$\delta V \approx A(r) \cdot \delta r.$$

The total volume is the summation of all shell volumes:

$$V \approx \sum A(r) \cdot r.$$

In the limit as δr approaches zero this equation becomes:

$$V = \int_0^r A(r')dr'.$$

Substitute V:

$$\frac{4}{3}\pi r^3 = \int_0^r A(r')dr'.$$

Differentiating both sides of this equation with respect to r yields A as a function of r:

$$4\pi r^2 = A(r).$$

Which is generally abbreviated as:

$$A = 4\pi r^2.$$

Alternatively, the area element on the sphere is given in spherical coordinates by $dA = r^2 \sin\theta\, d\theta\, d\varphi$. In Cartesian coordinates, the area element is

$$dS = \frac{r}{\sqrt{r^2 - \sum_{i \neq k} x_i^2}} \prod_{i \neq k} dx_i, \; \forall k.$$

The total area can thus be obtained by integration:

$$A = \int_0^{2\pi} \int_0^{\pi} r^2 \sin\theta\, d\theta\, d\varphi = 4\pi r^2.$$

Enclosed Volume

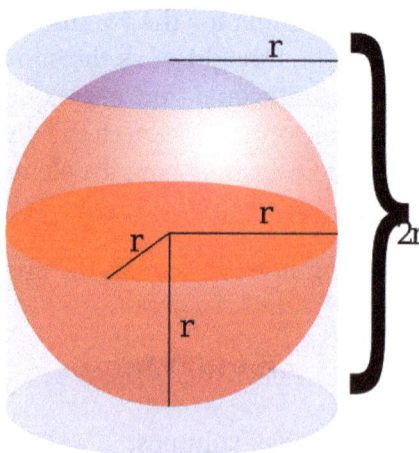

Circumscribed cylinder to a sphere

In three dimensions, the volume inside a sphere (that is the volume of a ball) is derived to be

$$V = \frac{4}{3}\pi r^3$$

where r is the radius of the sphere. Archimedes first derived this formula, which shows that the volume inside a sphere is 2/3 that of a circumscribed cylinder. (This assertion follows from Cavalieri's principle.) In modern mathematics, this formula can be derived using integral calculus, i.e. disk integration to sum the volumes of an infinite number of circular disks of infinitesimally small thickness stacked centered side by side along the x-axis from $x = 0$ where the disk has radius r (i.e. $y = r$) to $x = r$ where the disk has radius 0 (i.e. $y = 0$).

At any given x, the incremental volume (δV) equals the product of the cross-sectional area of the disk at x and its thickness (δx):

$$\delta V \approx \pi y^2 \cdot \delta x.$$

The total volume is the summation of all incremental volumes:

$$V \approx \sum \pi y^2 \cdot \delta x.$$

In the limit as δx approaches zero this equation becomes:

$$V = \int_{-r}^{r} \pi y^2 dx.$$

At any given x, a right-angled triangle connects x, y and r to the origin; hence, applying the Pythagorean theorem yields:

$$y^2 = r^2 - x^2.$$

Thus, substituting y with a function of x gives:

$$V = \int_{-r}^{r} \pi (r^2 - x^2) dx.$$

Which can now be evaluated as follows:

$$V = \pi \left[r^2 x - \frac{x^3}{3} \right]_{-r}^{r} = \pi \left(r^3 - \frac{r^3}{3} \right) - \pi \left(-r^3 + \frac{r^3}{3} \right) = \frac{4}{3}\pi r^3.$$

Therefore the volume of a sphere is:

$$V = \frac{4}{3}\pi r^3.$$

Alternatively this formula is found using spherical coordinates, with volume element

$$dV = r^2 \sin\theta \, dr \, d\theta \, d\varphi$$

so

$$V = \int_0^{2\pi}\int_0^{\pi}\int_0^r r'^2 \sin\theta \, dr' \, d\theta \, d\varphi = \frac{4}{3}\pi r^3$$

For most practical purposes, the volume inside a sphere inscribed in a cube can be approximated as 52.4% of the volume of the cube, since $\pi/6 \approx 0.5236$. For example, a sphere with diameter 1 meter has 52.4% the volume of a cube with edge length 1 meter, or about 0.524 m³.

In higher dimensions, the analog of a sphere is called a hypersphere, which encloses an n-ball. General recursive and non-recursive formulas exist for the volume of an n-ball.

Equations in Three-dimensional Space

In analytic geometry, a sphere with center (x_0, y_0, z_0) and radius r is the locus of all points (x, y, z) such that

$$(x-x_0)^2 + (y-y_0)^2 + (z-z_0)^2 = r^2.$$

The points on the sphere with radius r can be parameterized via

$$x = x_0 + r\cos\theta \sin\varphi$$
$$y = y_0 + r\sin\theta \sin\varphi \qquad (0 \leq \theta \leq 2\pi \text{ and } 0 \leq \varphi \leq \pi)$$
$$z = z_0 + r\cos\varphi$$

A sphere of any radius centered at zero is an integral surface of the following differential form:

$$x\,dx + y\,dy + z\,dz = 0.$$

This equation reflects that position and velocity vectors of a point traveling on the sphere are always orthogonal to each other.

The sphere has the smallest surface area of all surfaces that enclose a given volume, and it encloses the largest volume among all closed surfaces with a given surface area. The sphere therefore appears in nature: for example, bubbles and small water drops are roughly spherical because the surface tension locally minimizes surface area.

The surface area relative to the mass of a sphere is called the specific surface area and can be expressed from the above stated equations as

$$\text{SSA} = \frac{A}{V\rho} = \frac{3}{r\rho},$$

where ρ is the density (the ratio of mass to volume).

A sphere can also be defined as the surface formed by rotating a circle about any diameter. Replacing the circle with an ellipse rotated about its major axis, the shape becomes a prolate spheroid; rotated about the minor axis, an oblate spheroid.

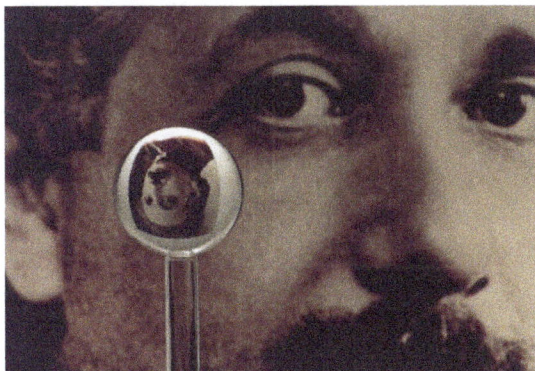

An image of one of the most accurate human-made spheres, as it refracts the image of Einstein in the background. This sphere was a fused quartz gyroscope for the Gravity Probe B experiment, and differs in shape from a perfect sphere by no more than 40 atoms (less than 10 nanometers) of thickness. It was announced on 1 July 2008 that Australian scientists had created even more nearly perfect spheres, accurate to 0.3 nanometers, as part of an international hunt to find a new global standard kilogram.

Terminology

Pairs of points on a sphere that lie on a straight line through the sphere's center are called antipodal points. A great circle is a circle on the sphere that has the same center and radius as the sphere and consequently divides it into two equal parts. The shortest distance along the surface between two distinct non-antipodal points on the surface is on the unique great circle that includes the two points. Equipped with the great-circle distance, a great circle becomes the Riemannian circle.

If a particular point on a sphere is (arbitrarily) designated as its *north pole*, then the corresponding antipodal point is called the *south pole*, and the equator is the great circle that is equidistant to them. Great circles through the two poles are called lines (or meridians) of longitude, and the line connecting the two poles is called the axis of rotation. Circles on the sphere that are parallel to the equator are lines of latitude. This terminology is also used for such approximately spheroidal astronomical bodies as the planet Earth.

Hemisphere

Any plane that includes the center of a sphere divides it into two equal hemispheres. Any two intersecting planes that include the center of a sphere subdivide the sphere into four lunes or biangles, the vertices of which all coincide with the antipodal points lying on the line of intersection of the planes.

The antipodal quotient of the sphere is the surface called the real projective plane, which can also be thought of as the northern hemisphere with antipodal points of the equator identified.

The round hemisphere is conjectured to be the optimal (least area) filling of the Riemannian circle.

The circles of intersection of any plane not intersecting the sphere's center and the sphere's surface are called *spheric sections*.

Generalizations

Dimensionality

Spheres can be generalized to spaces of any number of dimensions. For any natural number n, an "n-sphere," often written as S^n, is the set of points in $(n + 1)$-dimensional Euclidean space that are at a fixed distance r from a central point of that space, where r is, as before, a positive real number. In particular:

- S^0: a 0-sphere is a pair of endpoints of an interval $[-r, r]$ of the real line

- S^1: a 1-sphere is a circle of radius r

- S^2: a 2-sphere is an ordinary sphere

- S^3: a 3-sphere is a sphere in 4-dimensional Euclidean space.

Spheres for $n > 2$ are sometimes called hyperspheres.

The n-sphere of unit radius centered at the origin is denoted S^n and is often referred to as "the" n-sphere. Note that the ordinary sphere is a 2-sphere, because it is a 2-dimensional surface (which is embedded in 3-dimensional space).

The surface area of the $(n - 1)$-sphere of radius 1 is

$$\frac{2\pi^{n/2}}{\Gamma\left(\dfrac{n}{2}\right)}$$

where $\Gamma(z)$ is Euler's gamma function.

Another expression for the surface area is

$$\begin{cases} \dfrac{(2\pi)^{n/2} r^{n-1}}{2 \cdot 4 \cdots (n-2)}, & \text{if } n \text{ is even;} \\[4ex] \dfrac{2(2\pi)^{(n-1)/2} r^{n-1}}{1 \cdot 3 \cdots (n-2)}, & \text{if } n \text{ is odd.} \end{cases}$$

and the volume is the surface area times r/n or

$$\begin{cases} \dfrac{(2 \ )^{n/2} \ ^n}{2 \cdot 4 \ }, & \text{if } \text{ is even;} \\[4ex] \dfrac{2(2 \ )^{(n \ 1)/2} \ ^n}{1 \cdot 3 \ }, & \text{if } \text{ is odd.} \end{cases}$$

Metric Spaces

More generally, in a metric space (E,d), the sphere of center x and radius $r > 0$ is the set of points y such that $d(x,y) = r$.

If the center is a distinguished point that is considered to be the origin of E, as in a normed space, it is not mentioned in the definition and notation. The same applies for the radius if it is taken to equal one, as in the case of a unit sphere.

Unlike a ball, even a large sphere may be an empty set. For example, in Z^n with Euclidean metric, a sphere of radius r is nonempty only if r^2 can be written as sum of n squares of integers.

Topology

In topology, an n-sphere is defined as a space homeomorphic to the boundary of an $(n + 1)$-ball; thus, it is homeomorphic to the Euclidean n-sphere, but perhaps lacking its metric.

- A 0-sphere is a pair of points with the discrete topology.

- A 1-sphere is a circle (up to homeomorphism); thus, for example, (the image of) any knot is a 1-sphere.

- A 2-sphere is an ordinary sphere (up to homeomorphism); thus, for example, any spheroid is a 2-sphere.

The n-sphere is denoted S^n. It is an example of a compact topological manifold without boundary. A sphere need not be smooth; if it is smooth, it need not be diffeomorphic to the Euclidean sphere.

The Heine–Borel theorem implies that a Euclidean n-sphere is compact. The sphere is the inverse image of a one-point set under the continuous function $||x||$. Therefore, the sphere is closed. S^n is also bounded; therefore it is compact.

Remarkably, it is possible to turn an ordinary sphere inside out in a three-dimensional space with possible self-intersections but without creating any crease, in a process called sphere eversion.

Spherical Geometry

Great circle on a sphere

The basic elements of Euclidean plane geometry are points and lines. On the sphere, points are defined in the usual sense. The analogue of the "line" is the geodesic, which is a great circle; the defining characteristic of the latter is that the plane containing all its points also passes through the center of the sphere. Measuring by arc length shows that the shortest path between two points lying entirely on the sphere is a segment of the great circle that includes the points.

Many theorems from classical geometry hold true for spherical geometry as well, but not all do because the sphere fails to satisfy some of classical geometry's postulates, including the parallel postulate. In spherical trigonometry, angles are defined between great circles. Thus spherical trigonometry differs from ordinary trigonometry in many respects. For example, the sum of the interior angles of a spherical triangle exceeds 180 degrees. Also, any two similar spherical triangles are congruent.

Eleven Properties of the Sphere

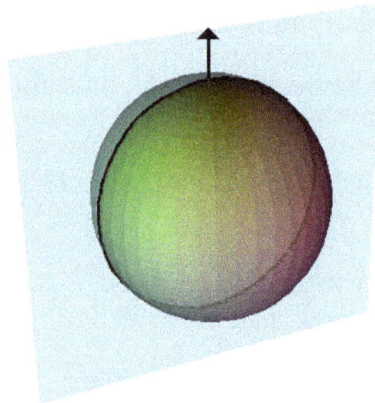

A normal vector to a sphere, a normal plane and its normal section. The curvature of the curve of intersection is the sectional curvature. For the sphere each normal section through a given point will be a circle of the same radius, the radius of the sphere. This means that every point on the sphere will be an umbilical point.

In their book *Geometry and the Imagination* David Hilbert and Stephan Cohn-Vossen describe eleven properties of the sphere and discuss whether these properties uniquely determine the sphere. Several properties hold for the plane, which can be thought of as a sphere with infinite radius. These properties are:

1. *The points on the sphere are all the same distance from a fixed point. Also, the ratio of the distance of its points from two fixed points is constant.*

    The first part is the usual definition of the sphere and determines it uniquely. The second part can be easily deduced and follows a similar result of Apollonius of Perga for the circle. This second part also holds for the plane.

2. *The contours and plane sections of the sphere are circles.*

    This property defines the sphere uniquely.

3. *The sphere has constant width and constant girth.*

The width of a surface is the distance between pairs of parallel tangent planes. Numerous other closed convex surfaces have constant width, for example the Meissner body. The girth of a surface is the circumference of the boundary of its orthogonal projection on to a plane. Each of these properties implies the other.

4. *All points of a sphere are umbilics.*

At any point on a surface a normal direction is at right angles to the surface because the sphere these are the lines radiating out from the center of the sphere. The intersection of a plane that contains the normal with the surface will form a curve that is called a *normal section,* and the curvature of this curve is the *normal curvature.* For most points on most surfaces, different sections will have different curvatures; the maximum and minimum values of these are called the principal curvatures. Any closed surface will have at least four points called *umbilical points.* At an umbilic all the sectional curvatures are equal; in particular the principal curvatures are equal. Umbilical points can be thought of as the points where the surface is closely approximated by a sphere.

For the sphere the curvatures of all normal sections are equal, so every point is an umbilic. The sphere and plane are the only surfaces with this property.

5. *The sphere does not have a surface of centers.*

For a given normal section exists a circle of curvature that equals the sectional curvature, is tangent to the surface, and the center lines of which lie along on the normal line. For example, the two centers corresponding to the maximum and minimum sectional curvatures are called the *focal points,* and the set of all such centers forms the focal surface.

For most surfaces the focal surface forms two sheets that are each a surface and meet at umbilical points. Several cases are special:

- o   For channel surfaces one sheet forms a curve and the other sheet is a surface

- o   For cones, cylinders, tori and cyclides both sheets form curves.

- o   For the sphere the center of every osculating circle is at the center of the sphere and the focal surface forms a single point. This property is unique to the sphere.

6. *All geodesics of the sphere are closed curves.*

Geodesics are curves on a surface that give the shortest distance between two points. They are a generalization of the concept of a straight line in the plane. For the sphere the geodesics are great circles. Many other surfaces share this property.

7. *Of all the solids having a given volume, the sphere is the one with the smallest surface area; of all solids having a given surface area, the sphere is the one having the greatest volume.*

It follows from isoperimetric inequality. These properties define the sphere uniquely and can be seen in soap bubbles: a soap bubble will enclose a fixed volume, and surface tension minimizes its surface area for that volume. A freely floating soap bubble therefore approximates a sphere (though such external forces as gravity will slightly distort the bubble's shape).

8. *The sphere has the smallest total mean curvature among all convex solids with a given surface area.*

The mean curvature is the average of the two principal curvatures, which is constant because the two principal curvatures are constant at all points of the sphere.

9. *The sphere has constant mean curvature.*

The sphere is the only imbedded surface that lacks boundary or singularities with constant positive mean curvature. Other such immersed surfaces as minimal surfaces have constant mean curvature.

10. *The sphere has constant positive Gaussian curvature.*

Gaussian curvature is the product of the two principal curvatures. It is an intrinsic property that can be determined by measuring length and angles and is independent of how the surface is embedded in space. Hence, bending a surface will not alter the Gaussian curvature, and other surfaces with constant positive Gaussian curvature can be obtained by cutting a small slit in the sphere and bending it. All these other surfaces would have boundaries, and the sphere is the only surface that lacks a boundary with constant, positive Gaussian curvature. The pseudosphere is an example of a surface with constant negative Gaussian curvature.

11. *The sphere is transformed into itself by a three-parameter family of rigid motions.*

Rotating around any axis a unit sphere at the origin will map the sphere onto itself. Any rotation about a line through the origin can be expressed as a combination of rotations around the three-coordinate axis. Therefore, a three-parameter family of rotations exists such that each rotation transforms the sphere onto itself; this family is the rotation group SO(3). The plane is the only other surface with a three-parameter family of transformations (translations along the x- and y-axes and rotations around the origin). Circular cylinders are the only surfaces with two-parameter families of rigid motions and the surfaces of revolution and helicoids are the only surfaces with a one-parameter family.

Cubes in Relation to Spheres

For every sphere there are multiple cuboids that may be inscribed within the sphere. The largest cuboid which can be inscribed within a sphere is a cube.

References

- Büeler, B.; Enge, A.; Fukuda, K. (2000). "Exact Volume Computation for Polytopes: A Practical Study". Polytopes — Combinatorics and Computation. p. 131. doi:10.1007/978-3-0348-8438-9_6. ISBN 978-3-7643-6351-2.

- Hilbert, David; Cohn-Vossen, Stephan (1952). Geometry and the Imagination (2nd ed.). Chelsea. ISBN 0-8284-1087-9.

- Dunham, William. The Mathematical Universe: An Alphabetical Journey Through the Great Proofs, Problems and Personalities. pp. 28, 226. ISBN 0-471-17661-3..

Understanding Trigonometry

The branch of mathematics that studies the relation between lengths and angles is termed as trigonometry. Some of the fields that use trigonometry are astronomy, electronics, chemistry, seismology and oceanography. Trigonometry in today's times has immense number of purposes.

Trigonometry

Trigonometry is a branch of mathematics that studies relationships involving lengths and angles of triangles. The field emerged in the Hellenistic world during the 3rd century BC from applications of geometry to astronomical studies.

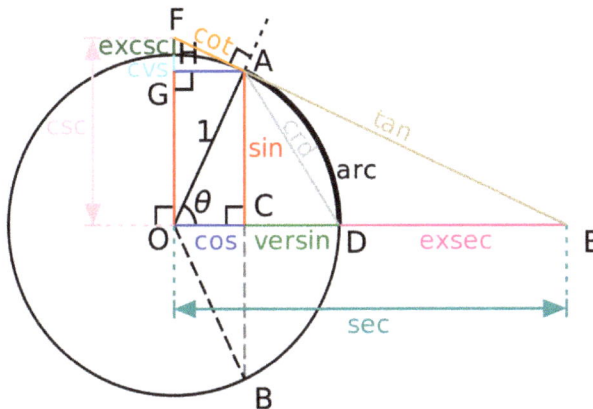

All of the trigonometric functions of an angle θ can be constructed geometrically in terms of a unit circle centered at O.

Trigonometry

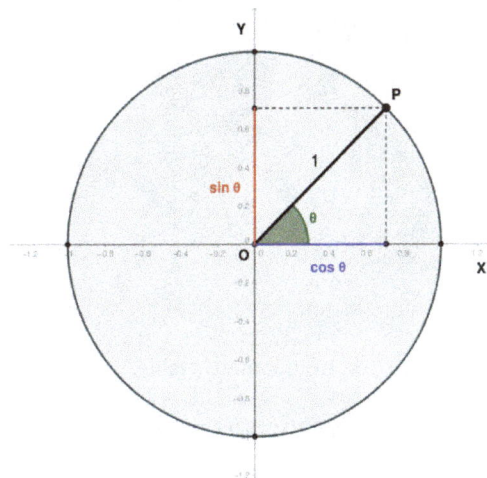

The 3rd-century astronomers first noted that the lengths of the sides of a right-angle triangle and the angles between those sides have fixed relationships: that is, if at least the length of one side and the value of one angle is known, then all other angles and lengths can be determined algorithmically. These calculations soon came to be defined as the trigonometric functions and today are pervasive in both pure and applied mathematics: fundamental methods of analysis such as the Fourier transform, for example, or the wave equation, use trigonometric functions to understand cyclical phenomena across many applications in fields as diverse as physics, mechanical and electrical engineering, music and acoustics, astronomy, ecology, and biology. Trigonometry is also the foundation of surveying.

Trigonometry is most simply associated with planar right-angle triangles (each of which is a two-dimensional triangle with one angle equal to 90 degrees). The applicability to non-right-angle triangles exists, but, since any non-right-angle triangle (on a flat plane) can be bisected to create two right-angle triangles, most problems can be reduced to calculations on right-angle triangles. Thus the majority of applications relate to right-angle triangles. One exception to this is spherical trigonometry, the study of triangles on spheres, surfaces of constant positive curvature, in elliptic geometry (a fundamental part of astronomy and navigation). Trigonometry on surfaces of negative curvature is part of hyperbolic geometry.

Trigonometry basics are often taught in schools, either as a separate course or as a part of a pre-calculus course.

History

Sumerian astronomers studied angle measure, using a division of circles into 360 degrees. They, and later the Babylonians, studied the ratios of the sides of similar triangles and discovered some properties of these ratios but did not turn that into a systematic method for finding sides and angles of triangles. The ancient Nubians used a similar method.

Hipparchus, credited with compiling the first trigonometric table, is known as "the father of trigonometry".

In the 3rd century BC, Hellenistic mathematicians such as Euclid and Archimedes studied the properties of chords and inscribed angles in circles, and they proved theorems that are equivalent to modern trigonometric formulae, although they presented them geometrically rather than algebraically. In 140 BC, Hipparchus (from Nicaea, Asia Minor) gave the first tables of chords, analogous to modern tables of sine values, and used them to solve problems in trigonometry and spherical trigonometry. In the 2nd century AD, the Greco-Egyptian astronomer Ptolemy (from Alexandria, Egypt) printed detailed trigonometric tables (Ptolemy's table of chords) in Book 1, chapter 11 of his Almagest. Ptolemy used chord length to define his trigonometric functions, a minor difference from the sine convention we use today. (The value we call sin(θ) can be found by looking up the chord length for twice the angle of interest (2θ) in Ptolemy's table, and then dividing that value by two.) Centuries passed before more detailed tables were produced, and Ptolemy's treatise remained in use for performing trigonometric calculations in astronomy throughout the next 1200 years in the medieval Byzantine, Islamic, and, later, Western European worlds.

The modern sine convention is first attested in the *Surya Siddhanta*, and its properties were further documented by the 5th century (AD) Indian mathematician and astronomer Aryabhata. These Greek and Indian works were translated and expanded by medieval Islamic mathematicians. By the 10th century, Islamic mathematicians were using all six trigonometric functions, had tabulated their values, and were applying them to problems in spherical geometry. At about the same time, Chinese mathematicians developed trigonometry independently, although it was not a major field of study for them. Knowledge of trigonometric functions and methods reached Western Europe via Latin translations of Ptolemy's Greek Almagest as well as the works of Persian and Arabic astronomers such as Al Battani and Nasir al-Din al-Tusi. One of the earliest works on trigonometry by a northern European mathematician is *De Triangulis* by the 15th century German mathematician Regiomontanus, who was encouraged to write, and provided with a copy of the Almagest, by the Byzantine Greek scholar cardinal Basilios Bessarion with whom he lived for several years. At the same time, another translation of the Almagest from Greek into Latin was completed by the Cretan George of Trebizond. Trigonometry was still so little known in 16th-century northern Europe that Nicolaus Copernicus devoted two chapters of *De revolutionibus orbium coelestium* to explain its basic concepts.

Driven by the demands of navigation and the growing need for accurate maps of large geographic areas, trigonometry grew into a major branch of mathematics. Bartholomaeus Pitiscus was the first to use the word, publishing his *Trigonometria* in 1595. Gemma Frisius described for the first time the method of triangulation still used today in surveying. It was Leonhard Euler who fully incorporated complex numbers into trigonometry. The works of the Scottish mathematicians James Gregory in the 17th century and Colin Maclaurin in the 18th century were influential in the development of trigonometric series. Also in the 18th century, Brook Taylor defined the general Taylor series.

Overview

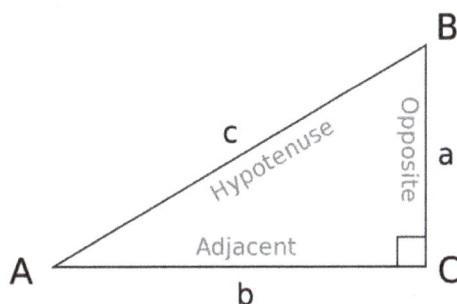

In this right triangle: $\sin A = a/c$; $\cos A = b/c$; $\tan A = a/b$.

If one angle of a triangle is 90 degrees and one of the other angles is known, the third is thereby fixed, because the three angles of any triangle add up to 180 degrees. The two acute angles therefore add up to 90 degrees: they are complementary angles. The shape of a triangle is completely determined, except for similarity, by the angles. Once the angles are known, the ratios of the sides are determined, regardless of the overall size of the triangle. If the length of one of the sides is known, the other two are determined. These ratios are given by the following trigonometric functions of the known angle A, where a, b and c refer to the lengths of the sides in the accompanying figure:

- Sine function (sin), defined as the ratio of the side opposite the angle to the hypotenuse.

$$\sin A = \frac{\text{opposite}}{\text{hypotenuse}} = \frac{a}{c}.$$

- Cosine function (cos), defined as the ratio of the adjacent leg to the hypotenuse.

$$\cos A = \frac{\text{adjacent}}{\text{hypotenuse}} = \frac{b}{c}.$$

- Tangent function (tan), defined as the ratio of the opposite leg to the adjacent leg.

$$\tan A = \frac{\text{opposite}}{\text{adjacent}} = \frac{a}{b} = \frac{a}{c} * \frac{c}{b} = \frac{a}{c} / \frac{b}{c} = \frac{\sin A}{\cos A}.$$

The hypotenuse is the side opposite to the 90 degree angle in a right triangle; it is the longest side of the triangle and one of the two sides adjacent to angle A. The adjacent leg is the other side that is adjacent to angle A. The opposite side is the side that is opposite to angle A. The terms perpendicular and base are sometimes used for the opposite and adjacent sides respectively. Many people find it easy to remember what sides of the right triangle are equal to sine, cosine, or tangent, by memorizing the word SOH-CAH-TOA.

The reciprocals of these functions are named the cosecant (csc or cosec), secant (sec), and cotangent (cot), respectively:

$$\csc A = \frac{1}{\sin A} = \frac{\text{hypotenuse}}{\text{opposite}} = \frac{c}{a},$$

$$\sec A = \frac{1}{\cos A} = \frac{\text{hypotenuse}}{\text{adjacent}} = \frac{c}{b},$$

$$\sec A = \frac{1}{\cos A} = \frac{\text{hypotenuse}}{\text{adjacent}} = \frac{c}{b},$$

The inverse functions are called the arcsine, arccosine, and arctangent, respectively. There are arithmetic relations between these functions, which are known as trigonometric identities. The cosine, cotangent, and cosecant are so named because they are respectively the sine, tangent, and secant of the complementary angle abbreviated to "co-".

With these functions, one can answer virtually all questions about arbitrary triangles by using the law of sines and the law of cosines. These laws can be used to compute the remaining angles and sides of any triangle as soon as two sides and their included angle or two angles and a side or three sides are known. These laws are useful in all branches of geometry, since every polygon may be described as a finite combination of triangles.

Extending the Definitions

The above definitions only apply to angles between 0 and 90 degrees (0 and $\pi/2$ radians). Using the unit circle, one can extend them to all positive and negative arguments. The trigonometric functions are periodic, with a period of 360 degrees or 2π radians. That means their values repeat at those intervals. The tangent and cotangent functions also have a shorter period, of 180 degrees or π radians.

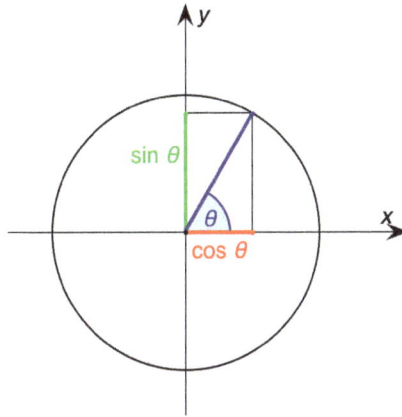

Fig. 1a – Sine and cosine of an angle θ defined using the unit circle.

The trigonometric functions can be defined in other ways besides the geometrical definitions above, using tools from calculus and infinite series. With these definitions the trigonometric functions can be defined for complex numbers. The complex exponential function is particularly useful.

$$e^{x+iy} = e^x (\cos y + i \sin y).$$

See Euler's and De Moivre's formulas.

Graphing process of $y = \sin(x)$ using a unit circle.

Graphing process of $y = \csc(x)$, the reciprocal of sine, using a unit circle.

Graphing process of $y = \tan(x)$ using a unit circle.

Mnemonics

A common use of mnemonics is to remember facts and relationships in trigonometry. For example, the *sine*, *cosine*, and *tangent* ratios in a right triangle can be remembered by representing them and their corresponding sides as strings of letters. For instance, a mnemonic is SOH-CAH-TOA:

Sine = Opposite ÷ Hypotenuse

Cosine = Adjacent ÷ Hypotenuse

Tangent = Opposite ÷ Adjacent

One way to remember the letters is to sound them out phonetically (i.e., *SOH-CAH-TOA*, which is pronounced 'so-kə-toe-uh' /soʊkəˈtoʊə/). Another method is to expand the letters into a sentence, such as "Some Old Hippie Caught Another Hippie Trippin' On Acid".

Calculating Trigonometric Functions

Trigonometric functions were among the earliest uses for mathematical tables. Such tables were incorporated into mathematics textbooks and students were taught to look up values and how to interpolate between the values listed to get higher accuracy. Slide rules had special scales for trigonometric functions.

Today, scientific calculators have buttons for calculating the main trigonometric functions (sin, cos, tan, and sometimes cis and their inverses). Most allow a choice of angle measurement methods: degrees, radians, and sometimes gradians. Most computer programming languages provide function libraries that include the trigonometric functions. The floating point unit hardware incorporated into the microprocessor chips used in most personal computers has built-in instructions for calculating trigonometric functions.

Applications of Trigonometry

Sextants are used to measure the angle of the sun or stars with respect to the horizon. Using trigonometry and a marine chronometer, the position of the ship can be determined from such measurements.

There is an enormous number of uses of trigonometry and trigonometric functions. For instance, the technique of triangulation is used in astronomy to measure the distance to nearby stars, in geography to measure distances between landmarks, and in satellite navigation systems. The sine and cosine functions are fundamental to the theory of periodic functions, such as those that describe sound and light waves.

Fields that use trigonometry or trigonometric functions include astronomy (especially for locating apparent positions of celestial objects, in which spherical trigonometry is essential) and hence navigation (on the oceans, in aircraft, and in space), music theory, audio synthesis, acoustics, optics, electronics, biology, medical imaging (CAT scans and ultrasound), pharmacy, chemistry, number theory (and hence cryptology), seismology, meteorology, oceanography, many physical sciences, land surveying and geodesy, architecture, image compression, phonetics, economics, electrical engineering, mechanical engineering, civil engineering, computer graphics, cartography, crystallography and game development.

Pythagorean Identities

The following identities are related to the Pythagorean theorem and hold for any value:

$$\sin^2 A + \cos^2 A = 1$$

$$\tan^2 A + 1 = \sec^2 A$$

$$\cot^2 A + 1 = \csc^2 A$$

Angle Transformation Formulae

$$\sin(A \pm B) = \sin A \cos B \pm \cos A \sin B$$

$$\cos(A \pm B) = \cos A \cos B \mp \sin A \sin B$$

$$\tan(A \pm B) = \frac{\tan A \pm \tan B}{1 \mp \tan A \tan B}$$

$$\cot(A \pm B) = \frac{\cot A \cot B \mp 1}{\cot B \pm \cot A}$$

Common Formulae

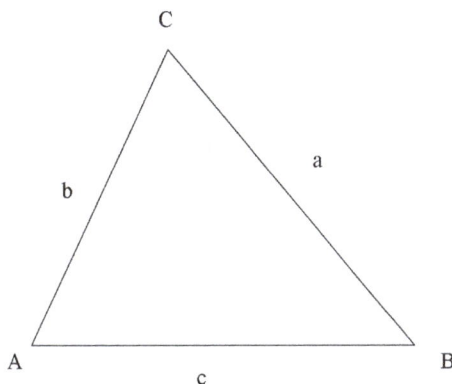

Triangle with sides a,b,c and respectively opposite angles A,B,C

Certain equations involving trigonometric functions are true for all angles and are known as *trigonometric identities*. Some identities equate an expression to a different expression involving the same angles. These are listed in List of trigonometric identities. Triangle identities that relate the sides and angles of a given triangle are listed below.

In the following identities, *A*, *B* and *C* are the angles of a triangle and *a*, *b* and *c* are the lengths of sides of the triangle opposite the respective angles (as shown in the diagram).

Law of Sines

The law of sines (also known as the "sine rule") for an arbitrary triangle states:

$$\frac{a}{\sin A} = \frac{b}{\sin B} = \frac{c}{\sin C} = 2R = \frac{abc}{2\Delta},$$

where Δ is the area of the triangle and *R* is the radius of the circumscribed circle of the triangle:

$$R = \frac{abc}{\sqrt{(a+b+c)(a-b+c)(a+b-c)(b+c-a)}}.$$

Another law involving sines can be used to calculate the area of a triangle. Given two sides *a* and *b* and the angle between the sides *C*, the area of the triangle is given by half the product of the lengths of two sides and the sine of the angle between the two sides:

$$\text{Area} = \Delta = \frac{1}{2}ab\sin C.$$

Law of Cosines

The law of cosines (known as the cosine formula, or the "cos rule") is an extension of the Pythagorean theorem to arbitrary triangles:

$$c^2 = a^2 + b^2 - 2ab\cos C,$$

or equivalently:

$$\cos C = \frac{a^2 + b^2 - c^2}{2ab}.$$

The law of cosines may be used to prove Heron's formula, which is another method that may be used to calculate the area of a triangle. This formula states that if a triangle has sides of lengths *a*, *b*, and *c*, and if the semiperimeter is

$$s = \frac{1}{2}(a+b+c),$$

then the area of the triangle is:

$$\text{Area} = \Delta = \sqrt{s(s-a)(s-b)(s-c)} = \frac{abc}{4R},$$

where R is the radius of the circumcircle of the triangle.

Law of Tangents

The law of tangents:

$$\frac{a-b}{a+b} = \frac{\tan\left[\frac{1}{2}(A-B)\right]}{\tan\left[\frac{1}{2}(A+B)\right]}$$

Euler's Formula

Euler's formula, which states that $e^{ix} = \cos x + i\sin x$, produces the following analytical identities for sine, cosine, and tangent in terms of e and the imaginary unit i:

$$\sin x = \frac{e^{ix} - e^{-ix}}{2i}, \qquad \cos x = \frac{e^{ix} + e^{-ix}}{2}, \qquad \tan x = \frac{i(e^{-ix} - e^{ix})}{e^{ix} + e^{-ix}}.$$

Trigonometric Functions

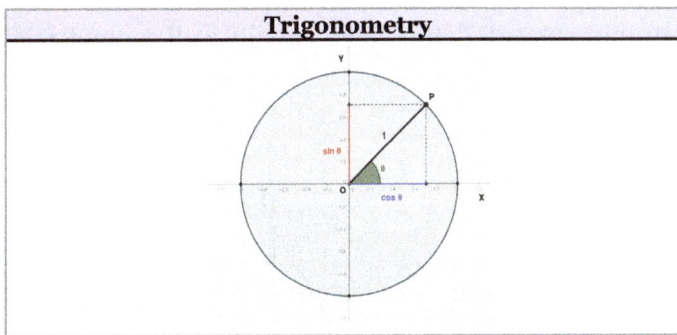

Trigonometry

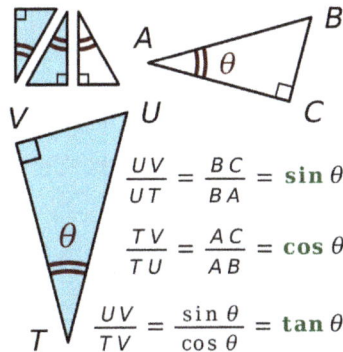

$$\frac{UV}{UT} = \frac{BC}{BA} = \sin\theta$$

$$\frac{TV}{TU} = \frac{AC}{AB} = \cos\theta$$

$$\frac{UV}{TV} = \frac{\sin\theta}{\cos\theta} = \tan\theta$$

Base of trigonometry: if two right triangles have equal acute angles, they are similar, so their side lengths are proportional. Proportionality constants are written within the image: $\sin\theta$, $\cos\theta$, $\tan\theta$, where θ is the common measure of five acute angles.

In mathematics, the trigonometric functions (also called the circular functions) are functions of an angle. They relate the angles of a triangle to the lengths of its sides. Trigonometric functions are important in the study of triangles and modeling periodic phenomena, among many other applications.

The most familiar trigonometric functions are the sine, cosine, and tangent. In the context of the standard unit circle (a circle with radius 1 unit), where a triangle is formed by a ray starting at the origin and making some angle with the x-axis, the sine of the angle gives the length of the y-component (the opposite to the angle or the rise) of the triangle, the cosine gives the length of the x-component (the adjacent of the angle or the run), and the tangent function gives the slope (y-component divided by the x-component). More precise definitions are detailed below. Trigonometric functions are commonly defined as ratios of two sides of a right triangle containing the angle, and can equivalently be defined as the lengths of various line segments from a unit circle. More modern definitions express them as infinite series or as solutions of certain differential equations, allowing their extension to arbitrary positive and negative values and even to complex numbers.

Trigonometric functions have a wide range of uses including computing unknown lengths and angles in triangles (often right triangles). In this use, trigonometric functions are used, for instance, in navigation, engineering, and physics. A common use in elementary physics is resolving a vector into Cartesian coordinates. The sine and cosine functions are also commonly used to model periodic function phenomena such as sound and light waves, the position and velocity of harmonic oscillators, sunlight intensity and day length, and average temperature variations through the year.

In modern usage, there are six basic trigonometric functions, tabulated here with equations that relate them to one another. Especially with the last four, these relations are often taken as the *definitions* of those functions, but one can define them equally well geometrically, or by other means, and then derive these relations.

Right-angled Triangle Definitions

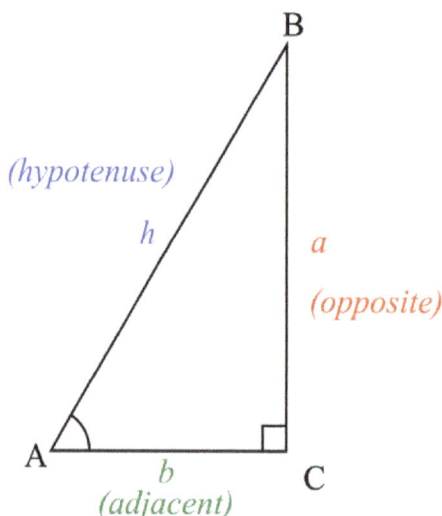

The notion that there should be some standard correspondence between the lengths of the sides of a triangle and the angles of the triangle comes as soon as one recognizes that similar triangles maintain the same ratios between their sides. That is, for any similar triangle the ratio of the hypotenuse (for example) and another of the sides remains the same. If the hypotenuse is twice as long, so are the sides. It is these ratios that the trigonometric functions express.

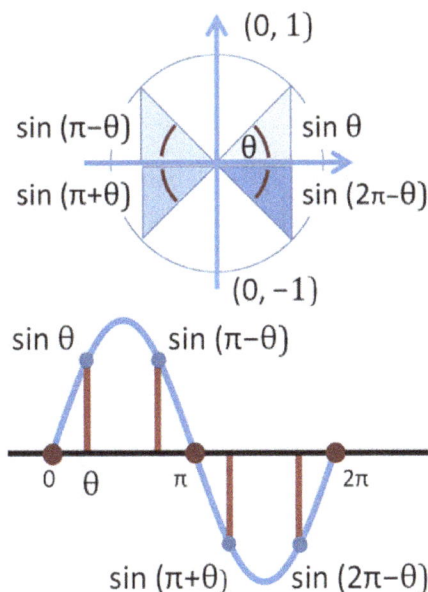

Top: Trigonometric function $\sin\theta$ for selected angles θ, $\pi - \theta$, $\pi + \theta$, and $2\pi - \theta$ in the four quadrants.
Bottom: Graph of sine function versus angle. Angles from the top panel are identified.

To define the trigonometric functions for the angle A, start with any right triangle that contains the angle A. The three sides of the triangle are named as follows:

- The *hypotenuse* is the side opposite the right angle, in this case side **h**. The hypotenuse is always the longest side of a right-angled triangle.

- The *opposite side* is the side opposite to the angle we are interested in (angle A), in this case side **a**.

- The *adjacent side* is the side having both the angles of interest (angle A and right-angle C), in this case side **b**.

In ordinary Euclidean geometry, according to the triangle postulate, the inside angles of every triangle total 180° (π radians). Therefore, in a right-angled triangle, the two non-right angles total 90° ($\pi/2$ radians), so each of these angles must be in the range of (0°,90°) as expressed in interval notation. The following definitions apply to angles in this 0° – 90° range. They can be extended to the full set of real arguments by using the unit circle, or by requiring certain symmetries and that they be periodic functions. For example, the figure shows $\sin(\theta)$ for angles θ, $\pi - \theta$, $\pi + \theta$, and $2\pi - \theta$ depicted on the unit circle (top) and as a graph (bottom). The value of the sine repeats itself apart from sign in all four quadrants, and if the range of θ is extended to additional rotations, this behavior repeats periodically with a period 2π.

The trigonometric functions are summarized in the following table and described in more detail below. The angle θ is the angle between the hypotenuse and the adjacent line – the angle at A in the accompanying diagram.

Function	Abbreviation	Description	Identities (using radians)
sine	sin	opposite/ hypotenuse	$\sin\theta = \cos\left(\dfrac{\pi}{2}-\theta\right) = \dfrac{1}{\csc\theta}$
cosine	cos	adjacent/ hypotenuse	$\cos\theta = \sin\left(\dfrac{\pi}{2}-\theta\right) = \dfrac{1}{\sec\theta}$
tangent	tan (or tg)	opposite/ adjacent	$\tan\theta = \dfrac{\sin\theta}{\cos\theta} = \cot\left(\dfrac{\pi}{2}-\theta\right) = \dfrac{1}{\cot\theta}$
cotangent	cot (or cotan or cotg or ctg or ctn)	adjacent/ opposite	$\cot\theta = \dfrac{\cos\theta}{\sin\theta} = \tan\left(\dfrac{\pi}{2}-\theta\right) = \dfrac{1}{\tan\theta}$
secant	sec	hypotenuse/ adjacent	$\sec\theta = \csc\left(\dfrac{\pi}{2}-\theta\right) = \dfrac{1}{\cos\theta}$
cosecant	csc (or cosec)	hypotenuse/ opposite	$\csc\theta = \sec\left(\dfrac{\pi}{2}-\theta\right) = \dfrac{1}{\sin\theta}$

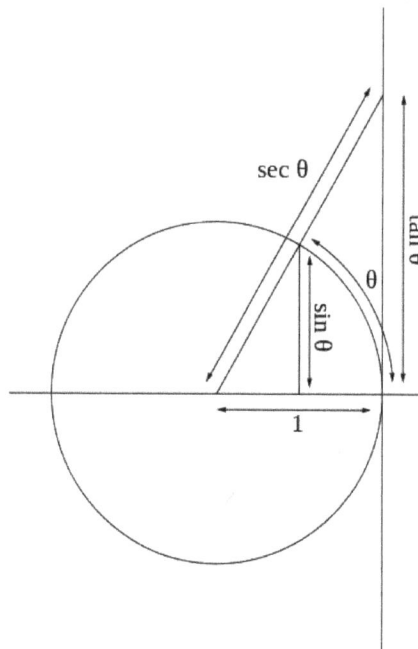

The sine, tangent, and secant functions of an angle constructed geometrically in terms of a unit circle. The number θ is the length of the curve; thus angles are being measured in radians. The secant and tangent functions rely on a fixed vertical line and the sine function on a moving vertical line. ("Fixed" in this context means not moving as θ changes; "moving" means depending on θ.) Thus, as θ goes from 0 up to a right angle, $\sin\theta$ goes from 0 to 1, $\tan\theta$ goes from 0 to ∞, and $\sec\theta$ goes from 1 to ∞.

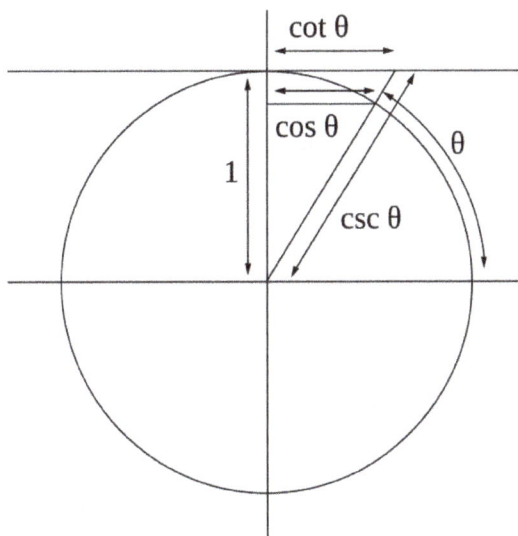

The cosine, cotangent, and cosecant functions of an angle θ constructed geometrically in terms of a unit circle. The functions whose names have the prefix *co-* use horizontal lines where the others use vertical lines.

Sine, Cosine and Tangent

The sine of an angle is the ratio of the length of the opposite side to the length of the hypotenuse. (The word comes from the Latin *sinus* for gulf or bay, since, given a unit circle, it is the side of the triangle on which the angle *opens*.) In our case

$$\sin A = \frac{\text{opposite}}{\text{hypotenuse}} = \frac{a}{h}.$$

This ratio does not depend on the size of the particular right triangle chosen, as long as it contains the angle A, since all such triangles are similar.

The cosine of an angle is the ratio of the length of the adjacent side to the length of the hypotenuse: so called because it is the sine of the complementary or co-angle. In our case

$$\cos A = \frac{\text{adjacent}}{\text{hypotenuse}} = \frac{b}{h}.$$

The tangent of an angle is the ratio of the length of the opposite side to the length of the adjacent side: so called because it can be represented as a line segment tangent to the circle, that is the line that touches the circle, from Latin *linea tangens* or touching line (cf. *tangere*, to touch). In our case

$$\tan A = \frac{\text{opposite}}{\text{adjacent}} = \frac{a}{b}.$$

The acronyms "SOH-CAH-TOA" ("soak-a-toe", "sock-a-toa", "so-kah-toa") and "OHSAHCOAT" are commonly used trigonometric mnemonics for these ratios.

Reciprocal Functions

The remaining three functions are best defined using the above three functions.

The cosecant csc(A), or cosec(A), is the reciprocal of sin(A); i.e. the ratio of the length of the hypotenuse to the length of the opposite side; so called because it is the secant of the complementary or co-angle:

$$\csc A = \frac{1}{\sin A} = \frac{\text{hypotenuse}}{\text{opposite}} = \frac{h}{a}.$$

The secant sec(A) is the reciprocal of cos(A); i.e. the ratio of the length of the hypotenuse to the length of the adjacent side:

$$\sec A = \frac{1}{\cos A} = \frac{\text{hypotenuse}}{\text{adjacent}} = \frac{h}{b}.$$

It is so called because it represents the line that *cuts* the circle.

The cotangent cot(A) is the reciprocal of tan(A); i.e. the ratio of the length of the adjacent side to the length of the opposite side; so called because it is the tangent of the complementary or co-angle:

$$\cot A = \frac{1}{\tan A} = \frac{\text{adjacent}}{\text{opposite}} = \frac{b}{a}.$$

Slope Definitions

Equivalent to the right-triangle definitions, the trigonometric functions can also be defined in terms of the *rise*, *run*, and *slope* of a line segment relative to horizontal. The slope is commonly taught as "rise over run" or rise/run. The three main trigonometric functions are commonly taught in the order sine, cosine, tangent. With a line segment length of 1 (as in a unit circle), the following mnemonic devices show the correspondence of definitions:

1. "Sine is first, rise is first" meaning that Sine takes the angle of the line segment and tells its vertical rise when the length of the line is 1.

2. "Cosine is second, run is second" meaning that Cosine takes the angle of the line segment and tells its horizontal run when the length of the line is 1.

3. "Tangent combines the rise and run" meaning that Tangent takes the angle of the line segment and tells its slope; or alternatively, tells the vertical rise when the line segment's horizontal run is 1.

This shows the main use of tangent and arctangent: converting between the two ways of telling the slant of a line, *i.e.*, angles and slopes.

While the length of the line segment makes no difference for the slope (the slope does not depend on the length of the slanted line), it does affect rise and run. To adjust and find the actual rise and run when the line does not have a length of 1, just multiply the sine and cosine by the line length. For instance, if the line segment has length 5, the run at an angle of 7° is $5\cos(7°)$

Unit-circle Definitions

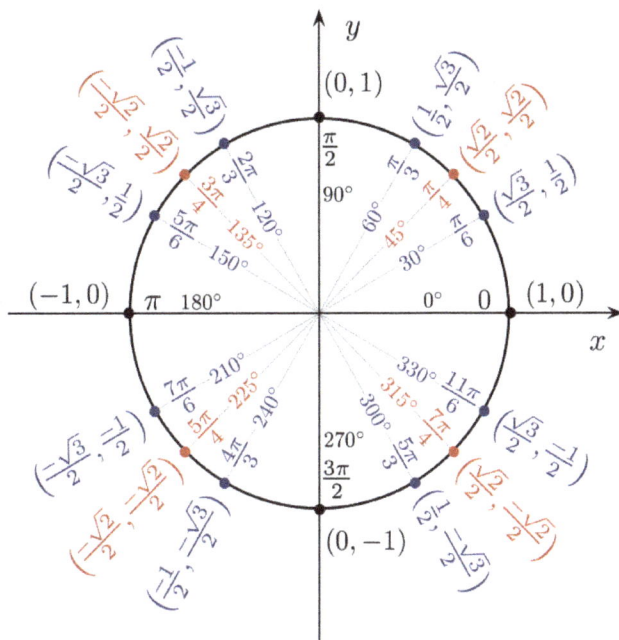

The unit circle, with some points labeled with their Cartesian coordinates and the corresponding angles in radians and degrees.

Signs of trigonometric functions in each quadrant. The mnemonic "all science teachers (are) crazy" lists the functions which are positive from quadrants I to IV. This is a variation on the mnemonic "All Students Take Calculus".

The six trigonometric functions can also be defined in terms of the unit circle, the circle of radius one centered at the origin. The unit circle relies on right triangles for most angles.

The unit circle definition does, however, permit the definition of the trigonometric functions for all positive and negative arguments, not just for angles between 0 and $\pi/2$ radians.

From the Pythagorean theorem the equation for the unit circle is

$$x^2 + y^2 = 1.$$

Let a line through the origin, making an angle of θ with the positive half of the x-axis. The line intersects the unit circle at a point P whose x- and y-coordinates are $\cos(\theta)$ and $\sin(\theta)$

Let us consider the right triangle whose vertexes are the point P, the center of the circle O, and the point H of the x-axis, that has the same x-coordinate as P. The radius of the circle is equal to the hypotenuse OP, and has length 1, so we have $\sin(\theta) = y/1$ and $\cos(\theta) = x/1$. The unit circle can be thought of as a way of looking at an infinite number of triangles by varying the lengths of their legs but keeping the lengths of their hypotenuses equal to 1.

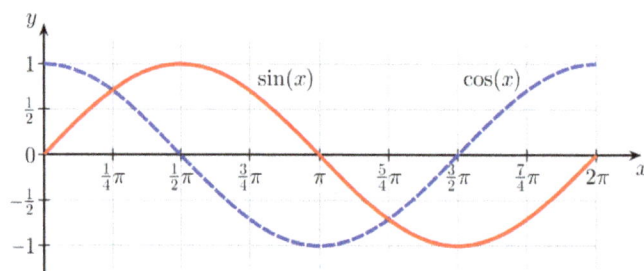

The sine and cosine functions graphed on the Cartesian plane.

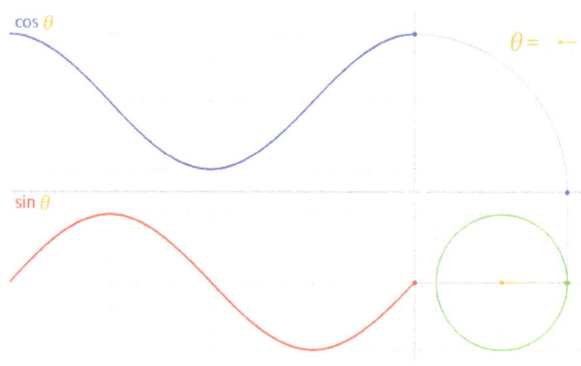

Animation showing the relationship between the unit circle and the sine and cosine functions.

For angles greater than 2π or less than -2π, one simply continues to rotate around the circle; sine and cosine are periodic functions with period 2π:

$$\sin\theta = \sin\left(\theta + 2\pi k\right),$$
$$\cos\theta = \cos\left(\theta + 2\pi k\right),$$

for any angle θ and any integer k.

The smallest positive period of a periodic function is called the primitive period of the function.

The primitive period of the sine or cosine is a full circle, i.e. 2π radians or 360 degrees.

Above, only sine and cosine were defined directly by the unit circle, but other trigonometric functions can be defined by:

$$\tan\theta = \frac{\sin\theta}{\cos\theta}, \cot\theta = \frac{\cos\theta}{\sin\theta} = \frac{1}{\tan\theta}$$

$$\sec\theta = \frac{1}{\cos\theta}, \csc\theta = \frac{1}{\sin\theta}$$

So :

- The primitive period of the secant, or cosecant is also a full circle, i.e. 2π radians or 360 degrees.

- The primitive period of the tangent or cotangent is only a half-circle, i.e. π radians or 180 degrees.

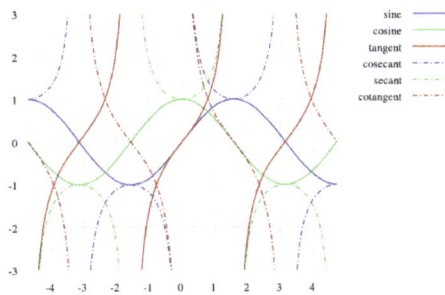

Trigonometric functions: Sine, Cosine, Tangent, Cosecant (dotted), Secant (dotted), Cotangent (dotted)

The image at right includes a graph of the tangent function.

- Its θ-intercepts correspond to those of $\sin(\theta)$ while its undefined values correspond to the θ-intercepts of $\cos(\theta)$.

- The function changes slowly around angles of $k\pi$, but changes rapidly at angles close to $(k + 1/2)\pi$.

- The graph of the tangent function also has a vertical asymptote at $\theta = (k + 1/2)\pi$, the θ-intercepts of the cosine function, because the function approaches infinity as θ approaches $(k + 1/2)\pi$ from the left and minus infinity as it approaches $(k + 1/2)\pi$ from the right.

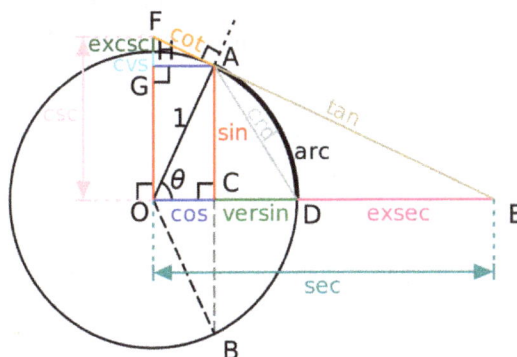

All of the trigonometric functions of the angle θ can be constructed geometrically in terms of a unit circle centered at O.

Alternatively, *all* of the basic trigonometric functions can be defined in terms of a unit circle centered at O (as shown in the adjacent picture), and similar such geometric definitions were used historically.

- In particular, for a chord AB of the circle, where θ is half of the subtended angle, $\sin(\theta)$ is AC (half of the chord), a definition introduced in India.

- $\cos(\theta)$ is the horizontal distance OC, and $\text{versin}(\theta) = 1 - \cos(\theta)$ is CD.

- $\tan(\theta)$ is the length of the segment AE of the tangent line through A, hence the word *tangent* for this function. $\cot(\theta)$ is another tangent segment, AF.

- $\sec(\theta) = OE$ and $\csc(\theta) = OF$ are segments of secant lines (intersecting the circle at two points), and can also be viewed as projections of OA along the tangent at A to the horizontal and vertical axes, respectively.

- DE is $\text{exsec}(\theta) = \sec(\theta) - 1$ (the portion of the secant outside, or *ex*, the circle).

- From these constructions, it is easy to see that the secant and tangent functions diverge as θ approaches $\pi/2$ (90° degrees) and that the cosecant and cotangent diverge as θ approaches zero. (Many similar constructions are possible, and the basic trigonometric identities can also be proven graphically.)

Algebraic Values

The algebraic expressions for $\sin(0°)$, $\sin(30°)$, $\sin(45°)$, $\sin(60°)$ and $\sin(90°)$ are

$$0, \quad \frac{1}{2}, \quad \frac{\sqrt{2}}{2}, \quad \frac{\sqrt{3}}{2}, \quad 1,$$

respectively. Such simple expressions generally do not exist for other angles.

For an angle which, measured in degrees, is a multiple of three, the sine and the cosine may be expressed in terms of square roots, as shown below. These values of the sine and the cosine may thus be constructed by ruler and compass.

For an angle of an integer number of degrees, the sine and the cosine may be expressed in terms of square roots and the cube root of a non-real complex number. Galois theory allows to prove that, if the angle is not a multiple of 3°, non-real cube roots are unavoidable.

For an angle which, measured in degrees, is a rational number, the sine and the cosine are algebraic numbers, which may be expressed in terms of nth roots. This results from the fact that the Galois groups of the cyclotomic polynomials are cyclic.

For an angle which, measured in degrees, is not a rational number, then either the angle or both the sine and the cosine are transcendental numbers. This is a corollary of Baker's theorem, proved in 1966.

Explicit Values

Algebraic expressions for 15°, 18°, 36°, 54°, 72° and 75° are as follows:

$$\sin 15° = \cos 75° = \frac{\sqrt{6} - \sqrt{2}}{4}$$

$$\sin 18° = \cos 72° = \frac{\sqrt{5} - 1}{4}$$

$$\sin 36° = \cos 54° = \frac{\sqrt{10 - 2\sqrt{5}}}{4}$$

$$\sin 54° = \cos 36° = \frac{\sqrt{5} + 1}{4}$$

$$\sin 72° = \cos 18° = \frac{\sqrt{10 + 2\sqrt{5}}}{4}$$

$$\sin 75° = \cos 15° = \frac{\sqrt{6} + \sqrt{2}}{4}.$$

From these, the algebraic expressions for all multiples of 3° can be computed. For example:

$$\sin 3° = \cos 87° = \frac{2(1 - \sqrt{3})\sqrt{5 + \sqrt{5}} + (1 + \sqrt{3})(\sqrt{10} - \sqrt{2})}{16}$$

$$\sin 6° = \cos 84° = \frac{\sqrt{30 - 6\sqrt{5}} - \sqrt{5} - 1}{8}$$

$$\sin 9° = \cos 81° = \frac{\sqrt{10} + \sqrt{2} - 2\sqrt{5 - \sqrt{5}}}{8}$$

$$\sin 84° = \cos 6° = \frac{\sqrt{10 - 2\sqrt{5}} + \sqrt{3} + \sqrt{15}}{8}$$

$$\sin 87° = \cos 3° = \frac{2\left(1 + \sqrt{3}\right)\sqrt{5 + \sqrt{5}} - \left(1 - \sqrt{3}\right)\left(\sqrt{10} - \sqrt{2}\right)}{16}.$$

Algebraic expressions can be deduced for other angles of an integer number of degrees, for example,

$$\sin 1° = \frac{\sqrt[3]{z} - \dfrac{1}{\sqrt[3]{z}}}{2i},$$

where $z = a + ib$, and a and b are the above algebraic expressions for, respectively, cos 3° and sin 3°, and the principal cube root (that is, the cube root with the largest real part) is to be taken.

Series Definitions

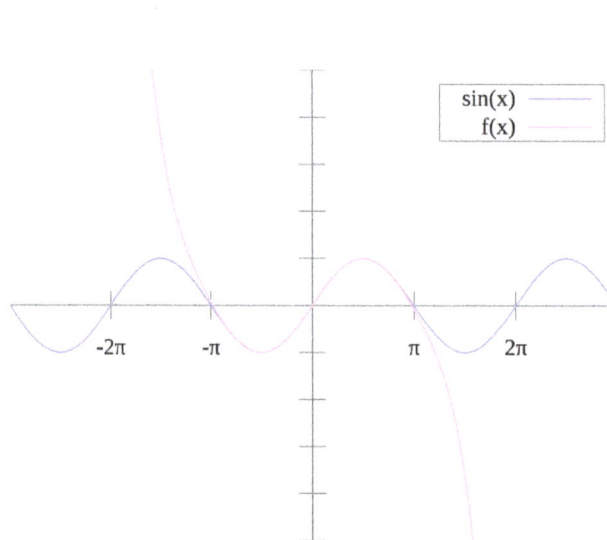

The sine function (blue) is closely approximated by its Taylor polynomial of degree 7 (pink) for a full cycle centered on the origin.

Trigonometric functions are analytic functions. Using only geometry and properties of limits, it can be shown that the derivative of sine is cosine and the derivative of cosine is the negative of sine. (Here, and generally in calculus, all angles are measured in radians.) One can then use the theory of Taylor series to show that the following identities hold for all real numbers x:

$$\sin x = x - \frac{x^3}{3!} + \frac{x^5}{5!} - \frac{x^7}{7!} + \cdots$$
$$= \sum_{n=0}^{\infty} \frac{(-1)^n x^{2n+1}}{(2n+1)!},$$
$$\cos x = 1 - \frac{x^2}{2!} + \frac{x^4}{4!} - \frac{x^6}{6!} + \cdots$$
$$= \sum_{n=0}^{\infty} \frac{(-1)^n x^{2n}}{(2n)!}.$$

The infinite series appearing in these identities are convergent in the whole complex plane and are often taken as the definitions of the sine and cosine functions of a complex variable. Another standard (and equivalent) definition of the sine and the cosine as functions of a complex variable is through their differential equation, below.

Other series can be found. For the following trigonometric functions:

U_n is the nth up/down number,

B_n is the nth Bernoulli number, and

E_n (below) is the nth Euler number.

Tangent

$$\tan x = \sum_{n=0}^{\infty} \frac{U_{2n+1} x^{2n+1}}{(2n+1)!}$$

$$= \sum_{n=1}^{\infty} \frac{(-1)^{n-1} 2^{2n} (2^{2n} - 1) B_{2n} x^{2n-1}}{(2n)!}$$

$$= x + \frac{1}{3} x^3 + \frac{2}{15} x^5 + \frac{17}{315} x^7 + \cdots, \qquad \text{for } |x| < \frac{\pi}{2}.$$

When this series for the tangent function is expressed in a form in which the denominators are the corresponding factorials, the numerators, called the "tangent numbers", have a combinatorial interpretation: they enumerate alternating permutations of finite sets of odd cardinality. The series itself can be found by a power series solution of the aforementioned differential equation.

Cosecant

$$\csc x = \sum_{n=0}^{\infty} \frac{(-1)^{n+1} 2 (2^{2n-1} - 1) B_{2n} x^{2n-1}}{(2n)!}$$

$$= x^{-1} + \frac{1}{6} x + \frac{7}{360} x^3 + \frac{31}{15120} x^5 + \cdots, \qquad \text{for } 0 < |x| < \pi.$$

Secant

$$\sec x = \sum_{n=0}^{\infty} \frac{U_{2n} x^{2n}}{(2n)!} = \sum_{n=0}^{\infty} \frac{(-1)^n E_{2n} x^{2n}}{(2n)!}$$

$$= 1 + \frac{1}{2} x^2 + \frac{5}{24} x^4 + \frac{61}{720} x^6 + \cdots, \qquad \text{for } |x| < \frac{\pi}{2}.$$

When this series for the secant function is expressed in a form in which the denominators are the corresponding factorials, the numerators, called the "secant numbers", have a combinatorial interpretation: they enumerate alternating permutations of finite sets of even cardinality.

Cotangent

From a theorem in complex analysis, there is a unique analytic continuation of this real function to the domain of complex numbers. They have the same Taylor series, and so the trigonometric functions are defined on the complex numbers using the Taylor series above.

$$\cot x = \sum_{n=0}^{\infty} \frac{(-1)^n 2^{2n} B_{2n} x^{2n-1}}{(2n)!}$$

$$= x^{-1} - \frac{1}{3}x - \frac{1}{45}x^3 - \frac{2}{945}x^5 - \cdots, \qquad \text{for } 0 < |x| < \pi.$$

There is a series representation as partial fraction expansion where just translated reciprocal functions are summed up, such that the poles of the cotangent function and the reciprocal functions match:

$$\pi \cdot \cot(\pi x) = \lim_{N \to \infty} \sum_{n=-N}^{N} \frac{1}{x+n}.$$

This identity can be proven with the Herglotz trick. Combining the $(-n)$th with the nth term lead to absolutely convergent series:

$$\pi \cdot \cot(\pi x) = \frac{1}{x} + \sum_{n=1}^{\infty} \frac{2x}{x^2 - n^2}, \qquad \frac{\pi}{\sin(\pi x)} = \frac{1}{x} + \sum_{n=1}^{\infty} \frac{(-1)^n 2x}{x^2 - n^2}.$$

Relationship to Exponential Function and Complex Numbers

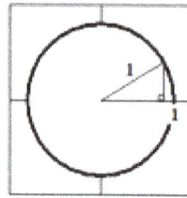

Unit Circle

Euler's formula illustrated with the three dimensional helix, starting with the 2D orthogonal components of the unit circle, sine and cosine (using $\theta = t$).

It can be shown from the series definitions that the sine and cosine functions are respectively the imaginary and real parts of the exponential function of a purely imaginary argument. That is, if x is real, we have

$$\cos x = \text{Re}(e^{ix}), \qquad \sin x = \text{Im}(e^{ix}),$$

and

$$e^{ix} = \cos x + i \sin x.$$

The latter identity, although primarily established for real x, remains valid for every complex x, and is called Euler's formula.

Euler's formula can be used to derive most trigonometric identities from the properties of the exponential function, by writing sine and cosine as:

$$\sin x = \frac{e^{ix} - e^{-ix}}{2i}, \qquad \cos x = \frac{e^{ix} + e^{-ix}}{2}.$$

It is also sometimes useful to express the complex sine and cosine functions in terms of the real and imaginary parts of their arguments.

$$\sin(x+iy) = \sin x \cosh y + i \cos x \sinh y, \cos(x+iy) = \cos x \cosh y - i \sin x \sinh y.$$

This exhibits a deep relationship between the complex sine and cosine functions and their real (sin, cos) and hyperbolic real (sinh, cosh) counterparts.

Complex Graphs

In the following graphs, the domain is the complex plane pictured, and the range values are indicated at each point by color. Brightness indicates the size (absolute value) of the range value, with black being zero. Hue varies with argument, or angle, measured from the positive real axis. (more)

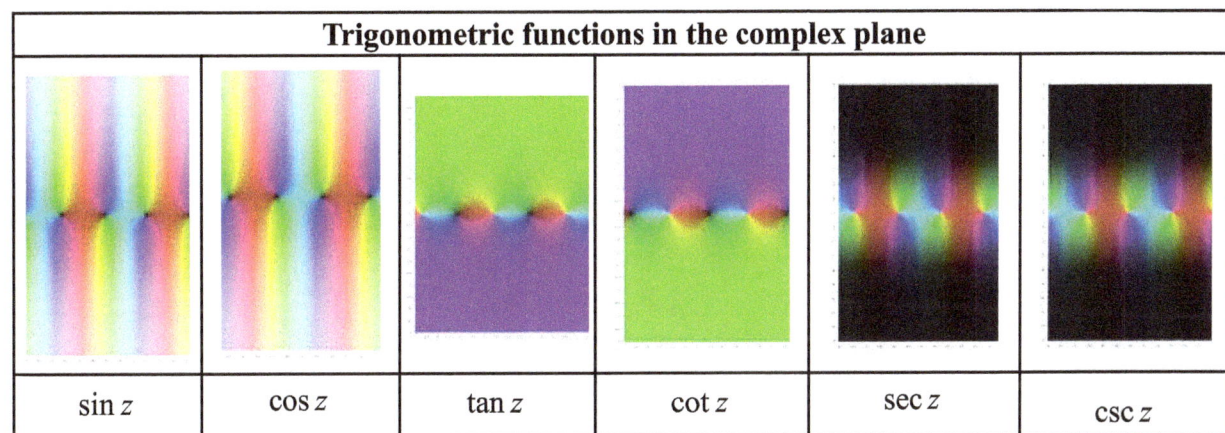

Trigonometric functions in the complex plane					
$\sin z$	$\cos z$	$\tan z$	$\cot z$	$\sec z$	$\csc z$

Definitions via Differential Equations

Both the sine and cosine functions satisfy the differential equation:

$$y'' = -y.$$

That is to say, each is the additive inverse of its own second derivative. Within the 2-dimensional function space V consisting of all solutions of this equation,

- the sine function is the unique solution satisfying the initial condition $\left(y'(0), y(0)\right) = (1, 0)$ and

- the cosine function is the unique solution satisfying the initial condition $\left(y'(0), y(0)\right) = (0, 1)$..

Since the sine and cosine functions are linearly independent, together they form a basis of V. This method of defining the sine and cosine functions is essentially equivalent to using Euler's formula. It turns out that this differential equation can be used not only to define the sine and cosine functions but also to prove the trigonometric identities for the sine and cosine functions.

Further, the observation that sine and cosine satisfies $y'' = -y$ means that they are eigenfunctions of the second-derivative operator.

The tangent function is the unique solution of the nonlinear differential equation

$$y' = 1 + y^2$$

satisfying the initial condition $y(0) = 0$. There is a very interesting visual proof that the tangent function satisfies this differential equation.

The Significance of Radians

Radians specify an angle by measuring the length around the path of the unit circle and constitute a special argument to the sine and cosine functions. In particular, only sines and cosines that map radians to ratios satisfy the differential equations that classically describe them. If an argument to sine or cosine in radians is scaled by frequency,

$$f(x) = \sin(kx),$$

then the derivatives will scale by amplitude.

$$f'(x) = k \cos(kx).$$

Here, k is a constant that represents a mapping between units. If x is in degrees, then

$$k = \frac{\pi}{180^\circ}.$$

This means that the second derivative of a sine in degrees does not satisfy the differential equation

$$y'' = -y$$

but rather

$$y'' = -k^2 y.$$

The cosine's second derivative behaves similarly.

This means that these sines and cosines are different functions, and that the fourth derivative of sine will be sine again only if the argument is in radians.

Identities

Many identities interrelate the trigonometric functions. Among the most frequently used is the Pythagorean identity, which states that for any angle, the square of the sine plus the square of the cosine is 1. This is easy to see by studying a right triangle of hypotenuse 1 and applying the Pythagorean theorem. In symbolic form, the Pythagorean identity is written

$$\sin^2 x + \cos^2 x = 1$$

which is standard shorthand notation for

$$(\sin x)^2 + (\cos x)^2 = 1.$$

Other key relationships are the sum and difference formulas, which give the sine and cosine of the sum and difference of two angles in terms of sines and cosines of the angles themselves. These can be derived geometrically, using arguments that date to Ptolemy. One can also produce them algebraically using Euler's formula.

Sum

$$\sin(x+y) = \sin x \cos y + \cos x \sin y,$$
$$\cos(x+y) = \cos x \cos y - \sin x \sin y,$$

Difference

$$\sin(x-y) = \sin x \cos y - \cos x \sin y,$$
$$\cos(x-y) = \cos x \cos y + \sin x \sin y.$$

These in turn lead to the following three-angle formulae:

$$(x+y+z) = \sin x \cos y \cos z + \sin y \cos z \cos x + \sin z \cos y \cos x - \sin x \sin y \sin z,$$
$$\cos(x+y+z) = \cos x \cos y \cos z - \cos x \sin y \sin z - \cos y \sin x \sin z - \cos z \sin x \sin y,$$

When the two angles are equal, the sum formulas reduce to simpler equations known as the double-angle formulae.

$$\sin(2x) = 2 \sin x \cos x,$$
$$\cos(2x) = \cos^2 x - \sin^2 x = 2\cos^2 x - 1 = 1 - 2\sin^2 x.$$

When three angles are equal, the three-angle formulae simplify to

$$\sin(3x) = 3 \sin x - 4 \sin^3 x.$$
$$\cos(3x) = 4 \cos^3 x - 3 \cos x.$$

These identities can also be used to derive the product-to-sum identities that were used in antiquity to transform the product of two numbers into a sum of numbers and greatly speed operations, much like the logarithm function.

Calculus

For integrals and derivatives of trigonometric functions, see the relevant sections of Differentiation of trigonometric functions, Lists of integrals and List of integrals of trigonometric functions.

Below is the list of the derivatives and integrals of the six basic trigonometric functions. The number C is a constant of integration.

$f(x)$	$f'(x)$	$\int f(x)dx$
$\sin x$	$\cos x$	$-\cos x + C$
$\cos x$	$-\sin x$	$\sin x + C$
$\tan x$	$\sec^2 x = 1 + \tan^2 x$	$-\ln|\cos x| + C$
$\cot x$	$-\csc^2 x = -(1 + \cot^2 x)$	$\ln|\sin x| + C$
$\sec x$	$\sec x \tan x$	$\ln|\sec x + \tan x| + C$
$\csc x$	$-\csc x \cot x$	$-\ln|\csc x + \cot x| + C$

Definitions Using Functional Equations

In mathematical analysis, one can define the trigonometric functions using functional equations based on properties like the difference formula. Taking as given these formulas, one can prove that only two real functions satisfy those conditions. Symbolically, we say that there exists exactly one pair of real functions – sin and cos – such that for all real numbers x and y, the following equation holds:

$$\cos(x - y) = \cos x \cos y + \sin x \sin y$$

with the added condition that

$$0 < x \cos x < \sin x < x \text{ for } 0 < x < 1.$$

Other derivations, starting from other functional equations, are also possible, and such derivations can be extended to the complex numbers. As an example, this derivation can be used to define trigonometry in Galois fields.

Computation

The computation of trigonometric functions is a complicated subject, which can today be avoided by most people because of the widespread availability of computers and scientific calculators that provide built-in trigonometric functions for any angle. This section, however, describes details of their computation in three important contexts: the historical use of trigonometric tables, the modern techniques used by computers, and a few "important" angles where simple exact values are easily found.

The first step in computing any trigonometric function is range reduction—reducing the given angle to a "reduced angle" inside a small range of angles, say 0 to $\pi/2$, using the periodicity and symmetries of the trigonometric functions.

Prior to computers, people typically evaluated trigonometric functions by interpolating from a detailed table of their values, calculated to many significant figures. Such tables have been available for as long as trigonometric functions have been described, and were typically generated by repeated application of the half-angle and angle-addition identities starting from a known value (such as sin(π/2) = 1).

Modern computers use a variety of techniques. One common method, especially on higher-end processors with floating point units, is to combine a polynomial or rational approximation (such as Chebyshev approximation, best uniform approximation, and Padé approximation, and typically for higher or variable precisions, Taylor and Laurent series) with range reduction and a table look-up—they first look up the closest angle in a small table, and then use the polynomial to compute the correction. Devices that lack hardware multipliers often use an algorithm called CORDIC (as well as related techniques), which uses only addition, subtraction, bitshift, and table lookup. These methods are commonly implemented in hardware floating-point units for performance reasons.

For very high precision calculations, when series expansion convergence becomes too slow, trigonometric functions can be approximated by the arithmetic-geometric mean, which itself approximates the trigonometric function by the (complex) elliptic integral.

Finally, for some simple angles, the values can be easily computed by hand using the Pythagorean theorem, as in the following examples. For example, the sine, cosine and tangent of any integer multiple of π/60 radians (3°) can be found exactly by hand.

Consider a right triangle where the two other angles are equal, and therefore are both π/4 radians (45°). Then the length of side b and the length of side a are equal; we can choose $a = b = 1$. The values of sine, cosine and tangent of an angle of π/4 radians (45°) can then be found using the Pythagorean theorem:

$$c = \sqrt{a^2 + b^2} = \sqrt{2}.$$

Therefore:

$$\sin\frac{\pi}{4} = \sin 45° = \cos\frac{\pi}{4} = \cos 45° = \frac{1}{\sqrt{2}} = \frac{\sqrt{2}}{2},$$

$$\tan\frac{\pi}{4} = \tan 45° = \frac{\sin\dfrac{\pi}{4}}{\cos\dfrac{\pi}{4}} = \frac{1}{\sqrt{2}}\cdot\frac{\sqrt{2}}{1} = \frac{\sqrt{2}}{\sqrt{2}} = 1.$$

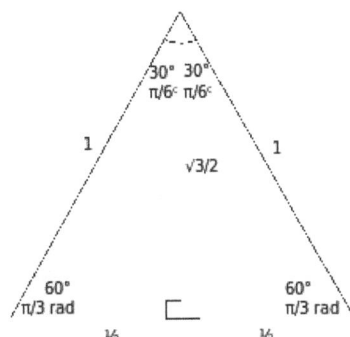

Computing trigonometric functions from an equilateral triangle

To determine the trigonometric functions for angles of $\pi/3$ radians (60°) and $\pi/6$ radians (30°), we start with an equilateral triangle of side length 1. All its angles are $\pi/3$ radians (60°). By dividing it into two, we obtain a right triangle with $\pi/6$ radians (30°) and $\pi/3$ radians (60°) angles. For this triangle, the shortest side is 1/2, the next largest side is $\sqrt{3}/2$ and the hypotenuse is 1. This yields:

$$\sin\frac{\pi}{6} = \sin 30° = \cos\frac{\pi}{3} = \cos 60° = \frac{1}{2},$$

$$\cos\frac{\pi}{6} = \cos 30° = \sin\frac{\pi}{3} = \sin 60° = \frac{\sqrt{3}}{2},$$

$$\tan\frac{\pi}{6} = \tan 30° = \cot\frac{\pi}{3} = \cot 60° = \frac{1}{\sqrt{3}} = \frac{\sqrt{3}}{3}.$$

Special values in Trigonometric Functions

There are some commonly used special values in trigonometric functions, as shown in the following table.

Radian	0	$\frac{\pi}{12}$	$\frac{\pi}{8}$	$\frac{\pi}{6}$	$\frac{\pi}{4}$	$\frac{\pi}{3}$	$\frac{5\pi}{12}$	$\frac{\pi}{2}$
Degree	0°	15°	22.5°	30°	45°	60°	75°	90°
sin	0	$\frac{\sqrt{6}-\sqrt{2}}{4}$	$\frac{\sqrt{2-\sqrt{2}}}{2}$	$\frac{1}{2}$	$\frac{\sqrt{2}}{2}$	$\frac{\sqrt{3}}{2}$	$\frac{\sqrt{6}+\sqrt{2}}{4}$	1
cos	1	$\frac{\sqrt{6}+\sqrt{2}}{4}$	$\frac{\sqrt{2+\sqrt{2}}}{2}$	$\frac{\sqrt{3}}{2}$	$\frac{\sqrt{2}}{2}$	$\frac{1}{2}$	$\frac{\sqrt{6}-\sqrt{2}}{4}$	0
tan	0	$2-\sqrt{3}$	$\sqrt{2}-1$	$\frac{\sqrt{3}}{3}$	1	$\sqrt{3}$	$2+\sqrt{3}$	∞
cot	∞	$2+\sqrt{3}$	$\sqrt{2}+1$	$\sqrt{3}$	1	$\frac{\sqrt{3}}{3}$	$2-\sqrt{3}$	0
sec	1	$\sqrt{6}-\sqrt{2}$	$\sqrt{2}\sqrt{2-\sqrt{2}}$	$\frac{2\sqrt{3}}{3}$	$\sqrt{2}$	2	$\sqrt{6}+\sqrt{2}$	∞
csc	∞	$\sqrt{6}+\sqrt{2}$	$\sqrt{2}\sqrt{2+\sqrt{2}}$	2	$\sqrt{2}$	$\frac{2\sqrt{3}}{3}$	$\sqrt{6}-\sqrt{2}$	1

The symbol ∞ here represents the point at infinity on the projectively extended real line, the limit on the extended real line is $+\infty$ on one side and $-\infty$ on the other.

Inverse Functions

The trigonometric functions are periodic, and hence not injective, so strictly they do not have an inverse function. Therefore, to define an inverse function we must restrict their domains so that the trigonometric function is bijective. In the following, the functions on the left are *defined* by the equation on the right; these are not proved identities. The principal inverses are usually defined as:

Function	Definition	Value Field
$\arcsin x = y$	$\sin y = x$	$-\dfrac{\pi}{2} \leq y \leq \dfrac{\pi}{2}$
$\arccos x = y$	$\cos y = x$	$0 \leq y \leq \pi$
$\arctan x = y$	$\tan y = x$	$-\dfrac{\pi}{2} < y < \dfrac{\pi}{2}$
$\text{arccot}\, x = y$	$\cot y = x$	$0 < y < \pi$
$\text{arcsec}\, x = y$	$\sec y = x$	$0 \leq y \leq \pi, y \neq \dfrac{\pi}{2}$
$\text{arccsc}\, x = y$	$\csc y = x$	$-\dfrac{\pi}{2} \leq y \leq \dfrac{\pi}{2}, y \neq 0$

The notations $\sin^{-1}$ and $\cos^{-1}$ are often used for arcsin and arccos, etc. When this notation is used, the inverse functions could be confused with the multiplicative inverses of the functions. The notation using the "arc-" prefix avoids such confusion, though "arcsec" for arcsecant can be confused with "arcsecond".

Just like the sine and cosine, the inverse trigonometric functions can also be defined in terms of infinite series. For example,

$$\arcsin z = z + \left(\frac{1}{2}\right)\frac{z^3}{3} + \left(\frac{1\cdot3}{2\cdot4}\right)\frac{z^5}{5} + \left(\frac{1\cdot3\cdot5}{2\cdot4\cdot6}\right)\frac{z^7}{7} + \cdots.$$

These functions may also be defined by proving that they are antiderivatives of other functions. The arcsine, for example, can be written as the following integral:

$$\arcsin z = \int_0^z \frac{1}{\sqrt{1-x^2}}\, dx, \quad |z| < 1.$$

Analogous formulas for the other functions can be found at inverse trigonometric functions. Using the complex logarithm, one can generalize all these functions to complex arguments:

$$\arcsin z = -i \log\left(iz + \sqrt{1 - z^2}\right),$$

$$\arccos z = -i \log\left(z + \sqrt{z^2 - 1}\right),$$

$$\arctan z = \frac{1}{2} i \log\left(\frac{1 - iz}{1 + iz}\right).$$

Connection to the Inner Product

In an inner product space, the angle between two non-zero vectors is defined to be

$$\mathrm{angle}(x, y) = \arccos \frac{\langle x, y \rangle}{\| x \| \| y \|}.$$

Properties and Applications

The trigonometric functions, as the name suggests, are of crucial importance in trigonometry, mainly because of the following two results.

Law of Sines

The law of sines states that for an arbitrary triangle with sides a, b, and c and angles opposite those sides A, B and C:

$$\frac{\sin A}{a} = \frac{\sin B}{b} = \frac{\sin C}{c} = \frac{2\Delta}{abc},$$

where Δ is the area of the triangle, or, equivalently,

$$\frac{a}{\sin A} = \frac{b}{\sin B} = \frac{c}{\sin C} = 2R,$$

where R is the triangle's circumradius.

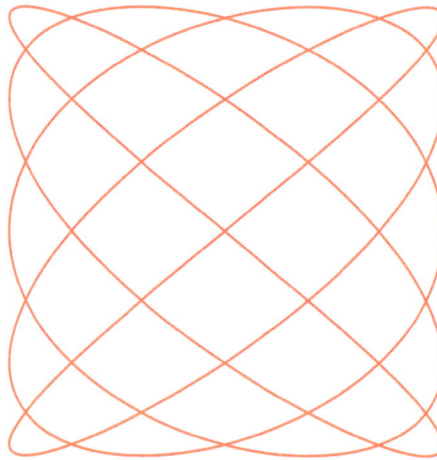

A Lissajous curve, a figure formed with a trigonometry-based function.

It can be proven by dividing the triangle into two right ones and using the above definition of sine. The law of sines is useful for computing the lengths of the unknown sides in a triangle if two angles and one side are known. This is a common situation occurring in *triangulation*, a technique to determine unknown distances by measuring two angles and an accessible enclosed distance.

Law of Cosines

The law of cosines (also known as the cosine formula or cosine rule) is an extension of the Pythagorean theorem:

$$c^2 = a^2 + b^2 - 2ab\cos C,$$

or equivalently,

$$\cos C = \frac{a^2 + b^2 - c^2}{2ab}.$$

In this formula the angle at C is opposite to the side c. This theorem can be proven by dividing the triangle into two right ones and using the Pythagorean theorem.

The law of cosines can be used to determine a side of a triangle if two sides and the angle between them are known. It can also be used to find the cosines of an angle (and consequently the angles themselves) if the lengths of all the sides are known.

Law of Tangents

The following all form the law of tangents

$$\frac{\tan\dfrac{A-B}{2}}{\tan\dfrac{A+B}{2}} = \frac{a-b}{a+b}; \qquad \frac{\tan\dfrac{A-C}{2}}{\tan\dfrac{A+C}{2}} = \frac{a-c}{a+c}; \qquad \frac{\tan\dfrac{B-C}{2}}{\tan\dfrac{B+C}{2}} = \frac{b-c}{b+c}$$

The explanation of the formulae in words would be cumbersome, but the patterns of sums and differences; for the lengths and corresponding opposite angles, are apparent in the theorem.

Law of Cotangents

If

$$\zeta = \sqrt{\frac{1}{s}(s-a)(s-b)(s-c)} \quad \text{(the radius of the inscribed circle for the triangle)}$$

and

$$s = \frac{a+b+c}{2} \quad \text{(the semi-perimeter for the triangle),}$$

then the following all form the law of cotangents

$$\cot\frac{A}{2} = \frac{s-a}{\zeta}; \qquad \cot\frac{B}{2} = \frac{s-b}{\zeta}; \qquad \cot\frac{C}{2} = \frac{s-c}{\zeta}$$

It follows that

$$\frac{\cot\dfrac{A}{2}}{s-a} = \frac{\cot\dfrac{B}{2}}{s-b} = \frac{\cot\dfrac{C}{2}}{s-c}.$$

In words the theorem is: the cotangent of a half-angle equals the ratio of the semi-perimeter minus the opposite side to the said angle, to the inradius for the triangle.

Periodic Functions

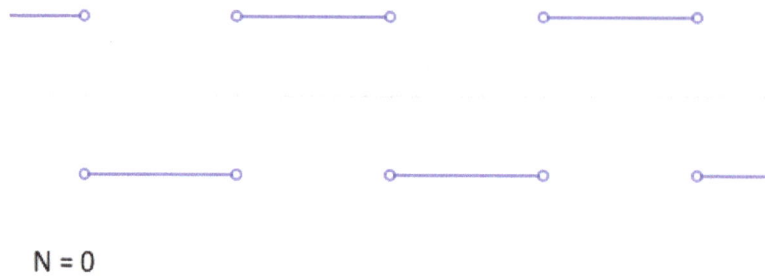

N = 0

An animation of the additive synthesis of a square wave with an increasing number of harmonics

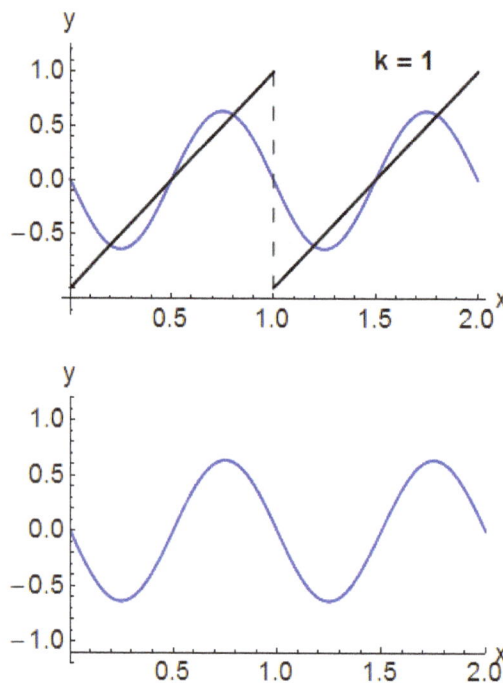

Sinusoidal basis functions (bottom) can form a sawtooth wave (top) when added. All the basis functions have nodes at the nodes of the sawtooth, and all but the fundamental ($k = 1$) have additional nodes. The oscillation seen about the sawtooth when k is large is called the Gibbs phenomenon

The trigonometric functions are also important in physics. The sine and the cosine functions, for example, are used to describe simple harmonic motion, which models many natural phenomena, such as the movement of a mass attached to a spring and, for small angles, the pendular motion of a mass hanging by a string. The sine and cosine functions are one-dimensional projections of uniform circular motion.

Trigonometric functions also prove to be useful in the study of general periodic functions. The characteristic wave patterns of periodic functions are useful for modeling recurring phenomena such as sound or light waves.

Under rather general conditions, a periodic function $f(x)$ can be expressed as a sum of sine waves or cosine waves in a Fourier series. Denoting the sine or cosine basis functions by φ_k, the expansion of the periodic function $f(t)$ takes the form:

$$f(t) = \sum_{k=1}^{\infty} c_k \varphi_k(t).$$

For example, the square wave can be written as the Fourier series

$$f_{\text{square}}(t) = \frac{4}{\pi} \sum_{k=1}^{\infty} \frac{\sin\big((2k-1)t\big)}{2k-1}.$$

In the animation of a square wave at top right it can be seen that just a few terms already produce a fairly good approximation. The superposition of several terms in the expansion of a sawtooth wave are shown underneath.

History

While the early study of trigonometry can be traced to antiquity, the trigonometric functions as they are in use today were developed in the medieval period. The chord function was discovered by Hipparchus of Nicaea (180–125 BC) and Ptolemy of Roman Egypt (90–165 AD).

The functions sine and cosine can be traced to the *jyā* and *koti-jyā* functions used in Gupta period Indian astronomy (*Aryabhatiya, Surya Siddhanta*), via translation from Sanskrit to Arabic and then from Arabic to Latin.

All six trigonometric functions in current use were known in Islamic mathematics by the 9th century, as was the law of sines, used in solving triangles. al-Khwārizmī produced tables of sines, cosines and tangents. They were studied by authors including Omar Khayyám, Bhāskara II, Nasir al-Din al-Tusi, Jamshīd al-Kāshī (14th century), Ulugh Beg (14th century), Regiomontanus (1464), Rheticus, and Rheticus' student Valentinus Otho.

Madhava of Sangamagrama (c. 1400) made early strides in the analysis of trigonometric functions in terms of infinite series.

The terms *tangent* and *secant* were first introduced in 1583 by the Danish mathematician Thomas Fincke in his book *Geometria rotundi*.

The first published use of the abbreviations *sin*, *cos*, and *tan* is by the 16th century French mathematician Albert Girard.

In a paper published in 1682, Leibniz proved that $\sin x$ is not an algebraic function of x.

Leonhard Euler's *Introductio in analysin infinitorum* (1748) was mostly responsible for establishing the analytic treatment of trigonometric functions in Europe, also defining them as infinite series and presenting "Euler's formula", as well as the near-modern abbreviations *sin., cos., tang., cot., sec.,* and *cosec.*

A few functions were common historically, but are now seldom used, such as the chord ($\text{crd}(\theta) = 2\sin(\theta/2)$), the versine ($\text{versin}(\theta) = 1 - \cos(\theta) = 2\sin^2(\theta/2)$) (which appeared in the earliest tables), the haversine ($\text{haversin}(\theta) = 1/2\text{versin}(\theta) = \sin^2(\theta/2)$), the exsecant ($\text{exsec}(\theta) = \sec(\theta) - 1$) and the excosecant ($\text{excsc}(\theta) = \text{exsec}(\pi/2 - \theta) = \csc(\theta) - 1$). Many more relations between these functions are listed in the article about trigonometric identities.

Etymology

The word *sine* derives from Latin *sinus*, meaning "bend; bay", and more specifically "the hanging fold of the upper part of a toga", "the bosom of a garment", which was chosen as the translation of what had been interpreted as the Arabic word *jaib*, meaning "pocket" or "fold" in twelfth-century European translations of works by Al-Battani and al-Khwārizmī. The choice was based on a mis-reading of the Arabic written form *j-y-b* (جيب), which itself originated as a transliteration from Sanskrit *jīvā*, which along with its synonym *jyā* (the standard Sanskrit term for the sine) translates to "bowstring".

The word *tangent* comes from Latin *tangens* meaning "touching", since the line *touches* the circle of unit radius, whereas *secant* stems from Latin *secans* — "cutting" — since the line *cuts* the circle.

The prefix "co-" (in "cosine", "cotangent", "cosecant") is found in Edmund Gunter's *Canon triangulorum* (1620), which defines the *cosinus* as an abbreviation for the *sinus complementi* (sine of the complementary angle) and proceeds to define the *cotangens* similarly.

References

- Grattan-Guinness, Ivor (1997). The Rainbow of Mathematics: A History of the Mathematical Sciences. W.W. Norton. ISBN 0-393-32030-8.

- Otto Neugebauer (1975). A history of ancient mathematical astronomy. 1. Springer-Verlag. pp. 744–. ISBN 978-3-540-06995-9.

- Robert E. Krebs (2004). Groundbreaking Scientific Experiments, Inventions, and Discoveries of the Middle Ages and the Renaissance. Greenwood Publishing Group. pp. 153–. ISBN 978-0-313-32433-8.

- William Bragg Ewald (2007). From Kant to Hilbert: a source book in the foundations of mathematics. Oxford University Press US. p. 93. ISBN 0-19-850535-3

- A sentence more appropriate for high schools is "'Some Old Horse Came A"Hopping Through Our Alley". Foster, Jonathan K. (2008). Memory: A Very Short Introduction. Oxford. p. 128. ISBN 0-19-280675-0.

- Peterson, John C. (2004). Technical Mathematics with Calculus (illustrated ed.). Cengage Learning. p. 856. ISBN 978-0-7668-6189-3.

- Abramowitz, Milton and Irene A. Stegun, Handbook of Mathematical Functions with Formulas, Graphs, and Mathematical Tables, Dover, New York. (1964). ISBN 0-486-61272-4.

- Joseph, George G., The Crest of the Peacock: Non-European Roots of Mathematics, 2nd ed. Penguin Books, London. (2000). ISBN 0-691-00659-8.

- Maor, Eli, Trigonometric Delights, Princeton Univ. Press. (1998). Reprint edition (February 25, 2002): ISBN 0-691-09541-8.

Permissions

Index

www.ingramcontent.com/pod-product-compliance
Lightning Source LLC
Chambersburg PA
CBHW061320190326
41458CB00011B/3845